S0-ATH-848

Design and Implementation of Intelligent Manufacturing Systems

From Expert Systems, Neural Networks, to Fuzzy Logic

PRENTICE HALL SERIES ON ENVIRONMENTAL AND INTELLIGENT MANUFACTURING SYSTEMS
M. Jamshidi, Editor

Volume 1: ***ROBOTICS AND REMOTE SYSTEMS FOR HAZARDOUS ENVIRONMENTS***, edited by M. Jamshidi and P. J. Eicker.

Volume 2: ***FUZZY LOGIC AND CONTROL: Software and Hardware Applications***, edited by M. Jamshidi, N. Vadiee, and T. J. Ross.

Volume 3: ***ARTIFICIAL INTELLIGENCE IN OPTIMAL DESIGN AND MANUFACTURING***, edited by Z. Dong.

Volume 4: ***SOFT COMPUTING: Fuzzy Logic, Neural Networks, and Distributed Artificial Intelligence***, edited by F. Aminzadeh and M. Jamshidi.

Volume 5: ***NONLINEAR CONTROL OF ROBOTIC SYSTEMS FOR ENVIRONMENTAL WASTE AND RESTORATION***, by D. Dawson, M. Bridges, and Z. Qu.

Volume 6: ***DESIGN AND IMPLEMENTATION OF INTELLIGENT MANUFACTURING SYSTEMS: From Expert Systems, Neural Networks, to Fuzzy Logic***, by H. R. Parsaei and M. Jamshidi

Design and Implementation of Intelligent Manufacturing Systems

From Expert Systems, Neural Networks, to Fuzzy Logic

Editors

Hamid R. Parsaei
University of Louisville, USA

M. Jamshidi
University of New Mexico, USA

Prentice Hall P T R
Upper Saddle River, New Jersey 07458

UNIVERSITY OF FLORIDA LIBRARIES

Library of Congress Cataloging-in-Publication Data

Design and implementation of intelligent manufacturing systems : from expert systems, neural networks, to fuzzy logic / editors Hamid R. Parsaei, Mo Jamshidi.
p. cm. -- (Prentice-Hall series on environmental and intelligent manufacturing systems)
Includes bibliographical references (p.) and index.
ISBN 0-13-458217-9
1. CAD/CAM systems 2. Manufacturing processes--Automation.
3. Expert systems (Computer science)--Industrial applications.
I. Parsaei, H. R. II. Jamshidi, Mohammad. III. Series.
TS155.6.D472 1995
670'.285--dc20 95-4068
CIP

Editorial/production supervision: *Patti Guerrieri*
Cover art: *Aren Graphics*
Manufacturing buyer: *Alexis R. Heydt*
Acquisitions editor: *Bernard Goodwin*
Editorial assistant: *Diane Spina*

©1995 by Prentice Hall PTR
Prentice-Hall Inc.
A Simon & Schuster Company
Upper Saddle River, NJ 07458

The publisher offers discounts on this book when ordered in bulk quantities. For more information, contact: Corporate Sales Department, Prentice Hall PTR, One Lake Street, Upper Saddle River, NJ 07458, Phone: 800-382-3419, Fax: 201-236-7141, e-mail: corpsales@prenhall.com

All rights reserved. No part of this book may be reproduced, in any form or by any means, without permission in writing from the publisher.

Printed in the United States of America
10 9 8 7 6 5 4 3 2 1

ISBN 0-13-458217-9

Prentice-Hall International (UK) Limited, *London*
Prentice-Hall of Australia Pty. Limited, *Sydney*
Prentice-Hall of Canada Inc., *Toronto*
Prentice-Hall Hispanoamericana, S.A., *Mexico*
Prentice-Hall of India Private Limited, *New Delhi*
Prentice-Hall of Japan, Inc., *Tokyo*
Simon & Schuster Asia Pte. Ltd., *Singapore*
Editora Prentice-Hall do Brasil, Ltda., *Rio de Janeiro*

TO OUR FAMILIES

Farah, Shadi, and Boback

Jila, Ava, and Nima

TABLE OF CONTENTS

Series Editor's Foreword

The world communities, for the most part, are searching for ways to cooperate with each other in order to solve their emerging problems, which affects not only their own generation but future generations as well. The concerns about our environment, energy supplies (the shortage of them), water resources, food, and agriculture are some of the most urgent issues facing every nation of the world. These issues will need to be continuously dealt with in every conceivable form.

This series was created three years ago to address the paramount issues of national economic security and respect for our environment (air, water, soil, lakes, etc.) and the emergence of artificial intelligence as an integral attribute of today's and tomorrow's products. This series attempts to build two bridges. The first is between manufacturing and environment, making environmental issues an integral part of any manufacturing design process leading to environmentally friendly manufacturing systems. The second is between manufacturing and artificial intelligence, leading to what is now being called MIQ (Machine Intelligence Quotient Systems). Today, AI techniques of expert systems, fuzzy logic, neural networks, and evolutionary algorithms have found their ways, to different degrees, in a wide spectrum of manufactured products and manufacturing systems.

We invite concerned authors from all parts of the world and all walks of science, engineering, and technology to come forth with proposals for this series of books, with the above concerns as a motivating factor. Together we can disseminate the most relevant information to the world community for the betterment of mankind.

The current volume, superbly put together by Professor Hamid R. Parsaei of the University of Louisville, is a step toward our goal. This volume attempts to close the gap between AI and manufacturing. I would like to thank Professor Parsaei and all the fine authors of this volume for making a difference with their research and opening new avenues for future generations.

Mohammad Jamshidi, Series Editor
Prentice Hall Series on:
Environmental and Intelligent Manufacturing Systems

PREFACE

The introduction of microprocessors and computer-controlled production tools into industry has given a new perspective to manufacturing processes both in the United States and abroad. Computer integrated manufacturing systems (CIMS), flexible manufacturing systems (FMS), group technology (GT), cellular manufacturing (CM), computer aided design (CAD), and computer aided manufacturing (CAM) have been considered by many as viable tools for reducing direct and indirect manufacturing costs and improving product quality and production flexibility. However, many of these new manufacturing techniques have not been able to become significantly efficient due to the inadequacy of the existing computer hardware and software.

The 1980's produced a remarkable growth in the area of the artificial intelligence (AI) which is viewed by many to be one of the most promising technologies for the factories of the future. In recent years, many papers and research results have reported the successful applications of artificial intelligence to engineering design and manufacturing planning and control.

The chapters in this book address the topics of intelligent manufacturing from a variety of theoretical, empirical, design, and implementation perspectives. The contributions to this volume represent a foundation for subsequent research work that could be done in this area that eventually would lead to efficient and intelligent manufacturing systems.

This volume consists of 15 chapters. In the first chapter, Jana and Auslander present a workcell programming environment to model and implement prototype intelligent manufacturing systems. In the second chapter, Reimann and Sarkis review a general conceptual system for computer aided process planning (CAPP) that was developed by the Consortium for Advanced Manufacturing - International (CAM-I). This framework specifies the structure and defines the functions necessary for automating the process planning of machined prismatic parts. The third chapter, by Ali *et al.*, proposes an artificial intelligence approach to metal forming processes and demonstrates the use of an intelligent hybrid tool to design and control the pack rolling process. The forth chapter, by Graham and Guan, describes a system for the monitoring and diagnosis of manufacturing systems which can best be described as a knowledge-based diagnosis system that uses a hybrid combination of symptom-based and functional reasoning. An efficient real time control method for part routing using fuzzy parameters is introduced by Ben-Arieh and Lee in the fifth chapter. Yeralan and Tan, in the sixth chapter, focus on high-level control language for embedded microcontroller applications. Their study investigates the suitability of fuzzy-logic control as a general-purpose high-level language for programming embedded microcontroller applications. An application of neural networks to the monitoring and recognition of both periodic and aperiodic process signals that may be encountered in manufacturing processes is presented in chapter 7 by Hou and Lin. In the eighth chapter, Indurkhya *et al.* report the results of study on modeling of Electro-Discharge Machining (EDM) and Wire Electro-Discharge Machining (WEDM) processes through an artificial neural network. The ninth chapter, by Shashikumar and Kamrani, develops a knowledge-based expert system for selection of industrial robots using a LEVEL5 expert system shell. Chapter 10, by Chung and Vassiliadis, reports on the development of a case-based knowledge system called ShootDem-Ks which effectively troubleshoots product failures without the need of specially trained personnel. Kamel and Ghenniwa, in chapter 11, introduce a general model of multi-machine systems that is able to represent different multi-machine environments including partially-overlapped systems. This model is based on the system's structure and its processing characteristics. The use of an object-oriented paradigm as a means for the design and implementation of an intelligent process planning system is proposed by Usher in chapter 12. Dong *et al.*, in chapter 13, present a prototype feature-based automated process planning (FBAPP) system to avoid costly traditional design and manufacturing processes later, and to make most product development decisions during the early stage of design. Chapter 14, by Nnaji and Kang, discusses the role of the computer-aided design for an automatic machine programming system. In the last chapter of this volume, Troxell *et al.* develop a framework to perform computer-assisted fault diagnosis of manufacturing processes.

We are indebted to our authors and reviewers for their outstanding contribution and assistance in preparation of this volume. We would also like to thank Dr. Herman R. Leep of the University of Louisville for his invaluable support and advice. Special word of thanks are due to Dongke An for providing exceptional help to make this endeavor possible. Finally, We would like to express our deepest gratitude to Bernard Goodwin and Michael Hays of Prentice-Hall for giving us the opportunity to initiate this project.

Hamid R. Parsaei
Mo Jamshidi
December, 1994

1

WORKCELL PROGRAMMING ENVIRONMENT FOR INTELLIGENT MANUFACTURING SYSTEMS

A. Jana, Southern University

D.M. Auslander, University of California at Berkeley

1.1 INTRODUCTION

A workcell programming environment to model and implement prototype intelligent manufacturing systems is being developed for education and research. The core of the integrated manufacturing system is the control software, which monitors and controls all of the activities of the workcell on real time basis. It is extremely difficult to generalize the manufacturing equipment required in a workcell, but for the purpose of training as well as research a set of typical equipment in benchtop versions may be used to create a test bed for intelligent manufacturing systems research. The system uses standard equipment such as robots, milling machines, material handling equipment, vision system etc. for training and research. This chapter discusses some of the independent machine control software, along with the communication protocols, the interfacing methodologies, and the control structures which are being used in order to implement a real time intelligent manufacturing system.

Extensive research is going on to develop a generalized control structure to implement intelligent manufacturing system, but the full structure has yet to emerge. The main difficulty in this kind of development is the dynamic nature of the manufacturing problem. This can be attributed to the diversity of products, variety of manufacturing equipment, numerous types of production processes, wide range of manufacturing hardware and software suppliers, and also to the variations in market demand. Expensive test-beds are needed for implementation and research with standard industrial quality integrated manufacturing systems. Although these test-beds are essential for industry and research, they are too complicated to maintain and use for academic environments. In this work, an effort is being made to develop a low-cost manufacturing workcell and a workcell programming environment to support implementation of intelligent manufacturing systems for education and research. The workcell programming environment consists of various independent software design and implementation tools developed in the C programming language which can be integrated to model intelligent manufacturing systems on a real time basis.

The work described here provides a supporting environment for the implementation of intelligent manufacturing systems by creating a software environment for carrying out complex manufacturing procedures. An integrated intelligent manufacturing system is defined by Rao and Luxhoj [1991] to be a large knowledge integration environment that consists of several symbolic reasoning systems and numerical computation packages, which are controlled through a meta-system for timely implementation. The main objective here is to use the software environment to explore the application of complex decision making processes related to real time manufacturing systems using low cost test-beds in academic environment. By providing a relatively easy-to-use software development environment, students can implement such systems in ways that would be impossible with standard industrial equipment.

In this chapter, first we discuss the manufacturing workcell, then we discuss the real time machine control software which has been developed, we then introduce a design methodology suitable for machine control software, and finally we present the control network and communication aspects of this work with future directions of our research.

1.2 Manufacturing Workcell

Researchers are attempting to develop manufacturing integration frameworks that seamlessly tie together people, software systems, machines, and activities across a large, geographically dispersed manufacturing enterprise. A framework can improve manufacturing flexibility by enhancing the functionality and inter-operability of the component applications in various significant ways [Glicksman et al, 1991]. In general, intelligent manufacturing frameworks are designed in two levels. A high level consisting of a manufacturing knowledge base that can be viewed as a meta-CIM system, and a low level that deals with equipment interfaces, operations and control. As an example, in a CIM environment the detection of process error is done in the low level whereas the error recovery process is initiated in the high level.

Real time considerations are normally relatively fast (in milliseconds) in comparison to high level operations (seconds and even minutes). This chapter deals mostly with low level implementation methodology.

A workcell is needed for testing and implementation of intelligent, integrated manufacturing. In academic settings, a low-cost test bed can be based on benchtop versions of robots, CNC machines, vision systems, and material handling equipment. In addition to being low-cost, these machines are easier to use and safer than their industrial counterparts, and take up much less space. Because of the closed nature of industrial CNCs (computer-numerical controllers), they are not suitable for the kind of low level work students need to do to fully understand the operation of intelligent manufacturing systems. The educational system being constructed for the work described here is programmed in a standard language (C) and allows students access to all of its low-level functionality.

The workcell will consist of four robots, two milling machines, one stamping machine, and seven personal computers. The material handling equipment consists of a conveyor, a rotary table, and a model cart, which can be used for part transfer. A computer vision system various electronic interface modules, a multi-axis motor controller (six axes to control the milling machine, conveyor, and rotary table) and data acquisition systems have been provided to model sensor-based, real-time manufacturing systems. The layout of the manufacturing workcell is shown in Figure 1. Most of this equipment has already been installed.

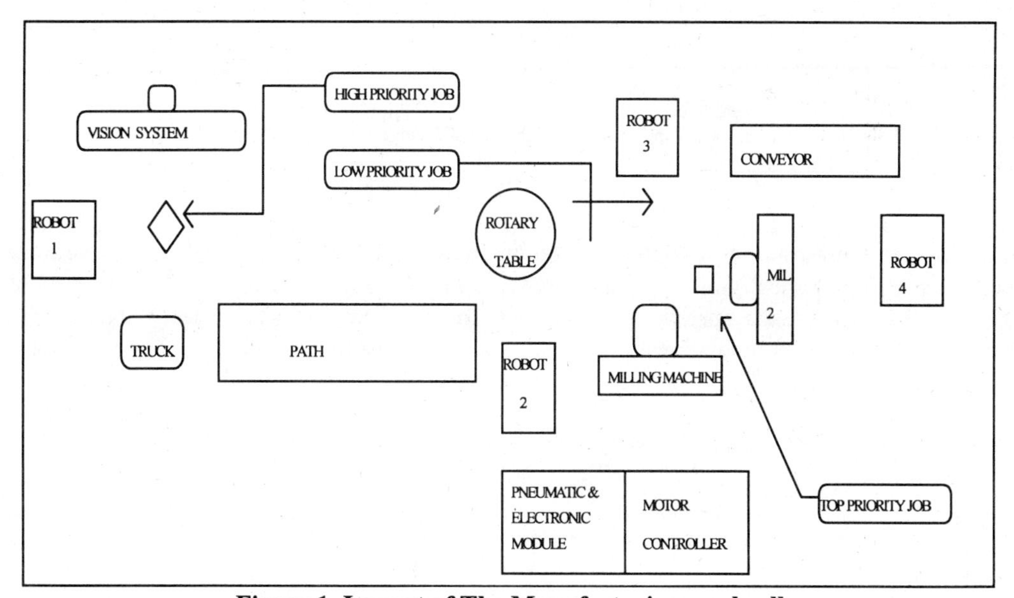

Figure 1. Layout of The Manufacturing workcell

1.3 Workcell Programming

The programming interface consists of different modules to support programming of individual components, world modeling, serial communication and also parallel communication through the printer port. Both the milling machine and the robot are connected to the serial port, and can be communicated with via a built-in command set that is interpreted by the robot as well as the milling machine controller. The following is the description of the different machine control software. The software models listed here are suitable for use of these elements in standard, synchronous programs. Modifications to them are suggested in a later section to make real time applications more effective.

1.3.1 Robot Programming

Most of the robots used here are five axis articulated manipulators, driven by stepper motors. An additional motor is provided for gripping. The functions used here are mainly for point to point control of the robot motion. The points in the robot space (robot coordinate frame or absolute coordinate) can be specified by the x,y,z coordinates (to move to a point with a fixed orientation), by a transformation matrix (x, y, z and the components of the orientation vectors), or by specifying the five joint angles. The most common low level routines, which are available for robot and gripper motion, and also for robot control are shown in Figure 2.

```
move_angle_to( speed, ang1,ang2,ang3,ang4,ang5,ang6)
move_x_y_z(speed, x, y, z)
move_increment_angle(speed, inc1, inc2,inc3,inc4,inc5)
move_trans_matrix(speed, trans_matrix[])
go_home(speed)
open_gripper(speed, opening)
grip_force(speed, force)
check_size_and_close(speed) /* returns the size of the object it is holding */
int_com1() /* to initialize serial communication */
```

Figure 2. Basic Robot Programming Routines

In addition to these routines, a high level routine is available to move the robot to a particular predefined point by specifying the point number, such as *go_to(speed, point1)*, where *point1* can be defined by a teach operation. In that case the robot motion will be faster than *move_..* commands, since less computational time is needed. The most basic routine for the robot motion is *move_angle_to();* the steps needed to execute this routine are given in Figure 3.

1. Use **@READ** command (controller specified) to find the current *motor steps*
2. Convert the *motor steps* to the corresponding *joint angles*. These are the *incremental angles* form the *home angles.*
3. Add *home angles* to calculate the *current angles*
4. Subtract *current angles* from the input angles to find the *increment in angles* needed to reach the *specified angles.*
5. Convert these *incremental angles* to the corresponding *motor steps.*
6. Use **@STEP** command (controller command) to move the necessary *motor steps.*

Figure 3. Steps to Execute *move_angle_to()* Routine

```
#include <mbot.h> /* header file containing robot control routines */
main()
{
float size;
        constant_angles(); /* set the values of the home angles */
        int_com1(); initialize();
        for(;;) {
        move_x_y_z(240,8.0.0.0,1.5); /* move to the top of the block */
        open_gripper(240,1.5); /* 1.5 inch to get 1.0 inch object */
        move_x_y_z(220,8.0,0.0,0.5); /* around the block */
        size = check_size_and_close(220);
        if ( size < 0.2) { printf("no block, going home\n");
                           go_home(240); }
        else {
              grip_force(200, 0.5); /* developing gripping force of 0.5 lbs */
              move_x_y_z(220, 8.0,0.0,2.5); /* go up by 2.5 inch */
              go_to(220, point[2]); /* point[]s are peviously saved points */
              go_to(220, point[3]); /* point just above the placing point */
              grip_force(220, -0.5); /* remove force */
              open_gripper(220, 0.5); /* the block will drop at the desired point */
              move_x_y_z(240, 8.0,4.0,2.5); /* same point as point[2] */
              go_home();
             }
        } /* for loop */
}
```

Figure 4. Robot Program

An example of a robot program using these software tools is shown in Figure 4. The task of the robot is to pick up a block (a cube of 1 inch side) from point 8,0,0 (x,y,z in inches) and to place it at 8,4,0 starting from home position. If the block is not available the robot will indicate it to the user and go home.

1.3.2 Milling Machine Programming

The milling machine is controlled by the six-axis motor controller, three of which are used to control the three axes x,y,z of the milling machine. The additional three axes can be used to drive another milling machine, or a lathe, or accessories such as the conveyor and rotary table. The motor controller can be connected to the computer through the RS-232 serial port. The controller defined commands like robot controller are also available to program the milling machine. A brief summary of the most common routines needed to program the milling machine is given in the Figure 5.

```
inches_to_step(inches, step);
mill_step(speed, j[]); /* rotate the mill motors by j[1],j[2],j[3] steps */
mill_spindle_on(); and ...._off(); /* spindle on and off */
current_x_y_z(c_pos[]); /* x,y, z of the current position and other status */
go_spindle_abs(speed, x, y, z); /* table x,y and spindle z */
go_spindle_inc(speed, ix,iy,iz); /* incremental motion */
spindle_go_home(speed); /* convenient position to enable loading and un loading */
```

Figure 5. Milling Machine Routines

Using these functions, a simple program to machine a rectangle (0.02 inch depth of cut) can be developed as shown in Figure 6.

1.3.3 Other Accessories

Other accessories like conveyor and rotary table can also be programmed using similar types of routines like *run_conveyor(speed, length)* and *rotate_rotary_table(speed, angle)* which are available in the header file **"cnc.h"**. The automatic job holding device located on the milling machine table is a pneumatic cylinder, and can be activated by the functions *mill_vice_on()* and *mill_vice_off()*. In the similar way the pneumatic cylinder of the stamping machine can also be activated by *stamping_on()* and ..._*off()*.

1.3.4 Cart and Sensors

The cart used in this manufacturing workcell is driven by a DC motor, and has been built from a commercial model building kit. This kit is equipped with various motors, limit switches, potentiometers, gears, wheels, electromagnets and photo electric sensors. The system can be controlled by the computer through a dedicated electronics interface box connected to the printer port. All together four motors, eight limits switches, and two photo sensors or two potentiometers can be connected to each interface. Various software tools have been developed to integrate this kit with the manufacturing workcell. Different routines related to this kit and a sample program are shown in Figure 7.

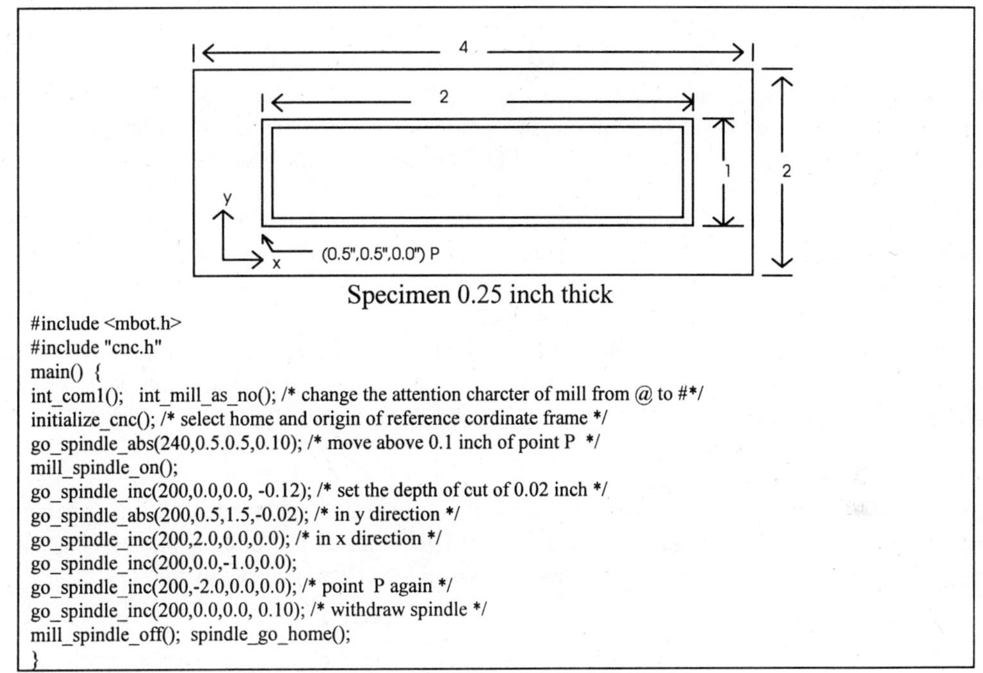


```
#include <mbot.h>
#include "cnc.h"
main() {
int_com1(); int_mill_as_no(); /* change the attention charcter of mill from @ to #*/
initialize_cnc(); /* select home and origin of reference cordinate frame */
go_spindle_abs(240,0.5.0.5,0.10); /* move above 0.1 inch of point P */
mill_spindle_on();
go_spindle_inc(200,0.0,0.0, -0.12); /* set the depth of cut of 0.02 inch */
go_spindle_abs(200,0.5,1.5,-0.02); /* in y direction */
go_spindle_inc(200,2.0,0.0,0.0); /* in x direction */
go_spindle_inc(200,0.0,-1.0,0.0);
go_spindle_inc(200,-2.0,0.0,0.0); /* point P again */
go_spindle_inc(200,0.0,0.0, 0.10); /* withdraw spindle */
mill_spindle_off(); spindle_go_home();
}
```

Figure 6. Milling Machine Programming

```
call_init(); /* initialize printer port communication */
run_motor_1(clockwise);
run_motor_2(counter_clock_wise);
run_motor_3(halt);
drive_cart(forward);
drive_cart(reverse);
drive_cart(stop);
switch_status(switch1); /* returns 1 for on 0 for off */
switch_status(switch2);
analog_in( line1); /* returns an integer value between 0 - 255 depending on the resistor value of
                the photo sensor or the potentiometer */
analog_in(line2);
/* sample program to run the cart as long as the switch number 1 is open */
#include <f_tek.h> /* header file for model building kit */
#include <f_tek.c> /* functions and extern variable declaration */
main()
{
call_init();
while( switch_status(switch1) == 0)
   {
    drive_cart(forward);
   }
drive_cart(stop);
}
```

Figure 7. Programming Sensors and Cart

1.4 Data Acquisition

Data acquisition systems have been implemented in this manufacturing cell to monitor the manufacturing processes, mainly the operation of milling machine, through the commercially available data acquisition cards like Metrabyte DAS08 & DAC 02, or Real Time Inc., ADA 2000 etc. To facilitate integration of sensors, an I/O interface box has been developed so that the general lab instruments can be connected easily for testing. Various low level routines such as *d_in()*; to read a digital value from the parallel input port, *d_out(integer)*, to write a value to the parallel port, *a2d(int channel_number)*; analog to digital conversion, and *d2a(int channel number, double volts)*, digital to analog conversion etc., have been developed for several standard low cost data acquisition boards. A sample program along with the setup for testing is shown in Figure 8. In this program an analog voltage is read through the I/O interface box, and the instantaneous value of this voltage is converted to the analog value and is compared with the input using an oscilloscope. Programs of this sort are used by the students as templates to ease development of their own programs.

We are in the processes of installing four load cells on the table of the each milling machine to monitor the machining process. These will be used to monitor the cutting process for use in diagnostics, for example, for worn tools, and in adjusting the cutting speed.

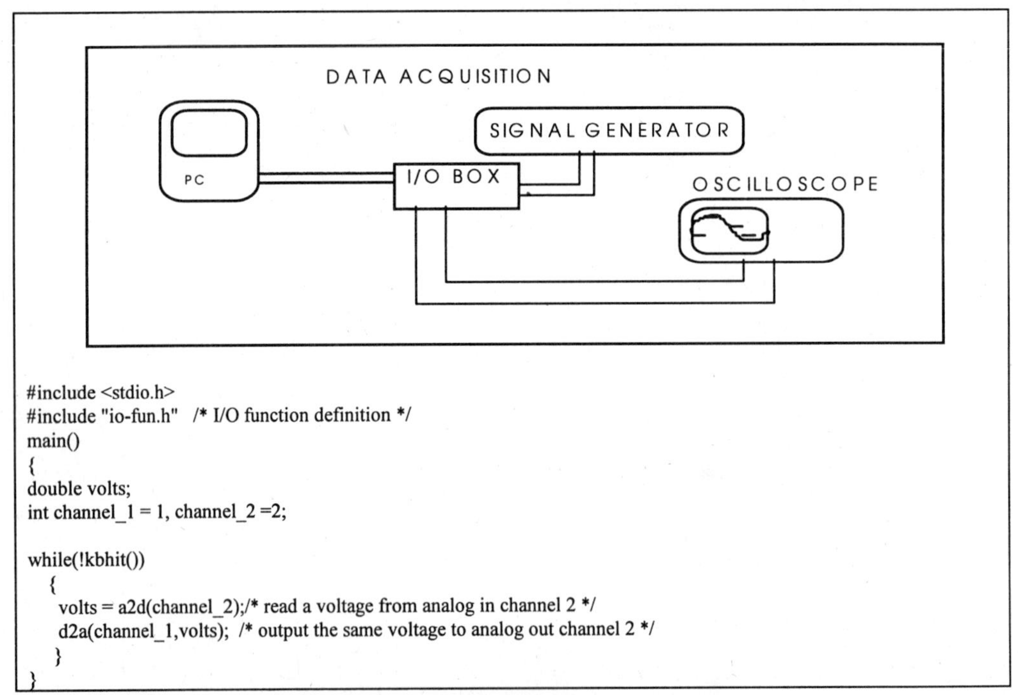


```
#include <stdio.h>
#include "io-fun.h"   /* I/O function definition */
main()
{
double volts;
int channel_1 = 1, channel_2 =2;

while(!kbhit())
   {
    volts = a2d(channel_2);/* read a voltage from analog in channel 2 */
    d2a(channel_1,volts);  /* output the same voltage to analog out channel 2 */
   }
}
```

Figure 8. Data Acquisition Programming

1.5 Real Time Programming

Real time computer systems should deliver the results in correct coordination with other systems operating asynchronously to the computer and to each other. A substantial methodology has grown up around the design and implementation of real time systems, but most of the available software is commercially proprietary and is only portable to the extent provided for by the manufacturer. Many simple real time problems can be solved by *synchronous* programming. Typical programs of this class have a section to initialize data, place physical devices in appropriate initial states and then run the program in an endless loop. All the programming examples discussed so far have been of this type. These programs are sequential or constructed with a *single-thread.*

However, to achieve a true multitasking environment in a time critical or in event driven situation, *asynchronous*, or *multi-threaded* programming is needed [Auslander and Tham, 1989]. Asynchronous programs are implemented by *interrupts,* which are hardware mechanisms in the computer that allow for the interruption of one thread of execution by another, higher priority thread. The terms *foreground* and *background* are often used in connection with the high and low priority sections of such programs. However, business use of the terms and engineering use are exactly contradictory, so we will avoid using them here.

Figure 9 shows a program template that could be used by students to build a data acquisition program or control program that requires an action taking place on a strict time schedule plus another activity that is not time critical. In a data acquisition program, for example, the time critical portion, the *interrupt service routine (isr)*, would be used to get the data from the relevant instruments. The *main()* function is the non-time-critical section. It might be used to get commands from the user as to when to start and stop the data acquisition. In a control program, the time-critical section would be used to implement a feedback loop (a PID control, for example) while the non-time-critical section might be used to get new setpoint commands from the user. Because it is connected to the interrupt mechanism, the interrupt service routine will preempt the CPU resource whenever the clock interrupt signal is present. Execution of the interrupt service routine will then continue until it is done, at which time execution of the non-time-critical section will resume.

Student programming efforts are aided by the *xignal* package, which takes care of all of the details of setting up the interrupt. The call to *xignal()* sets up a connection to the interrupt system so that the function *isr()* will be run whenever the clock interrupt occurs.

```
#include <stdio.h>
#include <8259.h>
#include <xignal.h>
#include <alarm.h>
#include "io-fun.h"  /* I/O function definition */
#define TIME  10.0  /* 10.0 millisec */
void isr();

main()
{
xignal(XIGTMR, isr);  /* setup interrupt service routine */
setalarm(TIME);       /* at an interval of 10.0 millises */

while(!kbhit())  /* Wait for user keyboard input to stop */
   {              /* non-time-critical process */
    /* Put code here that can be interrupted */
    disable();  /* Turn off the interrupt */
       /* Put code here that cannot be interrupted */
    enable();   /* Turn the interrupt back on again */
   }
/* User has given "done" signal -- put computer's interrupt and timing system
back to normal */
disable();
xignal(XIGALL,XIG_DFL);  /* Set interrupt vectors to default */
setalarm(-1.0);  /* Set clock back to default */
enable();
}
void isr(void)  /* Time-critical process */
{
 /* Put code here for the time-critical (interrupt-driven) task */
}
```

Figure 9. Asynchronous Implementation of Data Acquisition Program

1.6 State Transition Logic

The programming methods described above serve well for designing relatively simple systems, for example, a single machine using some form of process sensor. In that case, the sensor is connected and an interrupt task, and the non-interrupt portion of the program is used to operate the machine tool. This level of complexity provides an excellent introduction to the operation of basic manufacturing equipment.

When more complex problems are encountered, however, these software methods do not give enough organizational support to the design or implementation process. Further organizational support is needed for the production of software that is reliable and easily maintained, and, also, is portable across a variety of computing environments [Auslander et al, 1993]. The organization is accomplished by separating the engineering themes associated with the design from the computational implementation issues. Each of these is then separated into two levels:

Engineering: tasks, states
Computation: processes, threads

Tasks separate major activities of the control system. The tasks can run in parallel, so as many tasks as are relevant can be active at any given time. States specify the particular activity a task is currently engaged in. Thus, each active task is in one state at a time.

Processes describe computational modules that execute completely independently. In the manufacturing system control environment, they are most commonly defined for computational modules executing in independent computers, although independent processes can share the same computer. Threads are only semi-independent. Several threads can be resident in the same computer, and all threads in a process share a memory address space.

For the purposes of this presentation, we will concentrate on the definition of states since that follows most directly from the software environments already defined and is where the most complex organization is described. By separating the engineering from the computational implementation, the goals of specifying system operation in such a way that any engineers involved in the project can understand the detailed machine operation and strong modularization of the software are achieved. This, for example, enables substantive design review activities to take place; otherwise, with so much of the design expressed in a computer language that is not accessible to all engineering personnel, design reviews cannot probe deeply enough to be meaningful.

State transition logic is used to represent the internal task structure. It is widely used in synchronous sequential logic, where it represents a mathematical formalism. It is a specific case of a finite state machine. When applied to software development, it is used as a design formulation and documentation tool and does not carry the mathematical formalism it has in sequential logic. The *state* carries the major burden of the design process, and is an identifiable activity of the mechanical system. It is the basis of an engineering systems description of machine operation, not purely a software tool. Transitions occur to initiate new system activity by moving from one state to another. *Transitions* are usually caused by external events such as switch closures, voltage reaching predetermined limits, a clock running out, etc., or by internal events such as reaching a desired count. Transitions take the system to a new operating state. The notation to represent transition logic diagrams as well as a transition logic diagram to represent a simple interrupt task is shown in Figure 10.

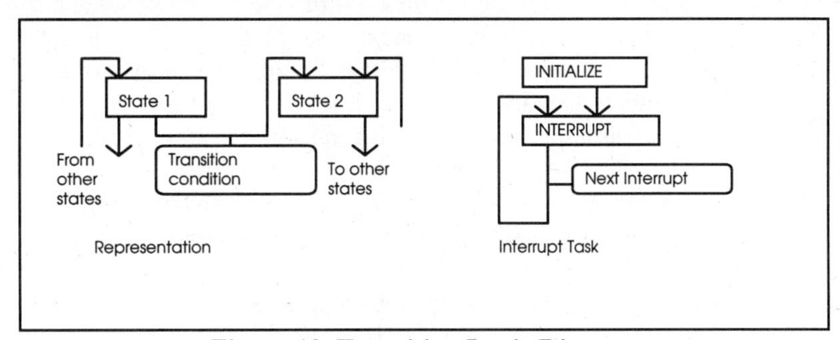

Figure 10. Transition Logic Diagram

The program code developed to implement transition logic is grouped by state, where each state has a specific set of functions associated with it:

Entry
Action
Transition test
Exit

The *entry* function is called when the state is first entered and is usually reserved for initialization of events which are to occur within the current state. The *action* function is called repeatedly, after the entry function, until the transition condition has been satisfied. The *transition* functions test for whether a specific transition should occur or not. There is a separate transition function for each possible transition associated with a state. In the case of multiple branch connections to other states the transition condition will provide unique information as to which branch will be taken. When a transition takes place the exit function associated with that transition is executed. These functions are transition specific, so can be used both for clean-up of activities related to that state that was executing and for initialization that are transition-specific rather than state-specific.

The programming is implemented by *scanning.* Each scan by the master program executes the set of functions associated with the current system state for each task. In order to make sure all possible transitions are considered, the state related functions must be of *non-blocking* type. This means that none of the functions can wait for events. They can test, but must not wait, Waiting is accomplished through the over all logic and is realized by multiple scans. The minimum scan rate depends on the critical time of the events coded. If the computer is capable of completing a scan in much less than critical time, more than one transition logic diagram can be implemented simultaneously or some other non-blocking task can be executed. If the scan is too slow for some high priority activities, more sophisticated computational environments can be use to move the high priority tasks into separate computing threads, or even separate processes. If separate processes are used, issues of communication must be faced in order to complete system implementation (see next section).

To meet the condition that all code be of non-blocking form, the machine tool and robot function calls listed above must be modified slightly. Instead of waiting for the operation (for example, a move) to complete, the functions return immediately. Additional functions are added to test for completion of the commanded operations. These functions are used in the transition test functions to determine when it is appropriate to make a transition to another state.

In order to implement real time control and monitoring of the manufacturing process, a set of transition conditions is being developed to implement between each operation (*state*) of the production machines, and also after each (or a block of) robot program step(s). While the non-interrupt task(s) is controlling the equipment, the interrupt (time-critical) task is updating all the possible manufacturing workcell system variables by scanning the relevant instruments. Each machine control computer updates the critical system variables, and the current status of all the variables are available to each computer. For example, if a task is to be terminated due to wrong machining (due to improper job location on the milling table, detected by the sensors after a while from the start of the operation) transition condition scan will find it by checking the status of the system variables and stop the process. Then, the error recovery state will start.

1.7 Communication and Control

The complete workcell is planned to have seven PCs. A distributed control architecture has been implemented to control the workcell activities. Figure 11 shows the complete network. All the computers will be connected in a standard network. The serial and parallel ports are used for connecting the computers to "smart" devices like robot and machine tool controllers. However, the robots used here are provided with two serial ports --*right* and *left*. The *right*

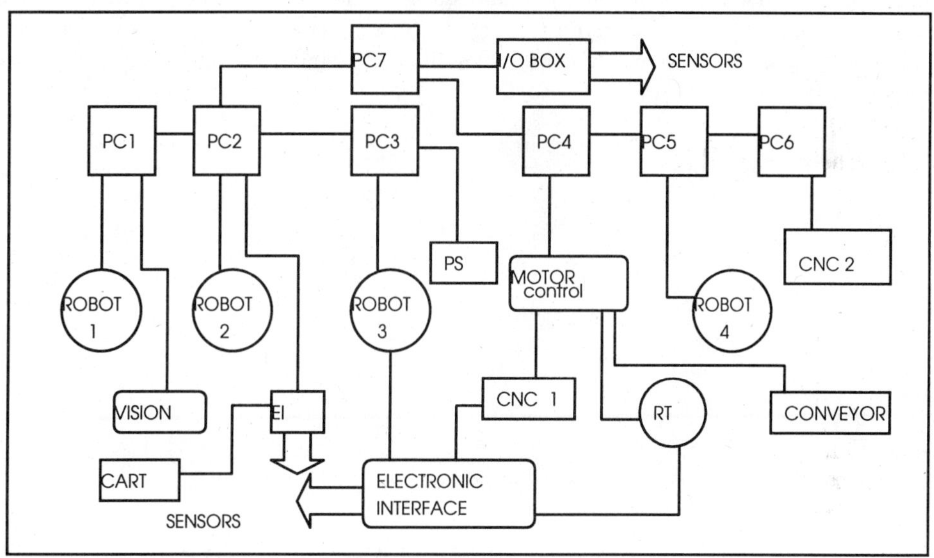

PC: Computer; PS: Photo Sensor; CNC: Milling; EI: Electronic Interface; RT: Rotary Table

Figure 11. Control Network

port is used to receive messages, but the robot controller passes any signal which is not sent for it through the left port. This facility can be used as a simple network to control several devices. The main communication objective is to update the critical system variables to implement transition logic in order to achieve real time control of the manufacturing cell. For this, software is being developed to facilitate programming of the data communication between computers. The software will work on the principle of pseudo-shared memory. That is, variables on different computers are used in the program as if they were in the same program, while the communications software keeps them updated across the network. In this method, some variables are defined as belonging to an (imaginary) memory that is supposed to be shared between all the machines. Whenever the variables value changes on one of the machines it gets updated on other machines within a certain period. Information in this software is passed back and forth in the form of data packets.

1.8 Implementation

The full integrated manufacturing system is built around jobs at three priority levels: high, medium and low. The first “products” to be made by the system will be lettered signs milled on plastic-coated plywood blocks. Different lettering jobs will be assigned to the available priority levels.

The jobs of different priority use separate raw-material entry points. The low priority job uses a gravity feeder, the medium priority job feeds through a vision system and uses a mobile cart for material transfer, and the high priority job triggers a photo-cell to indicate its presence. Use of these various raw material input mechanisms gives students a feeling for the variety of material handling methods that can be used in manufacturing systems. Figure 12 shows the transition logic used to describe the activities of a milling machine. This logic diagram considers the case of two priority levels. Students would start with a diagram of this sort and modify to increase the complexity.

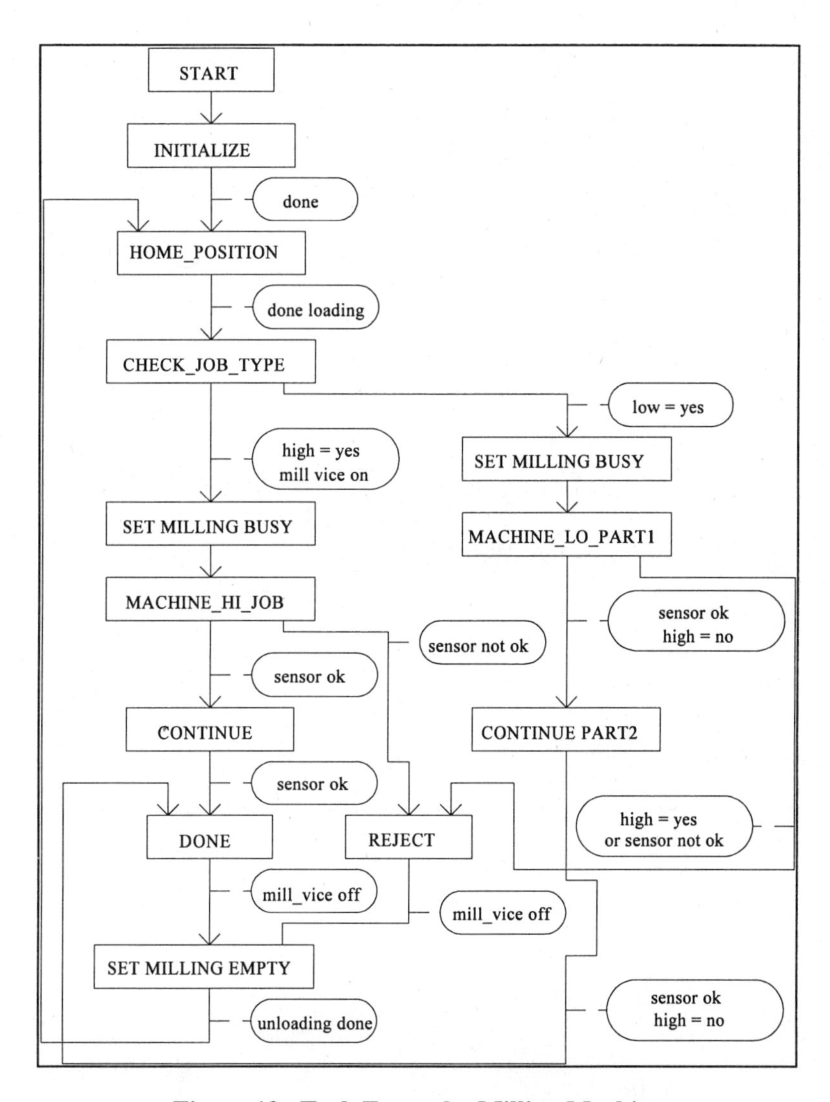

Figure 12. Task Example: Milling Machine

For each of the loading points, a different means of moving the blanks to the milling machines is used. For example, the medium priority part is first detected by the vision system, which measures its size and orientation. It signals this information to a robot, which places the part on the cart, in proper orientation. The part then moves from there to a rotary table, and finally to another robot for loading into the milling machine. Figure 13 shows a typical transition logic description for a robot control that is used for the material handling.

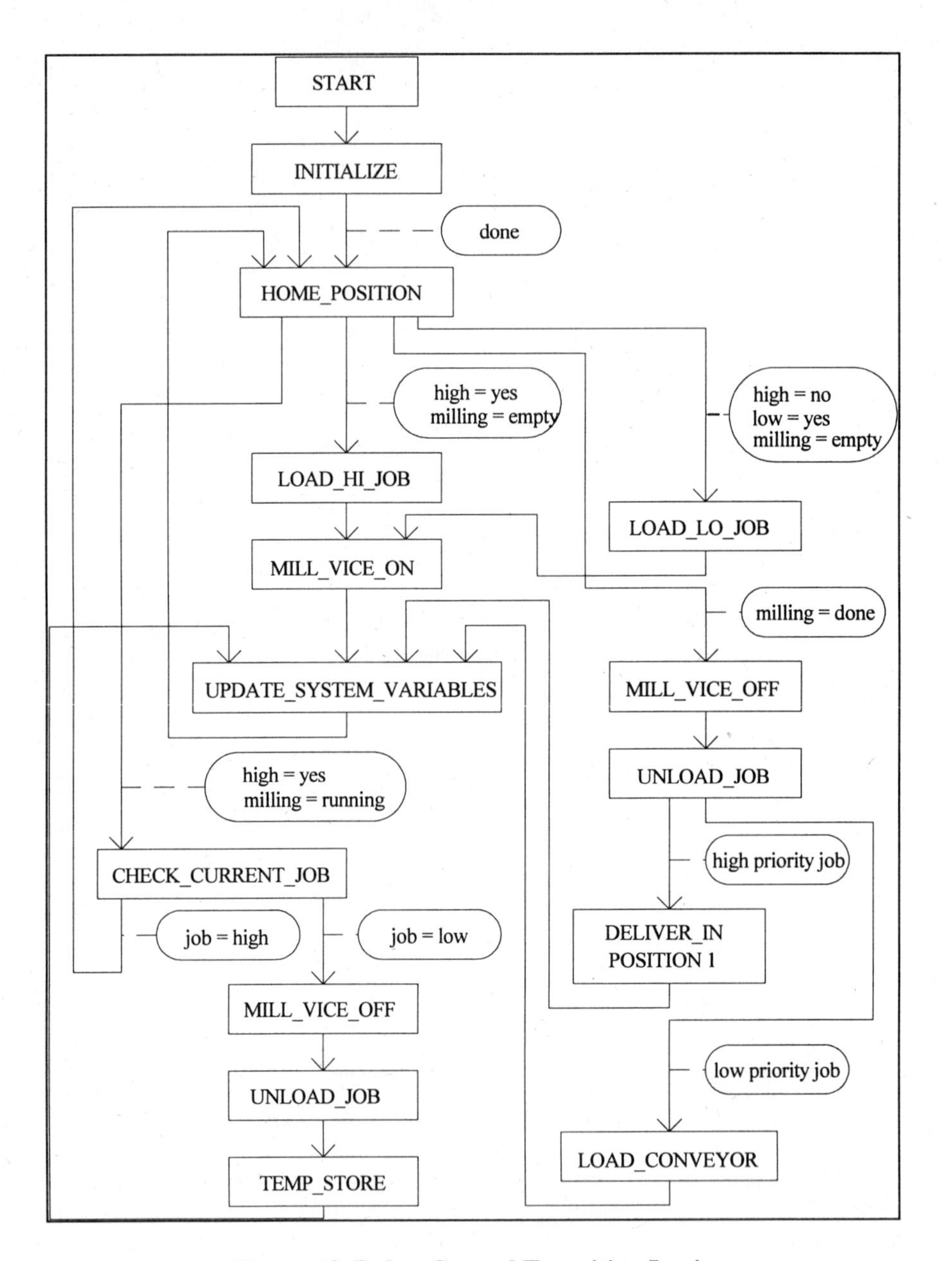

Figure 13. Robot Control Transition Logic

The finished parts are delivered through a conveyor to another vision system where they are inspected. Parts that pass inspection are placed in the delivery queue, while other parts are sent to a rejection area.

During the system operation, the transitions in the control tasks (as shown in the two figures, above) use information from system sensors to determine which transitions to take. This data comes from the sensor tasks which operate at high priority as interrupt-driven program elements.

1.9 Conclusion

Most of the equipment of the manufacturing workcell discussed here have been integrated in the teaching lab. The second milling machine and the fourth robot have not been installed yet. At this stage our main concentration is to develop communication software, as well as to refine and document the available software. Most of these work is related to the mechatronics and manufacturing curriculum development project supported by the Synthesis Coalition, a National Science Foundation Project. All the software and the related materials will be available through the National Engineering Education Delivery System (NEEDS).

1.10 Acknowledgments

The authors are grateful for the support from the NSF Synthesis Coalition and the IBM corporation. Undergraduate students Adrian Wallace, Randolph Williams of Southern University and graduate students Mark Lemkin and An-Chyau Huang of UC Berkeley are involved with various aspects of this project.

1.11 References

[1] Auslander, David M., and C.H. Tham, *Real-Time Software for Control,* Prentice Hall, Engle wood Cliffs, NJ, 1989

[2] Auslander, David M., Hanidu, G., Jana, A., Landsberger, S., Seif, S. and Young, Y.,"Mechatronics Curriculum in the Synthesis Coalition," *Proc. of 1992 IEEE/CHMT Int'l Electronic Manufacturing. Tech. Symp.*, Baltimore, MD, Sept. 1992 , pp. 165-168.

[3]Auslander, David M., Lemkin, M., Huang, A., "Control of Complex Mechanical Systems"., *IFAC 1993*, Australia

[4] "Research, Services, and Facilities", NIST Special Publication #817,National Institute of Standards and Technology, MD, August 1991.

[5] Ulsoy, Galip A., and DeVries R. Warren, *Microcomputer Applications in Manufacturing*, John Wiley & Sons, NY, 1989.

[6] Bonner, S., Shin, K. G., "Comparative Study of Robot Programming Languages", Computer, Vol.15, No. 12., Dec. 1982, pp 82-96.

[7] Glicksman, J., Hitson, B. L., Pan, J. C., Tenenbaum, J. M., "MKS: A Conceptually Centralized Knowledge Service for Distributed CIM Environments", *Journal of Intelligent Manufacturing*, Vol-2, No.1,February 1991, pp 27-42.

[8] Robotics, CNC, CAD/CAM, Computer Integrated Manufacturing , Catalogue No. A51, Feedback Incorporated., Hillsborough, NC.

[9] Koren, Y., *Computer Control of Manufacturing Systems*, McGraw-Hill Book Company, New York, 1983.

[10] Rembold, U., Nnaji, B. O., Storr, A., *Computer Integrated Manufacturing and Engineering*, Addison-Wesley, UK, 1993

[11] Jana, A., Cheng, S. W., Chehl,S. S.,"Development of a Flexible Manufacturing Workcell", *9th International CAD/CAM, Robotics and Factories of the Future Conference*, Newark, NJ, August 18-20, 1993

[12] Rao, M., Luxhoj, J.T., "Integration Framework for Intelligent Manufacturing Process", *Journal of Intelligent Manufacturing*, Vol-2, No.1,February 1991.pp 43-52.

2

AN INTELLIGENT SYSTEM FOR AUTOMATING THE INSPECTION OF MANUFACTURED PARTS

Michael D. Reimann, The University of Texas at Arlington
Joseph Sarkis, The University of Texas at Arlington

INTRODUCTION

Computer aided process planning (CAPP) is used in computer integrated manufacturing (CIM) environments to automate the linkage between the design and manufacturing phases of product development. A general conceptual functional framework for CAPP, that was developed by the Consortium for Advanced Manufacturing - International (CAM-I), will be described in this chapter. This general framework has been given the title of the CAM-I Advanced Numerical Control (ANC) [1] framework. The ANC framework specifies the structure and defines the functions necessary for automating the process planning of machined parismatic parts. The ANC framework also provides the conceptual foundation for developing an architecture for automating the inspection process planning activity.

There has been much work that has focused on CAPP in the research and practitioner literature [2,3,4], but the work that has focused on automating the inspection process planning function within the CIM environment has been much more limited. The major focus of this chapter is on the development of a functional framework for the automation and linkage of the inspection process planning activity (defined as the Expert Programming System-1 (EPS-1)). This activity has gained more attention now that quality, reliability and responsiveness can be used as a competitive advantage for manufacturing enterprises. Also with various total quality management (TQM) initiatives, the automation and integration of the quality function is even more imperative for successful TQM implementation.

To be able to show the conceptual development of the EPS-1 framework, we separate the chapter is into four major sections. First, the overall structure and functional components of the ANC framework will be described. The ANC framework which is made up of nine functional modules, has only been utilized as a conceptual model as of this time. The next section outlines a derivative of the ANC framework that is used for the EPS-1 framework. This provides a function to function mapping. The EPS-1 framework is compared to the basic ANC framework to describe its evolution. External components linking the automated inspection planning framework to a CIM environment are then presented. The final section summarizes major issues identified in this chapter and draws conclusions relative to practical implementations.

THE ANC FRAMEWORK

The objective for the development of the ANC framework is to functionally represent the automatic generation necessary for numerical control (NC) instruction sequence and supporting information for producing a finished parismatic part. Thus the ANC is intended to automatically link a computer-aided design (CAD) function with a computer-aided manufacturing (CAM) function in a CIM environment, or what has been defined as the CAPP function. In theory, all that should be required from the ANC user would be the identification of a part which exists in an external part model database. From the part definition, a complete process plan to machine a finished part is derived for the removal of volume from an initial raw stock. The structure and functional components of the ANC framework are illustrated by the $IDEF_0$ functional model shown in Figure 1.

$IDEF_0$ is a systems analysis and design technique that was developed by the Air Force's Integrated Computer Aided Manufacturing (ICAM) [5] program. IDEF stands for ICAM Definition and $IDEF_0$ is used to represent the *functional* framework of a system. There are five elements to a $IDEF_0$ functional model, the activity (or process) is represented by the boxes; inputs are represented by the arrows flowing into the left hand side of an activity box; outputs are represented by arrows flowing out the right hand side of an activity box; arrows flowing into the top of the box represent constraints or controls on the activities; the final element is the mechanism that performs the activity and these are represented by arrows flowing into the bottom of the activity box.

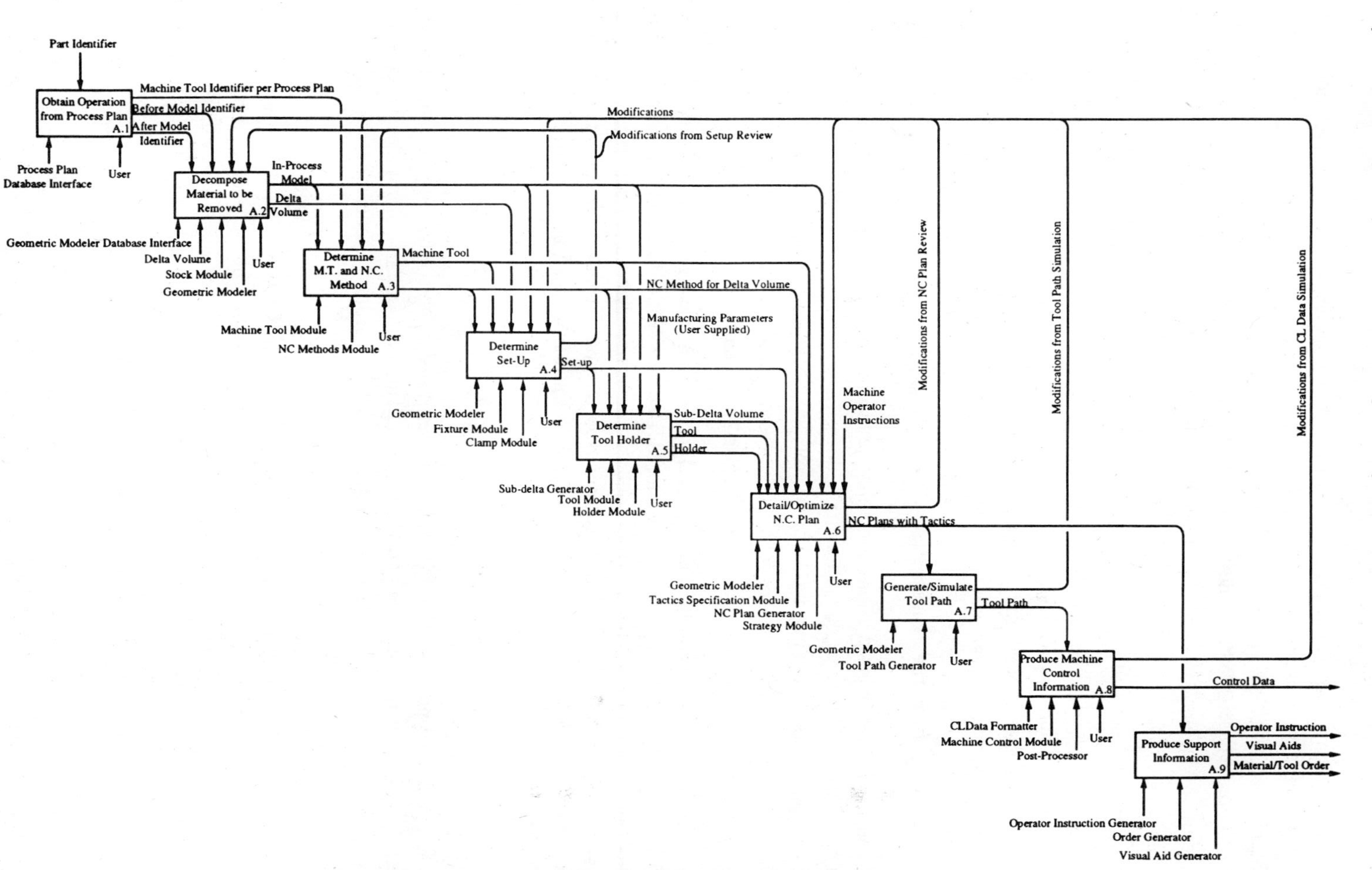

FIGURE 1: High Level $IDEF_0$ Diagram for ANC Framework

Each of nine functions within the ANC framework will be described in the remainder of this section. These functions include

- Obtain Operation from the Process Plan
- Decompose Material to be Removed
- Determine Machine Tool and NC Methods
- Determine Set-Up
- Determine Cutting Tool and Holder
- Detail/Optimize NC Plan
- Generate/Simulate Tool Path
- Produce Machine Control Information
- Produce Support Information.

Obtain Operation from the Process Plan

This function obtains basic information about a machined part from the ANC user. Data that are requested from the user by this function includes:

- The identification of the part model before the NC machine operation
- The identification of the part model after the NC machine operation
- A specification of a NC machine tool to be used (optional)

This information is used by subsequent functions in the ANC framework. The part identification numbers are used by the Material Decomposition function to retrieve various geometric models of the part and determine delta volumes necessary to produce intermediate and final part shapes. If the machine tool is known in advance, it can be specified at this time. Otherwise the Machine Tool and NC Tactics function must establish machine tool characteristics and select an appropriate machine tool.

Decompose Material to be Removed

Basic operational process plan information is generated by this function to establish models and delta volumes for intermediate part geometries. The technique of volume decomposition is employed to remove individual (delta) volumes of material from a model of stock material to produce an intermediate or final part model.

The Material Decomposition function uses "before" and "after" part model geometries, which are made available to the ANC framework, from an external part design database. The raw stock, from which the part is produced, is not modeled in the geometric part database. Specification for raw stock must be obtained from a Stock Definition library, which is external to the ANC. Various volume decomposition algorithms may be used by this function. These algorithms are also external to the ANC framework. Geometric models for the delta volume geometries are developed by an external geometric modeler.

As subsequent operations are performed in the ANC framework, adjustments in the results developed by this function may be needed. These requests for modifications are fed back as inputs to the Material Decomposition function if they occur.

Determine Machine Tool and NC Methods

In this function an appropriate machine tool and NC method to remove each delta volume is determined. A simplifying assumption made for this function is that only one machine tool may be used for the machining of the part. The machine tool can be optionally specified by the user when the models are initially identified. If the machine tool is predetermined, suitable NC methods for each delta volume are established by this function. When the machine tool is not pre-specified by the user, this function will first determine an appropriate NC method for each delta volume. Once the NC methods for all delta volumes are established, the most suitable machine tool is then selected.

The design for this function allows for either a manual or automatic mode for establishing machine tool and NC methods for the part machining process. This could potentially require the user to examine and assign NC methods for every delta volume. An intelligent decision tool would be needed to automate the decision process for determining an appropriate machine tool and NC methods. Such a decision aid would be external to the ANC framework.

Subsequent ANC framework operations may require adjustments to the results developed by this function. These requests for modifications are fed back as inputs to this Machine Tool and NC Methods function if they occur.

Determine Set-Up

The necessary setup characteristics for the part machining process are established by this function. Some of the specific issues that are addressed by this function include:

Orientation of the Part - Information from the previous functions is used to develop appropriate orientations for the part. Models of the part geometries were obtained by the Material Decomposition function. These models are used to determine the physical characteristics of the part that influence fixturing and clamping of the part. Constraints imposed by physical characteristics of the part, the NC methods employed, and the selected machine tool, are used to determine suitable part orientations. A geometric modeler is used to construct orientations for the part from which desired features can be machined. These orientations are compared to machine tool capabilities to insure compatibility with the machine tool.

Fixture Design - Geometry of the part, delta volume removal requirements, and part orientation are used to choose appropriate fixtures for holding the part. Features of the machined part are compared to the fixture feature capabilities of available fixtures. Based on this comparison, a suitable fixture is selected.

Clamp Design and Selection - Clamp design and selection uses part orientation and fixture information to choose clamps for holding the part. Candidate surfaces for clamping are compared to clamps defined in a clamp table. Clamp selection can be performed manually or by an automated decision process. Clamp selection data is fed back to the fixture design sub-function to insure compatibility between fixture and clamp selection.

Setup Evaluation - All support information and determinations made to this time are presented to the user for review and modification. Information that can be reviewed includes:

- Intermediate geometric part models
- NC methods
- Machine tool identification
- Part orientation
- Fixturing
- Clamping

Based on this review, the user can initiate modifications to previously obtained results.

Determine Cutting Tool and Holder

This function establishes the appropriate sub-delta material volumes, suitable cutting tool, and cutting tool holder for the removal of each sub-delta volume. Delta volumes are used to define the required volume to be removed from a "before" part model to produce a feature on an "after" part model. The actual types of machining steps (work elements) are defined by a specific NC method. Based on the total delta volume and the corresponding NC method, sub-delta volumes for each work element are established. Information obtained from previous ANC framework functions which are used by this function includes:

- Individual delta volumes
- NC methods for producing the delta volumes
- Manufacturing parameters and standards
- Machine tool setup data

Cutting tool selection requires sub-delta volume and NC method information to choose appropriate cutting tools for removing each desired sub-delta volume. Available cutting tools are specified in a cutting tool library. Individual cutting tools are compared against each required sub-delta volume in order to make a choice.

Tool holder selection uses setup and cutting tool information to choose the holder. The cutting tool and machine tool are compared with available holders. Compatibility must exist between the machine tool, cutting tool, cutting tool holder to result in holder selection. Special tool holders may be supported by this function.

Feedback from the NC Plan Review and Tool Path Simulation function may necessitate adjustments in this function.

Detail/Optimize NC Plan

The preparation of the detailed NC plan is carried out by this function. Many of the previously determined NC production characteristics are used by this function to carry out its objective. This information may include:

- Individual delta volumes for material removal from the before part model to produce the after part model

- Corresponding NC methods required to remove the delta volumes to produce the desired features
- Machine tool identification
- Part orientations, fixture and clamp specifications
- Sub-delta removal volumes
- Cutting tool and holder specifications

This function is comprised of the following four sub-functions:

Identification of Strategy - Identifies the overall strategy the NC process will employ to produce the part. Possible strategies may include the minimization of tool path travel lengths, minimization of the number of fixture changes, minimization of the number of part re-orientations or minimization of the number of cutting tool and holder changes. A number of decision steps are performed to determine the most appropriate strategy to be used by the NC Plan Generation sub-function.

NC Plan Generation - The chosen strategy is used to sequence the work elements in the NC plan. Information provided about each work element to this sub-function includes; setup data, NC method, cutting tool, and cutting tool holder specifications.. The result produced by this sub-function is a NC plan within which work elements have been ordered based on a designated hierarchy of strategies.

NC Tactics - The ANC user may review each work element in the NC plan through graphical techniques. NC tactics for the removal of sub-delta volumes can be altered by the ANC user through direct interaction with the system. Seven stages are associated with each NC tactic:

Approach - Movement from a clearance plane to an operative geometric position

Enstage - Movement from an operative geometric position to initial contact with part material

Entry - Movement from an initial part material contact to full part material contact

Machining - Movement through and continuous engagement with part material

Departure - Movement away from full part material contact to exit from material

Destage - Movement away from final contact with part material to an operative geometric position

Retract - Movement from an operative geometric position to a clearance plane

The ANC user accesses the geometric modeler in order to determine the cutting tool movements. A tactics table which includes coordinate points and other pertinent machine tool process information is developed by this sub-function.

Review of NC Plan - Once the NC plan and associated tactics have been determined, the user can review the part generations with the graphical interface. The geometric modeler can be used to display the position of the cutting tool and holder as they move in relation to the stationary part. Detection of proper coverage and collision avoidance can be observed with the NC Plan Review sub-function. In addition, interference with fixtures and clamps may be examined. Because this sub-function can detect conflict situations it will produce feedback to previously executed functions in order to rectify any problems.

Generate/Simulate Tool Path

The objective of this function is to generate tool path commands for each NC work element. Tactics records from the NC plan are used as an input to this function. Tactics specify the NC steps and characteristics which are necessary to produce the sub-delta volumes. Information in the tactics record is formatted for internal use by the ANC framework. A tool path generator sub-function converts a two dimensional representation into a three dimensional parametric equivalent. Graphical simulation of this approach enables the user view the tool path motion. The simulation uses the geometric modeler to produce the visual images.

Produce Machine Control Information

This function generates the appropriate cutter language (CL) instructions which will be used by a NC machine tool to machine the part. The tool path information that was generated in the previous function is converted into compatible CL data. This translation may use varied techniques that exist in the research literature, many of which utilize various forms of expert systems. Once these basic instructions are determined, graphical simulation will be used to represent the NC machining process. The generic CL instructions are then translated to machine dependent NC code.

Produce Support Information

The last function simply provides a user friendly interface, for the machine tool operator, to obtain information from the ANC framework. Components of this function include:

- Generation of operator instructions
- Visual aids
- Material orders

The finalized NC plan and associated tactics form the basis for information that is presented to the operator. This information is translated from an internal representation to an operator understandable form.

ANC Research Issues

A number of research opportunities can be identified in the ANC framework. Many of the functions which comprise this framework rely on some form of selection and sequencing decision process. Appropriate tools presently exist that would address many of these decision processes from an automation perspective [2,3,4,6,7,8]. Special algorithms and heuristics can be tested to determine which would be most efficient for each major set of functions. Additional research opportunities exist in regard to optimization of these decisions and selection and definition of appropriate database specifications and design. This area is also fertile for further development of each of the ANC steps and application these generic conceptual functions to various automated planning for machining, assembly, inspection and material handling processes. In the next section we show how the ANC process functions can be utilized to develop the instructions and plans for an automated inspection process planning system.

DEVELOPMENT OF AN AUTOMATED INSPECTION PROCESS PLANNING FRAMEWORK USING ADVANCED NUMERICAL CONTROL CONCEPTS

The ANC framework provides a general methodology for the automated development of process plans for numerically controlled machining equipment. This framework can be used as a foundation to automate other processes within a CIM environment. One of the more important manufacturing related processes that will be necessary from a more general competition perspective in manufacturing environments is the adoption of a total quality orientation and time based strategies. Both these competitive strategies (quality and time) can be met by integrating the inspection function into the CIM environment. Research in integrating automated inspection equipment, such as dimensional measuring equipment (DME) or coordinate measuring machines (CMM), with other components of CIM has received limited attention, until recently [9, 10, 11,12,13].

This section serves two purposes. First it demonstrates that the ANC framework is generalizable and can be used to develop other components in a CIM environment. It also describes a framework which is oriented toward quality functions and part inspection. This framework is referred to as the Expert Programming System - One (EPS-1) framework [14]. As we shall see in a later section, the EPS-1 is part of a larger operational environment that integrates inspection process planning with the remaining components of a CIM system.

The EPS-1 Architecture

The EPS-1 is a functional subset of the ANC framework. As such, the EPS-1 embodies many of the same characteristics as the ANC. Certain characteristics that are part of the ANC are not required in the EPS-1 framework. Only the differences between ANC and EPS-1 frameworks will be described in this paper. The major differences are primarily due to the underlying purpose of each processing framework. Whereas the ANC framework is primarily for machining purposes, the EPS-1 architecture is not a material transformation process, but an inspection process. Thus, many of the data and tool characteristics in the EPS-1 environment do not require material removal that are characteristic of the ANC framework.

The EPS-1 requires a part identification number and an overall process plan as inputs. The identification number is used to retrieve a geometric model and a dimensioning and tolerance model of a part. The DME NC plan generator uses the overall plan to produce Dimensional Measuring Interface Specification (DMIS) control data and support information for a specific DME. Figure 2 shows the relationships among nine functional modules that form the EPS-1 architecture. The remainder of this section describes each of these nine functional modules.

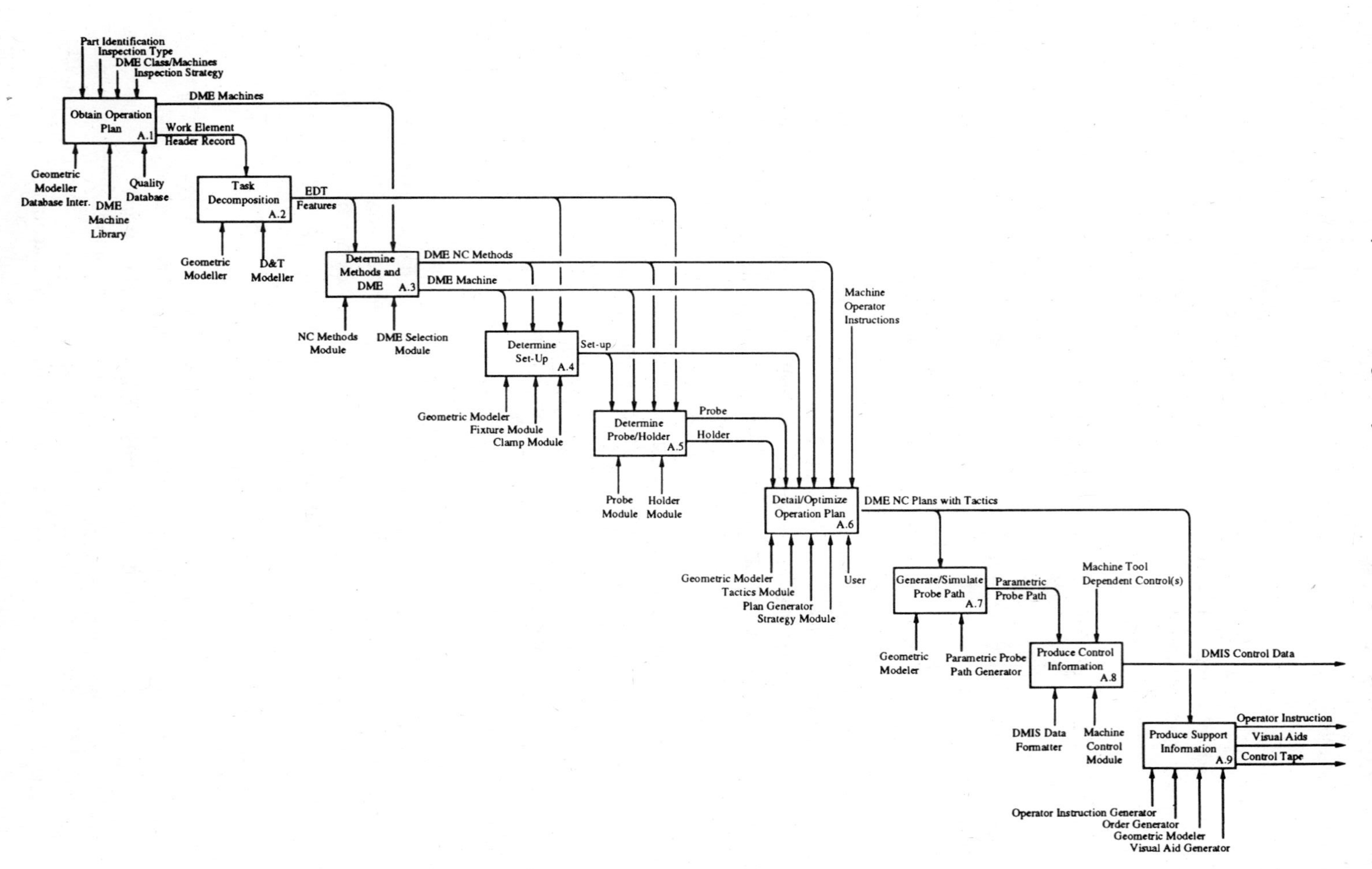

FIGURE 2: High Level IDEF$_0$ Diagram for EPS-1 Framework

Obtain Operation Plan

The EPS-1 user provides required data to the EPS-1 for initiating the automated inspection process. This function determines the scope of the inspection process to be performed on the indicated part. Data needed by the EPS-1 includes:

- A valid part identification number
- An inspection type
- A DME class or machine (optional)
- An inspection strategy (optional)

The inspection type establishes the scope of the inspection operation (i.e. courtesy inspection, in-process inspection, final inspection, etc.). The part identification number and the scope of the operation are used to determine the correct model in the geometric model database and the corresponding model in the D&T database. The output from this function includes work element identifiers that are passed on to the Task Decomposition function and the type of DME to be utilized which is used by the Determine NC Methods and DME function.

The EPS-1 must obtain the scope of the DME operation from the EPS-1 user, whereas the ANC obtained a machining process plan. Because the EPS-1 does not remove material from a part, it only requires the identification of a single part model. On the other hand, the ANC process requires the existence of before and after part models.

Task Decomposition

The Task Decomposition function determines the evaluated dimension and tolerance (EDT) features that are to be measured by the inspection process. The EPS-1 automatically determines these from information obtained by the Obtain Operational Plan function. The part identification and the inspection type uniquely identify the geometric and D&T models of the part. An entire D&T model is obtained through subroutine calls to the D&T modeler.

The D&T model is used by the Task Decomposition function to construct a skeletal list of the work elements necessary to perform the inspection process. For each node in the D&T database, an individual work element item is constructed. Subsequent functions in the EPS-1 use this information to determine other inspection process requirements and characteristics.

The EPS-1 only requires one part model identification and there is no product transformation feedback required. The output data representations also vary, whereas the EPS-1 data is a list of EDT features for the part to be inspected, the ANC outputs geometric models of the current state of the part and derived delta volumes. Another major difference is the lack of the need for a raw stock model in the EPS-1.

Determine Methods and DME

This function serves two purposes. First it determines the appropriate NC measurement methods for the inspection process. Then it identifies the most appropriate DME for taking the actual measurements. For the purposes of the EPS-1, the combination of tolerance class and sub-feature class, is used to establish individual NC measurement methods. Once the methods have been determined, a DME machine can be chosen. If a DME machine was specified during process initialization, selection of a DME will be bypassed. In this case, the NC measurement methods required for the inspection process are compared with those available from the specified DME, to ensure that the DME is capable of taking the required measurements. If a DME was not specified, then the EPS-1 compares the capabilities of available DME against the NC measurement methods required for the inspection process in order to choose a DME.

The requirements between the ANC and EPS-1 frameworks are similar for this function. Different methods of obtaining and transforming data are required since the characteristics of the equipment in each of the processes will utilize different internal representations of the data.

Determine Setup

Part setup generally applies to the part orientation, and fixture and clamp selection in a NC environment. In order to simplify the complexity of this process, fixtures and clamps were not incorporated into the initial specification of the EPS-1. Part orientation is accomplished by alignment of part surface normals with the DME measurement axes. Three potentially conflicting objectives are considered:

- Maximize the number of measurements in any given orientation
- Minimize the number of orientations to achieve the required inspections
- Eliminate possible collisions between the DME and the part due to improper orientation

The above strategy requires the geometric modeler to perform boolean algebra intersections on the part and probe models in order to detect potential collisions.

In the initial version of the EPS-1, fixtures and clamps are not required. Fixtures and clamps must be used in the ANC environment.

Determine Probe/Holder

Combinations of measurement probes and holders for the inspection process are determined by this function. Selection of a probe can become a complex task when the following factors are taken into consideration:

- DME technologies
- Probe relative orientation
- Multiple probe setups
- Probe shape

Due to this complexity, candidate probes must be identified through a process of elimination. The capabilities of the available probes are compared to required measurement characteristics of the inspection process. A probe is determined for each sub-feature on the part. Several probes may be required to measure the entire part. The strategy at this point may be to select probes based on minimizing the number of probes for the entire inspection process. Another probe selection strategy might be to maximize the number of measurements obtained from the individual probes.

There are four basic differences between ANC and EPS-1 frameworks relative to this function:

- The ANC sub-delta volume generator is replaced by the EDT sub-feature generator in the EPS-1
- The EPS-1 design does not require the user to supply manufacturing parameters
- The EPS-1 must select an appropriate probe instead of a cutting tool
- The EPS-1 does not determine a probe holder

Detail/Optimize Operation Plan

Inspection process information that was developed by previous functions, is used by this functions to complete the operation plan. The EPS-1 finalizes the sequence of work elements and determines the corresponding movement tactics for the inspection process. Three activities are performed by this function:

- NC plan generation
- Determination of NC tactics
- Review and modification of the NC plan

Sequencing of the individual work element items is determined by this function. The NC strategy previously specified by the user forms the basis for the ordering process. On the basis of this strategy, an appropriate set of logic is used to establish the specific sequencing of the work elements. The types of logic used depends on the sequencing strategy. Mixtures of heuristic, optimizing and decision table logic may be employed. Relative positioning of the probe is also determined for every sub-feature measurement to be made.

All the processing up to this point has been automatically carried out by the EPS-1. The work elements, NC plan, and NC tactics are now complete and the user can intervene to make any adjustments. The user may review a graphical representation of the inspection process sequence. This will indicate potential coverage and collision problems. The user can make the necessary corrections to eliminate any problems. Simple changes can be made by the user without any further intervention by the EPS-1.

The primary difference between the ANC and EPS-1 frameworks, for this function, is in the internal representation of the data and the decision making approach employed. These will be differ due to the need for more data items in the ANC framework.

Generate/Simulate Probe Path

This function serves two purposes. These are to generate locational and parametric probe path data and to visually simulate that data. The primary input to this function is the NC plan with tactics and the primary output is an internal representation of the probe path data.

The NC plan with tactics contains an ordered list of the work elements required to inspect the identified EDT features. The work elements are processed sequentially and contain the sub-feature name, the sub-features class, and the tolerance class for each entry.

The measurement process involves the generation of initial and departing probe positions and parametric probe path motion. Positioning of the probe for measurement is determined by generic routines that are geometry specific. EDT sub-feature class/tolerance class combinations dictate the appropriate routine. The routines use specific D&T information to establish actual probe position.

The parametric probe path data is mathematically determined for each measurement. The probe path data is composed of two curves that are parametrically synchronized. One of the curves represents the probe end, while the other represents the probe axis. Once the parametric probe path is created, it is used to graphically illustrate the movement of the probe in relation to the stationary part. The user can review the movements of the probe over the part to observe coverage and detect collisions.

The type of algorithms required by the ANC and EPS-1 frameworks for this function will necessarily have to differ. For example, there will need to be a greater diversity in the path patterns that are required for machining when compared to measurement operations. Generation of the probe path for taking dimensional measurements will incorporate different characteristics from that of a cutter tool path.

Produce Control Information

Numerically controlled DME inspection instructions are generated by this function. These instructions conform to the Dimensional Measuring Interface Specification (DMIS) [15]. The NC plan with tactics provides the necessary tolerancing information. The coordinates of the actual points for every movement and measurement are retrieved from an internal representation of the probe path. All of this information is combined to produce DMIS instructions. This function also allows DME independent control information to be interspersed with the DMIS instructions.

The major difference in this function between the ANC and EPS-1 frameworks is the type of output data. DMIS data will be output by the EPS-1, while APT compatible CL data will be output by the ANC framework.

Produce Support Information

Information that supports the actual NC inspection operation is developed by this function. It is defined in such a way that it can be tailored to support the specific environment of an individual implementation. Information for a DME operator is generated as well as for other users indirectly involved in the NC inspection operations. These individuals may include quality assurance engineers, shop supervisors, foremen, or tool crib operators.

The major difference between the ANC and EPS-1 frameworks for this function is that the EPS-1 will generate less support information.

OPERATIONAL ENVIRONMENT OF THE EPS-1

The EPS-1 interacts with various external components as shown in Figure 3. The major components in this operational environment include; A geometric modeler, the Dimensioning and Tolerancing (D&T) modeler, the Applications Interface Specification (AIS), the Dimensioning and Tolerancing Applications Interface Specification (DTAIS), the Dimensional Measuring Interface Specification (DMIS), and support databases. All of these components play an integral role in the EPS-1 framework, either as inputs, controls or mechanisms for processing. Each of these components will be described in the remainder of this section.

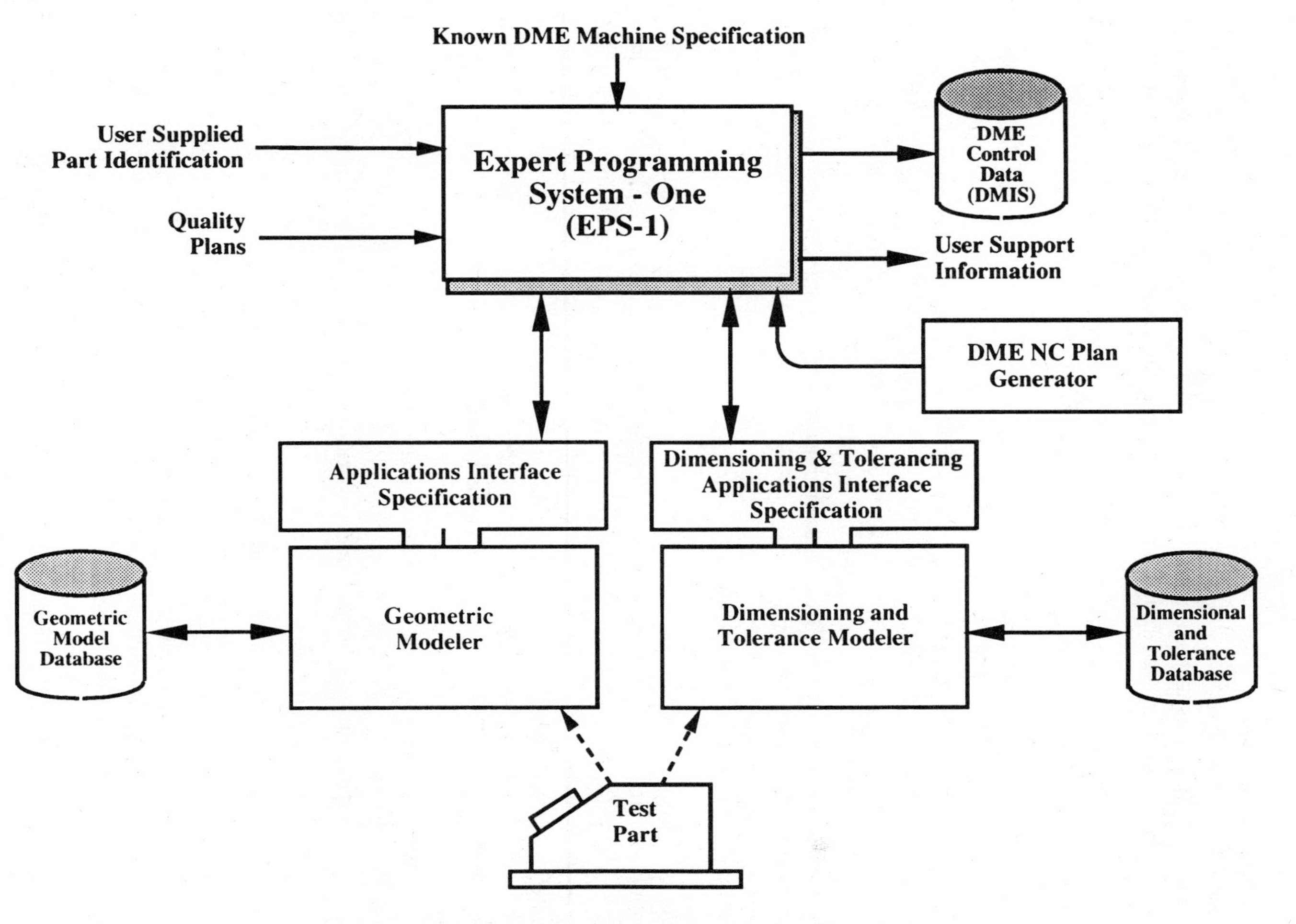

FIGURE 3: Operational Environment of the EPS-1

Geometric Modeler

A geometric modeler is used to delineate, identify and define the physical characteristics of a part. Various geometric entities can be described by their boundaries. The geometric modeler facilitates certain functions that are specific to inspection. In particular, collision between probe and part can be detected by performing a boolean intersection on geometric models of the part and probe.

Dimensioning and Tolerance Modeler

Inspection of physical objects by DME and CMMs requires specific data about the dimensions and tolerances for the geometric features of a part [16]. Many existing geometric modelers do not have the capability to produce the dimensioning and tolerancing information that is needed by inspection applications. Thus, a Dimension and Tolerancing (D&T) modeler [17] was designed and developed to fill this void. The D&T modeler defines various tolerance nodes and assigns these nodes to one or more geometric features. After geometric features have been added to a part model, the D&T modeler must be used to augment the geometries with specific dimensioning and tolerancing data.

Applications Interface Specification

Due to unique capabilities that are offered by various geometric modelers, an inspection application may require the use of several modelers. To avoid the necessity for developing unique interfaces for each combination of application and modeler, a standard interface similar to IGES has been developed. The Geometric Modeling Program of CAM-I has defined the Application Interface Specification (AIS) to satisfy this need.

The AIS is a dynamic interface between modelers and applications software. Modelers and applications that have been designed and implemented to conform to the AIS can interface directly with one another. The AIS defines a core set of capabilities that should be available from any geometric modeler. If a particular modeler has the capability to perform a specified function, then the interface acts only as a communications media. On the other hand, when the function is not supported by a modeler, the application interface would be required to perform that function.

Dimensioning and Tolerance Applications Interface Specification

The AIS allows a user to retrieve geometric data about a part and also enables applications to utilize the functional capabilities of the geometric modeler. The D&T model expands the geometric model of the part to include data that are necessary for dimensioning and tolerancing. Unfortunately, the AIS does not accommodate dimensioning and tolerancing data, nor does it enable external applications to access the functional capabilities of the D&T modeler. Therefore it was necessary to extend the AIS so that it could create, read, modify and delete nodes in the D&T model. The Dimensioning and Tolerancing Applications Interface Specification (DTAIS) [18] was created to satisfy this need. The DTAIS is similar in concept to the AIS in that it is used to construct the D&T model.

Dimensional Measuring Interface Specification

There are three classes of interface for linking CAD systems with DME [19].

> ***One-to-one*** interfaces require users to develop their own communication linkages that work with specific devices in their system
>
> ***One-to-many*** interfaces allow one CAD system to interface with numerous DME and CMMs
>
> ***Many-to-many*** interfaces allow the user to select the most appropriate CAD systems and DME which satisfy their particular needs

The many-to-many interface is the most desirable because it provides the greatest flexibility and has the highest potential for integration of inspection processes. The Dimensional Measuring Interface Specification (DMIS) [15, 20, 21] was recently adopted by ANSI as a National Standard for many-to-many interfaces. DMIS provides the linkage between the EPS-1 and DME as shown in Figure 3.

SUMMARY AND CONCLUSIONS

In this chapter, we have presented a review of the Advanced Numerical (ANC) framework developed by CAM-I's ANC Program. It was shown how the functional modules of an ANC framework could be used to develop a framework for automated quality and inspection control called Expert Programming System - One (EPS-1). To be competitive, manufacturers must produce their products to meet customer requirements in a timely fashion at a competitive price. The EPS-1 is an essential component for achieving these objectives in integrated manufacturing environments. The linkage of the EPS-1 in an operational environment was also described.

A description of how the EPS-1 framework is implemented to inspect representative parts is presented by Reimann and Sarkis [22]. Efforts addressing the issue of automating and integrating the inspection planning process have only recently emerged. This is not only due to competitive pressures, but the increased technological availability of various DME, specifically coordinate measuring machines which have shown increasing benefits [23].

Future research should focus on extending the ANC to include other aspects of automated planning within manufacturing environments. At this time, the ANC framework has only been conceptually developed, implementation of many of its elements is required. Development of decision tools to help in the automation of these steps. Numerous research possibilities in determining appropriate algorithms and knowledge bases for these decision tools exist for both the ANC and EPS-1.

REFERENCES

[1] *CAM-I Advanced Numerical Control Conceptual Design - Volume I Project Notebook & Status Report*, Computer Aided Manufacturing - International, Inc., Arlington, Texas, R-79-ANC-03, 1979.

[2] Evershiem, W., and Schneewind, J., "Computer-aided Process Planning - State of the Art and Future Development," *Operations Research/ Management Science*, Vol. 33, No. 5, pp. 573, 1993.

[3] Wang, H-P. and Jian-kang, L., *Computer Aided Process Planning*, Elsevier Publications, Amsterdam, 1991.

[4] Zhang, H.C., and Alting, L., "Computer Aided Process Planning: The State-of-the-Art Survey," *International Journal of Production Research*, Vol. 27, No. 4, pp. 553, 1989.

[5] Softech, Inc., Integrated Computer Aided Manufacturing (ICAM), Volume V, AFWAL-TR-82-4063, Wright Patterson Air Force Base, Dayton, Ohio, 1983.

[6] Dong, Z., and Hu, W., "Optimal Process Sequence Identification and Optimal Process Tolerance Assignment in Computer-Aided Process Planning," *Computers in Industry*, Vol. 17, No. 1, pp. 19, 1991.

[7] Gindy, N.N.Z., Huang, X., and Ratchev, T.M., "Feature-based Component Model for Computer-Aided Process Planning Systems," *Operations Research/Management Science*, Vol. 33, No. 5, pp. 579.

[8] Koshnevis, B., "A Real-Time Computer-Aided Process Planning System as a Support Tool for Economic Product Design," *Journal of Manufacturing Systems*, Vol. 12, No. 2, pp. 181, 1993.

[9] Brown, C.W., "IPPEX: An Automated Planning System for Dimensional Inspection," in the proceedings of *The 22nd CIRP International Seminar on Manufacturing Systems*, Enschede, The Netherlands, June 11, 1990.

[10] ElMaraghy, H.A., Gu, P.H., "Expert System for Inspection Planning," *Annals of the CIRP*, Vol. 36, No. 1, pp. 85-89, 1987.

[11] Menq, C. H, Yau, H.T., Lai, G.Y., "Automated Precision Measurement of Surface Profile in CAD-Directed Inspection," *IEEE Transactions on Robotics and Automation*, Vol. 8, No. 2, pp. 268-278, April, 1992.

[12] Menq, C.H., Yau, H.T., Wong, C.L., "An Intelligent Planning Environment for Automated Dimensional Inspection Using Coordinate Measuring Machines," *Journal of Engineering for Industry*, Vol. 114, pp. 222-230, May 1992.

[13] Reimann, M.D., Fowler, J.W., "The Expert Programming System - One (EPS-1)," in the proceedings of the *Fourth International Conference of the State-of-the-Art on Solids Modeling*, sponsored by CAD/CIM Alert and CAM-I, Boston, Massachusetts, May 1987.

[14] *The Expert Programming System - One (EPS-1), Task One Design Review*, Computer Aided Manufacturing - International, Inc., Arlington, Texas, R-87-ANC-01, 1987.

[15] *ANSI/CAMI 101-1990 Dimensional Measuring Interface Specification*, Computer Aided Manufacturing International, Arlington, Texas, 1990.

[16] *ANSI Y14.5M, Dimensioning and Tolerancing*, The American Society of Mechanical Engineers, ANSI Y14.5M - 1982.

[17] *CAM-I D&T Modeler Version 1.0 - Dimensioning and Tolerancing Feasibility Demonstration Final Report*, Computer Aided Manufacturing - International, Inc., Arlington, Texas, PS-86-ANC/GM01, Nov. 1986.

[18] Johnson, R.H., *Dimensioning and Tolerancing Final Report*, Computer Aided Manufacturing - International, Inc., Arlington, Texas, R-84-GM02.2, May, 1985.

[19] Zink, J.H., "Closing the CIM Loop With CMM", *Automation*, Vol. 36, No. 1, pp. 48- 50, 1989.

[20] Schreiber, R.R., "CMMs: Traits, Trends, Triumphs," *Manufacturing Engineering*, Vol. 104, No. 4, pp. 31-37, April 1990.

[21] Smith, L., "Quality too: CMM Smooths Mold Design," American Machinist, Vol.137, No. 4, pp.26-28, April, 1993.

[22] Reimann, M.D., and Sarkis, J., "An Architecture for Integrated Automated Quality Control," Journal of Manufacturing Systems, Vol. 12, No. 4, pp. 341-355, 1993.

[23] Stovicer, D., "CMMs Key to Plant With a Future," *Automation*, Vol. 37, No. 12, pp. 24-25, Dec. 1990.

3

AN INTELLIGENT HYBRID SYSTEM FOR SYNTHESIS AND CONTROL OF METAL FORMING PROCESSES

A.F. Ali, C.A. Vassiliadis, J.S. Gunasekera,
R. Anbajagane, and S. McDonald
Ohio University

3.1 INTRODUCTION

Design and control are central to any metal forming process. Process parameters involved in metal forming have to be controlled properly to achieve an optimum product without defects. Analytical models have become popular for analyzing forming processes, however, these models employ an iterative analysis procedure to arrive at solutions. Artificial Intelligence techniques such as expert systems and neural networks, provide a novel approach to design and control of a metal forming process and considerably reduce the time and effort to achieve a desirable product.

Use of expert systems and neural networks has seen tremendous growth in the past decade. Expert systems are a set of concepts, procedures and techniques that enable the computer to solve various problems in an intelligent way. Neural networks, on the other hand, are large networks of simple processing elements which process information dynamically. While both systems are tremendously successful, they have their limitations. The integration of the two techniques yields a more potent processing environment than the individual systems.

Several researchers have made efforts to establish a scientific approach to design [1, 2, 3, 4] which would improve product quality and aid in the development of new designs. Numerous expert system applications have been developed for diagnosis, monitoring and control, and inspection systems [5]. Right and Bourne [6] have employed expert systems for feature recognition and present a case study for an intelligent machining tool. Neural networks have been employed in various learning and optimizing tasks and for acquiring manufacturing knowledge [7]. Zarefar and Goulding [8] discussed a neural network-based hybrid CAD/CAM design environment for mechanical power transmissions. However, not enough inroads have been made into the design and control of metal forming problems. Researchers have made initial efforts into design and control of manufacturing processes [9, 10]. However, the approach presented in this paper is novel and the power of artificial intelligence tools is exploited for design and control of metal forming applications.

This study describes an artificial intelligence approach to metal forming processes and demonstrates the use of an intelligent hybrid tool to design and control the pack rolling process which is presented as an example. Pack rolling is a technique for rolling thin, hard and/or "difficult-to-form" materials. Proper design has become increasingly important with expanding global competition and stringent requirements of products and materials. The hybrid tool initially employs a neural network approach to establish a set of operating conditions for pack rolling. A set of input-output conditions for pack rolling which are employed by the neural network include: the starting and ending dimensions, the material strain, strain rate and temperature, the roll force and roll torque, etc. The neural network is interfaced with an expert system which firstly, serves as a user interface and secondly and more importantly, controls the solution methodology for the pack rolling process. Based on the machine limitations, the product requirements, the material under consideration, and the process conditions, the expert system establishes a processing path which most closely matches the output requirements using the neural network as a reasoning tool. The results are compared with actual experiments and presented to validate the strategy.

Manufacturing practices are generally established by trial and error. This process is not only cumbersome but expensive as well. At the same time, computer time to conduct numerical simulations has also become a cost effective factor in the competitive manufacturing environment. The outcome of designs, that is, the optimum solution, is dependent on a number of factors including the knowledge and experience of the designer. Expert systems use this idea of expert knowledge to achieve a workable, economical, and optimum solution of a problem in lesser time. This has become all the more important due to increased competition, greater demand, and stringent standards in industry.

Expert systems methodology represents an attempt to capture the expertise and knowledge of a design engineer in a computer program. This computer program can exploit the power of the existing analytical tools, CAD/CAM/CAE techniques for manufacturing, and heuristic rules to emulate tasks normally done by experts in this field. Neural networks, on the other hand, process information dynamically in response to external inputs and are thus classified as self adapting systems. An intelligent hybrid system, a combination of neural networks and expert systems, developed in this study envisages the implementation by coalescing computer aided engineering, computer aided design, material science theory, design theory, optimal process control, and AI techniques to achieve a robust and workable system. This hybrid system is capable of reasoning, drawing inference while performing specified tasks, as well as adjusting to new situations. The main motives in developing the hybrid system are to:

- Aid the user in arriving at workable solutions,
- Capture and imbed broad knowledge in a data base,
- Use intelligent computer techniques to optimize designs,
- Generate systems with self-adapting and self-learning capabilities, and
- Reduce the design time and relieve the engineer for other tasks.

Material processing operations are aimed at obtaining a desired geometry based on tolerance limits and microstructural attributes which may correspond to mechanical properties of the material. It is imperative that the processing conditions and behavior of the material be understood thoroughly. Even though computer models have been developed for analyzing material processes and have been found useful for optimizing the process, they are unable to generate new design concepts readily. Numerical techniques are widely being used in materials processing but most of them are traditionally based on an iterative cycle of "input, analysis of results, and modification of input data," as shown in Figure 3-1, for modeling the pack rolling process. The objective of the work envisaged is to employ an intelligent hybrid system to design and optimize the manufacturing process, rather than the conventional iterative procedure, mentioned above.

3.2 OPTIMAL DESIGN STRATEGY

The optimal design methodology presented in this paper is a generic methodology which can be applied to design and optimize metal forming processes. The salient steps to achieve the design are given below:

- Identification of Process Sequence,
- Conceptual Design,
- Process Models, and
- Process Optimization.

3.2.1 Identification of process sequence

Brain storming activities are part of the preliminary phase in any design philosophy. In the preliminary stage the metal forming process needs to be identified to achieve a desired product shape with the required material properties. In this stage, the goal variables or the desired output needs to be identified along with the resources and the set of initial or input conditions. In order to obtain the desired properties a sequence of manufacturing processes may have to be employed.

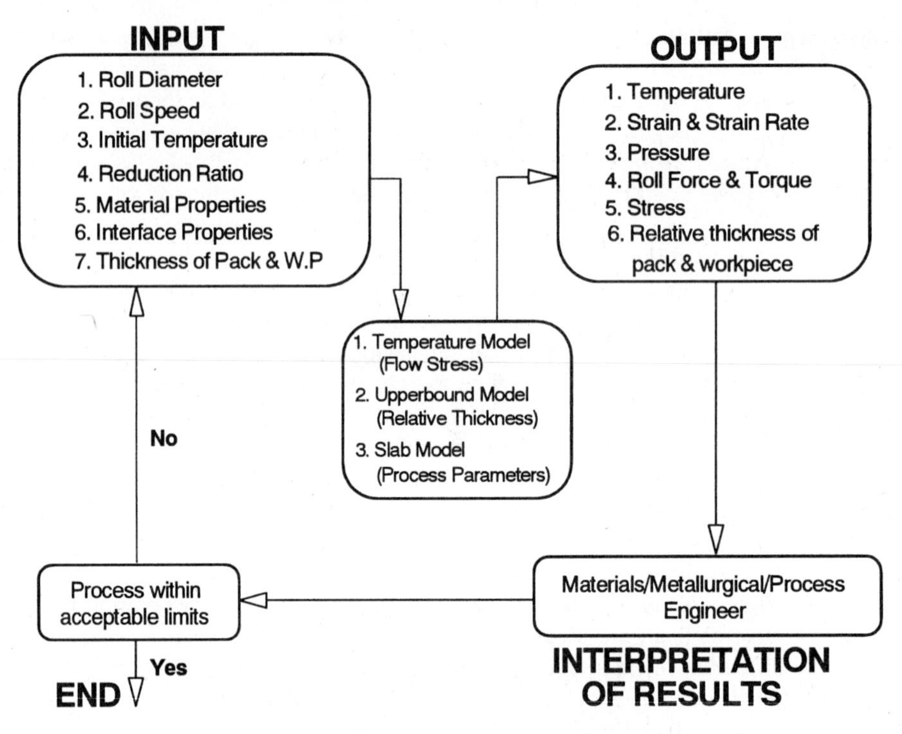

Figure 3-1: Iterative Procedure for Pack Rolling

The need for producing products of special purpose materials for aerospace and automotive applications requires proper material selection. The fundamental requirement in a part is a high strength-to-weight ratio. However, depending on the specific part a number of other requirements may also exist. The thermo-mechanical processing cycle to produce the final product with desired properties can be determined once the starting material and the manufacturing process is selected. There after, conceptual design, optimization and control techniques can be employed to optimize the process parameters and the process itself.

3.2.2 Conceptual design

Conceptual design is a critical activity which greatly determines the constraints and parameters to be further analyzed and optimized, and thereby significantly influences the life cycle performance and costs of products and processes. Conceptual design activities can be established through tools such as the axiomatic design approach developed by Suh [11]. The objective of this approach is to establish the important parameters - parameters that are required, the functional requirements {FRs}, and the controllable parameters or the design parameters {DPs}. A relationship may be established or mapped between the {FRs} and the {DPs} through a combination of empirical or analytical techniques. The axiomatic design philosophy establishes a first approximation for the design problem and provides further insight to the design situation.

3.2.3 Working of the Process

The working of the metal forming process has to be understood thoroughly. The intelligent hybrid system is linked closely to the actual metal forming process. The working of the process is available to the system as a set of analytical tools, empirical relationships, or heuristics derived from accumulated experience.

3.2.4 Optimization of the process

Manufacturing processes are aimed at obtaining desired geometry of the material such as the shape, size, and their tolerances and desired microstructural properties which translate into a desired strain, strain rate and temperature processing window. The outcome of the process variables is a function of the processing history of the variables such as roll or ram speeds, temperature, and reduction ratios. The intelligent optimization techniques identify the processing conditions which result in the desired geometry and material properties. The intelligent hybrid system invokes the thought process and provides the best processing conditions. However, expert knowledge and its involvement in the development of the intelligent system is crucial for any design process.

Specific tasks related to this study are shown in Table 3-1. As it is evident from the table, the task at hand is inter-disciplinary. The intelligent hybrid system brings together the knowledge domain from the various areas. The design methodology presented in this paper is generic and can be applied to any manufacturing process. A similar design methodology is developed and implemented for the pack rolling process and is presented in this paper as an example. A simple explanation of the pack rolling process is presented in the next section.

Table 3-1: Specific Tasks and Areas of Expertise for Work Described in this Paper

Task	Area / Expertise
• Optimal Processing Methodology • Identification of Process Sequence • Conceptual Design • Working of the Process • Optimization	Materials Processing Design Engineering CAD/CAM/CAE Intelligent Systems
• Constraints and Requirements • Material Data Base • Machine Requirements	Material Science Materials Processing

Table 3-1: Specific Tasks and Areas of Expertise for Work Described in this Paper

Task	Area / Expertise
• Intelligent Hybrid System • Development of the Expert System • Training of the Neural Network • Implementation of the ES-NN Interface • Optimization and Control	Materials Processing CAD/CAM/CAE Intelligent Systems
• Experimental Validation	Materials Processing Intelligent Systems

3.3 THE PACK ROLLING PROCESS

Pack rolling is a rolling technique for rolling thin, hard and/or "difficult-to-form" materials which otherwise would be difficult to roll [12] and [13]. This process involves the rolling of a harder material sandwiched or covered on both sides by a softer material, the "pack" - a sacrificial material, as shown in Figure 3-2. As presented in the figure, R is the radius, t is the thickness, V is the velocity. Subscripts p and w refer to the cover plate and the workpiece, respectively. This process considerably reduces the roll pressures required for deformation as the pack material being softer deforms more thereby setting up frictional tensile stresses beneficial to the deformation of the workpiece.

The objectives of the pack rolling design is to enable the processing of "difficult-to-form" high temperature alloys. The workpiece material is embedded or covered from all sides by a softer material of similar characteristics. The heat loss to the surrounding can be minimized by varying the thickness of the cover plate and/or the use of an insulation material between the workpiece and the cover plate. The inherent advantages of a design which meets the above criteria are:

- To protect the hot workpiece from the environment,
- To reduce heat loss for materials with a very narrow processing window,
- Ease of forming (i.e. reduce rolling loads),
- Lower roll forces,
- Tensile and frictional stresses help workpiece deformation,
- Multiple rolling of sheets, and
- To maintain the required high temperature of the workpiece.

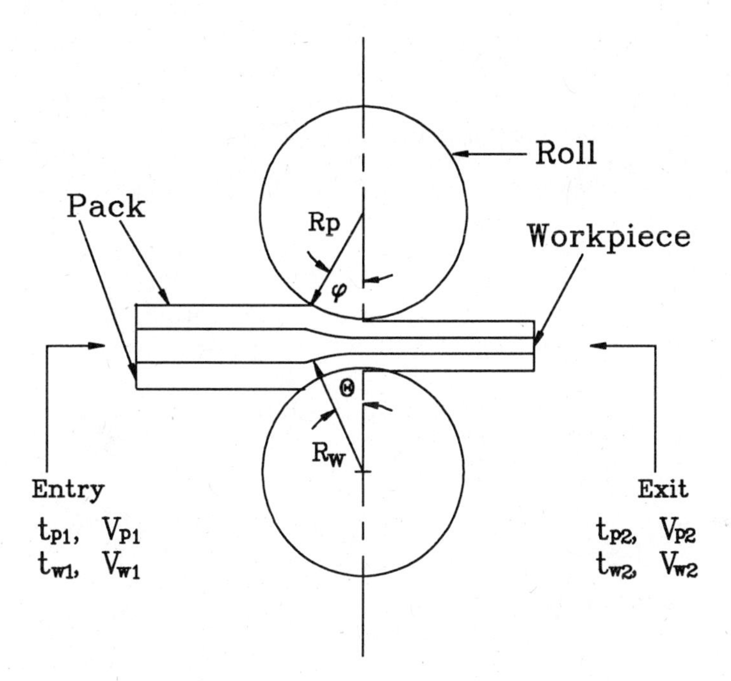

Figure 3-2: The Pack Rolling Process

Modeling the pack rolling process requires the determination of the relative velocity between the cover plate and the workpiece. The velocity distribution in pack rolling is essential because the two materials, the cover plate and the workpiece materials, deform in dissimilar modes due to the difference in the two material flow stresses and the frictional stresses acting on the interfaces. Because of the dissimilar nature of the materials, the workpiece and the cover plate undergo different deformations. The heat transfer for the process needs to be considered since the temperature combined with the strain rate (rate of deformation) and strain determines the stability of the material.

The functional requirements for the pack rolling model are:

- Ending thickness of the workpiece,
- Ending thickness of the cover plate,
- Temperature of the workpiece,
- Temperature of the cover plate,
- Ending length of the workpiece,
- Ending length of the cover plate,
- Mean stress,
- Strain rate, and
- Pressure.

The design parameters for the process are:

- Starting thickness of the workpiece,
- Starting thickness of the cover plate,
- Starting length of the workpiece,
- Starting length of the cover plate,
- Insulation between the cover plate and the workpiece,
- Roll diameter and speed,
- Number of pack rolling passes,
- Reduction ratio,
- Starting temperature of the cover plate and the workpiece, and
- Roll temperature.

In addition, the process has certain constraints driven primarily by the machine limitations, which are:

- Maximum machine force,
- Maximum machine torque,
- Maximum and minimum roll speed,
- Maximum furnace and roll temperature, and
- Maximum strain.

The objective of the design and optimization process is to achieve the functional requirements by appropriately setting the design parameters which also satisfy the machine constraints. The selection of the functional requirements, the design parameters and the optimization is discussed in a later section.

3.4 INTELLIGENT HYBRID SYSTEM

Knowledge is a valuable resource. This source has been connected with humans and their ability to decipher, comprehend and apply knowledge. An expert system is a computer program that attempts to capture the knowledge and experience of an expert in order to have the expertise readily available to aid in the decision making process. The expert system acts just like the human expert which has the knowledge and experience of the expert embedded in its data base. These systems are currently being used in a wide variety of fields. They, however, suffer from the fact that they rely solely upon the expert's knowledge in the knowledge base and cannot make decisions based on new situations which are not in the knowledge base. In other words, they are not self learning systems. Neural networks on the other hand, are large networks of processing elements or nodes which process information dynamically in response to external inputs. The nodes are simplified models of biological neurons. The knowledge in a neural network is distributed throughout the network in the form of internode connections and weighted links which form the inputs to the nodes. These models have been developed by inspiration from biological nervous systems and the ability of humans to learn. These systems can learn form examples but lack the ability of explaining decisions like the experts systems. Intelligent hybrid systems is a relatively new field of artificial intelligence which deals with the integration of expert systems with neural networks. This approach combines the functionality of an expert system shell with the learning capability of a neural network.

Expert systems and neural networks have both demonstrated their capabilities and are being exploited in a variety of fields. Both approaches have their advantages and limitations. Table 3-2 shows the similarities, differences, and strong points of both systems [14] and [15]. The integration of the two systems creates a hybrid tool which has the strong points of the two systems taken separately.

The ability to learn in any uncertain or fuzzy environments is an attractive component of any hybrid intelligent system. The ability to gather information, process it, enhance it, and the comprehension of new knowledge is crucial to the understanding of such systems. As such, these systems are able to gather information, learn in unsupervised situation, analyze results, and thereafter apply previous solutions, examples and experiences to form new solutions. The intelligent system formed by the integration of the two powerful techniques combine immense power and yield a system more powerful than either system operating alone [14].

TABLE 3-2: Comparison of Expert Systems and Neural Networks

Expert Systems	Neural Networks
They are domain specific.	They have a broad response.
They function well when the problem is well defined.	They have a capacity to provide general classification to a set of input conditions.
Implementation is a lengthy process depending on domain size.	Can capture a large number of cases quickly to provide acceptably accurate results.
Can communicate effectively with the user.	User interface may not be very friendly.
Can identify the data necessary to solve a problem.	They are not very productive when it comes to this feature.
Cannot learn from examples - depend on human experts for knowledge acquisition.	They are self-learning systems.
They are good in analysis and reasoning.	Neural networks can not explain their reasoning.
They have the ability of showing how the decision process was reached.	
Inability to synthesize new knowledge.	Ability to synthesize new knowledge.
Can not provide dynamic environments.	Provide dynamic knowledge by changing knowledge whenever necessary.

The integration of expert systems and neural networks, in general, requires a three stage process [14].

- Data Collection,
- Data Evaluation and Analysis, and
- Inferences and Conclusions.

3.4.1 Data Collection

Because of their ability to communicate with the user in an efficient and simple way, expert systems can be used to develop user friendly interfaces. This is accomplished by querying data bases, user interaction, and other programs. With the help of an expert system the data can be collected in an optimal fashion. Expert systems provide numerous advantages in communicating with a data base or a user. They provide flexibility and permit modification of data queries; they permit expandability in that additional data may be incorporated into an existing list of input conditions; and application of heuristic logic to the data collection process by making inferences on data made available from previous data. Expert systems can also explain why additional input parameters are required.

Knowledge acquisition entails the collection of sample cases with the output solutions. A small number of data points can be employed with some degree of reliability to train the neural network. This would however imply that a set of operating conditions or a set of feasible input-output data be available for training the neural network. The network will produce more accurate solutions if a larger number of training samples are used to represent the domain.

3.4.2 Data Evaluation and Analysis

Expert systems provide the capability of analyzing the set of input data and providing a solution. If the number of data points are relatively small the expert systems approach is adequate for analysis. However, if the knowledge base is incomplete, or the reasoning about the data is fuzzy, the neural network provides a viable solution by taking the data from the expert system and modifying it through learning. The transfer of information between the neural network and the expert system is bi-directional. Based on the modified information supplied from the neural network, the expert system performs the analysis by modifying, firing, and inferring new rules. The output values from the neural network can be analyzed by the expert system to determine the usefulness of the data and the reasoning process could be invoked to achieve a feasible solution. Conclusions and inferences are then drawn from these results.

3.4.3 Inferences and Conclusions

Conclusions are dependent on the goals set by the user. They indirectly depend on the input data, the output data, the threshold values of the parameters and the certainty factors. Based on the analysis of the input data, the goals can be determined by invoking the expert system.

3.5 DEVELOPMENT OF THE INTELLIGENT HYBRID SYSTEM

The overall control and optimization algorithm using the intelligent hybrid system is shown in Figure 3-3. This study, as opposed to the iterative analysis approach, is controlled and monitored by an expert system. The advantage of this approach is that it is a design process as opposed to an iterative analysis approach. Moreover, the expert system generates an optimal design.

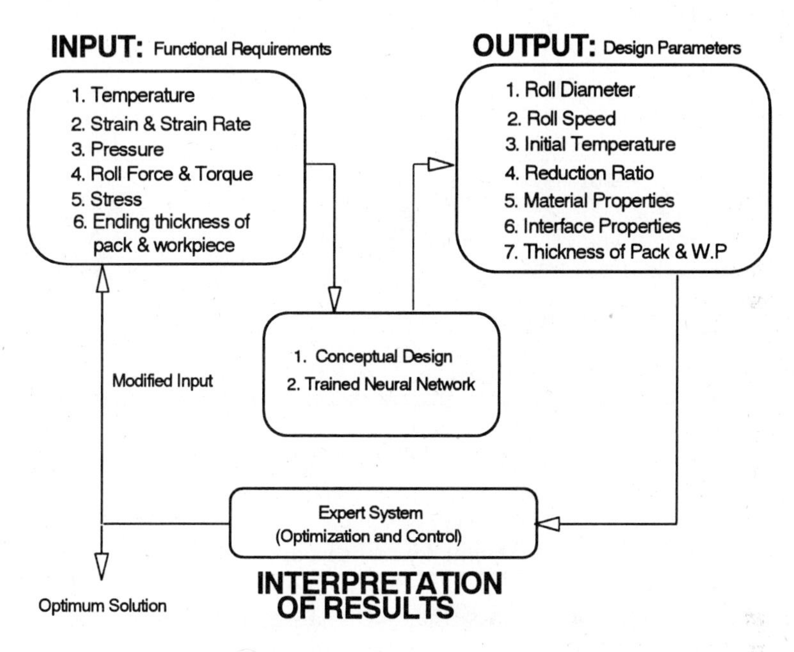

Figure 3-3: Intelligent hybrid system control flowchart for pack rolling

An overview of the intelligent hybrid system is shown in Figure 3-4. The user receives an input which is passed on to the dashed box which represents the expert system. An interface links the user with the expert system and receives the input design parameters and constraints, manipulates the design parameters, and generates an optimum solution which is sent back to the user. In addition, the expert system is linked to a material database. The database is an essential part of the system and contains process stability information for the tested materials. Process /variables such as strain, strain rate, and temperature, define the processing regimes for hot forming. The expert system is linked to the design tools, the neural network shell, and the CAD/CAM/CAE tools. Finally, the optimal solution is passed back to the user through the expert system interface.

3.5.1 Development of the Neural Network

The development of the neural network for the pack rolling process involved a four-step procedure which is outlined below.

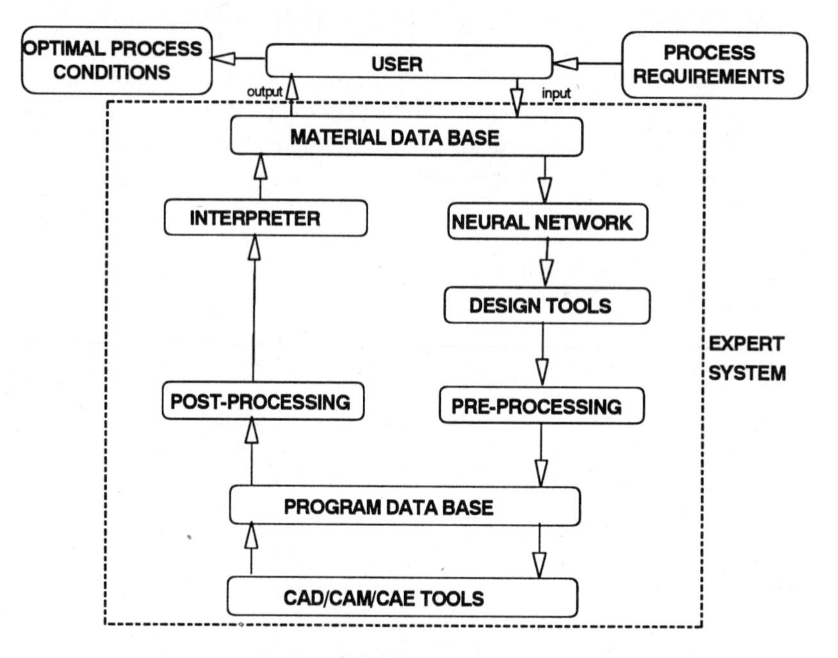

Figure 3-4: An Overview of the Intelligent Hybrid System

3.5.1.1 Conceptual Design: Identification of the input and output parameters

In this design phase the functional requirements and the design parameters for pack rolling are identified. The parameters of interest are chosen such that there is a corresponding relationship with the design parameters. According to the design axioms developed by Suh [11], the design parameters and the functional requirements can be reduced for simplicity.

The functional requirements for the process are:

- Output thickness of the workpiece,
- Output thickness of the cover plate,
- Temperature of the workpiece, and
- Strain rate.

The design parameters for the process are:

- Input thickness of the workpiece,
- Input thickness of the cover plate,
- Insulation between the cover plate and the workpiece,
- Roll speed,
- Reduction ratio,
- Initial temperature of the cover plate and the workpiece, and
- Roll temperature

The pack rolling constraints for the process are:

- Maximum machine force
- Maximum machine torque,
- Maximum and minimum roll speed,
- Maximum furnace and roll temperature, and
- Maximum strain.

The functional requirements and the design parameters are related by a Jacobian matrix according to the following relationship:

$$d\{ FR \} = [A] \; d\{ DP \} \qquad (1)$$

where,

FR is the function requirement,
DP is the design parameter,
d{ } is the differential, and
[A] is the Jacobian Matrix.

3.5.1.2 Data Collection: Design of Experiments

Data collection is of primary importance for both neural networks and expert systems. The generated data needs to be comprehensive to cover the spectrum of the operating conditions. Moreover, the data should be selected in an optimal fashion by conducting only a minimum number of experiments which would yield the maximum information. Pack rolling process data was generated using experimental design techniques, to achieve maximum information from a small number of experimental trials, and to gain a better understanding of the process. A 2^{7-3}_{IV} fractional factorial, resolution four design was selected for generating the data points for training the neural network. A resolution four design has main variables confounding with the three way interactions. The control variables or the design parameters were selected such that the variation of the parameters were linear in the processing regions. The settings of the control factors for generating the neural network data and the design matrix are shown in Tables 3-3a and b, respectively.

3.5.1.3 Building and Training the Neural Network

Neuroshell from Ward Systems Group, Inc., was used as the neural network software paradigm to address this design project. The paradigm is a back propagation neural network that learns a set of sample cases from valid data. Each case consists of the defining characteristics (the inputs), and the classifying characteristics (the outputs). Once the sample cases have been learned by the network, the system can process input data to produce results.

The pack-rolling network was trained with 32 pre-processed sample cases containing 8 inputs, and 7 outputs. There are several factors that can be manipulated in NeuroShell to increase the learning process and the probability for convergence, and therefore, insure accurate results.

Table 3-3a: Control Variable Settings

Number of runs - 16
Number of factors - 7
Design Resolution - 4

FACTORS		Low Level (-)	High Level (+)
T_w	- Starting Workpiece Thickness (in.)	0.50	0.60
T_c	- Starting Cover Plate Thickness (in.)	0.15	0.30
I	- Insulation	0	1
S	- Roll Speed (rpm)	3	15
R	- Reduction (%)	10	15
T	- Starting Pack Temperature (°F)	2000	2350
R_T	- Roll Temperature (°F)	80	200

Table 3-3b: Design Matrix

RUN	T_W	T_C	I	S	R	T	R_T
1	-	-	-	-	-	-	-
2	-	+	+	+	-	-	-
3	+	+	+	-	+	-	-
4	+	-	-	+	+	-	-
5	+	-	+	-	-	+	-
6	+	+	-	+	-	+	-
7	-	+	-	-	+	+	-
8	-	-	+	+	+	+	-
9	+	+	-	-	-	-	+
10	+	-	+	+	-	-	+
11	-	-	+	-	+	-	+
12	-	+	-	+	+	-	+
13	-	+	+	-	-	+	+
14	-	-	-	+	-	+	+
15	+	-	-	-	+	+	+
16	+	+	+	+	+	+	+

The first factor is the learning threshold. In this case, it was set to 0.0001. The learning process stops when the error for all of the cases falls below 0.0001. The threshold measures the sum of the squares of the errors in the sample cases. The system is capable of providing real time minimum error updates as learning progresses. A histogram shows what percentage of the training cases is within a particular error range. The system converges when the entire training set falls below the minimum error threshold.

The second factor is the number of hidden nodes. The hidden nodes are necessary to add one or more layers of nodes in addition to the input and output nodes, and enable the learning of more complex patterns. There is no set formula to choose the proper number of hidden nodes, but as the number of hidden nodes increases, so does the learning, but only to a certain point. As the number of hidden nodes increases past this point, the system's error rate does not decrease significantly, but the time it takes for the system to learn the input data, greatly increases. In this design, 10 hidden nodes proved to be sufficient.

The third factor is the learning rate. The weights leading to an output node are modified during the learning in the direction required to produce a smaller error the next time the same pattern is encountered. The amount of this weight modification is proportional to how much error is present. The greater the learning rate, the greater the weight changes, and the faster the learning. If the learning rate is relatively large, oscillation of the weight changes occurs, the learning is never complete, and the system converges to a non-optimum solution. For this project, a learning rate of 0.1 worked sufficiently.

The last factor is the momentum. The momentum is closely related to the learning rate because it determines the proportion of the last weight change that is added into the new weight change. This effect can allow faster learning without oscillation. In this project, the momentum was set to 0. This choice slows the learning process, but provides the system with better chance of converging.

The neural network was trained with the data generated from the fractional factorial design run of the pack rolling process. A pack rolling model, which was validated with actual pack rolling experiments, was used to generate the data for the neural network model. Due to the non-linear behavior of the problem, the training of the neural network was carried out in two sets of pack rolling passes.

3.5.1.4 Data Evaluation

Pack rolling output information from the trained neural network was validated with the validated pack rolling model. The validation was carried out for a multi-pass pack rolling process and is discussed in more detail in the validation section of this paper.

3.5.2 Development of the Expert System

The development of an expert system is a critical activity which is accomplished by utilizing techniques from the areas of software engineering (SE), database management (DBMS), operating systems (OS), analytical modeling (AM), and artificial intelligence (AI). An expert system is an intelligent system for prompting and aiding the design engineer in an effort to generate effective solutions to problems. The roots of an expert system, as discussed previously, are derived from the field of artificial intelligence. The use of AI is two fold:

- Developing a methodology for the manufacturing sequence of specified properties by using the inference process, and
- Retrieving, handling, recording, and processing information during the design, and implementing decisions.

The development of the expert system can become much easier if the intelligent system employs the following capabilities:

- Software Integration: for firing of rules and data flow algorithms and access information which the design engineer might have.
- Data Base Management: for storing and retrieving project information. Furthermore, the knowledge base contains the domain facts, beliefs, designers' perspectives, and analytical solution to problems.
- Operating System: to help integrate the various components and develop an interface monitor which will allow information exchange from the data base or other knowledge base libraries.

The expert system rules were developed using VP-EXPERT, from WordTech Systems, Inc., an inexpensive commercially available expert system package. The pertinent steps and phases in the development of the expert system are listed below:

3.5.2.1 Material Data Base Development - Generation of Stability Regions

To model and manufacture any material into a "net-shaped" part requires that the material behavior and the processing conditions are well understood. Special purpose alloys usually require complex sequence of operations and critical control of the process parameters. Constitutive models are analytical relations which describe the variation of material properties such as, flow stress or grain size with process variables namely, strain, strain rate and temperature. Constitutive equations are required for modeling material processes. These relations describe the linear or nonlinear behavior of the state variables such as, flow stress and grain size, with the processing variables. In order to fully optimize the pack rolling process, the constitutive relations need to be developed and stability regions established. The stability regions identify the good and bad processing areas for the material. In othèr words, processing in the unstable region will yield a product with defects and unsuitable for use. For hot forming, as is the present case, the stability region can be defined as a function of the material strain, strain rate (rate of deformation), and the temperature. Therefore, the three processing parameters need to be monitored and controlled at all times for the process to be in the stable domain.

3.5.2.2 Development of the User Interface

The role of the expert system in the control of the pack rolling process is to:

- Establish an interface with the user,
- Control the pack rolling process, and
- Incorporate the trained pack rolling neural network to model the process.

3.5.2.3 Incorporation of Constraints

The Expert system serves as a monitor for the pack rolling process and automatically adjusts the variables such that the pack rolling constraints are not violated. The pack rolling constraints are a combination of material and machine constraints. The user initially specifies the material and the machine limitations which generally do not change. The expert system control algorithm adjusts the pack rolling parameters automatically and adaptively changes the inputs to the trained pack rolling neural network. The output from the neural network is once again checked to match the constraints.

3.5.2.4 Interface with the Neural Network

The expert system is linked to a neural network through a software interface. As shown in Figure 3-3, the expert system furnishes a set of input data to the pack rolling neural network. The output from the neural network is processed back to the expert system which evaluates the constraints and the functional requirements. If the pack rolling constraints are not satisfied, the expert system adjusts the input conditions and again passes them on to the neural network for a new test. In this iterative manner a proper pack rolling solution is achieved.

3.5.3 Intelligent Control Algorithm

The strategy for the control of the pack rolling process using the Intelligent Hybrid System is shown in Figure 3-5. The desired properties supplied to the system by the user serve as a primary source for the design constraints. The parameters supplied by the user include:

- Material name,
- Desired output thickness of the workpiece,
- Strip width, and
- Available starting thickness of the cover plate and the workpiece.

The goal of the design is to reach the desired workpiece thickness within the optimum operating conditions. The expert system is linked to a material database and a machine database. The material database contains material properties and stability regions for user specified materials. The machine capabilities and limitations are generally fixed for a particular machine. The workability limit of the material establishes the number of passes and the number of annealing cycles required for the pack rolling process. The material and machine constraints dictate the selection of the process parameters or the design parameters. Generally, the constraint values lie within a certain range. The initial parameter settings made by the expert system lie within the center of these ranges. The parameters supplied to the trained pack rolling neural model are:

- Intermediate input thickness for the pass,
- Pass number,
- Maximum strain rate,
- Minimum billet temperature after rolling,
- Force,

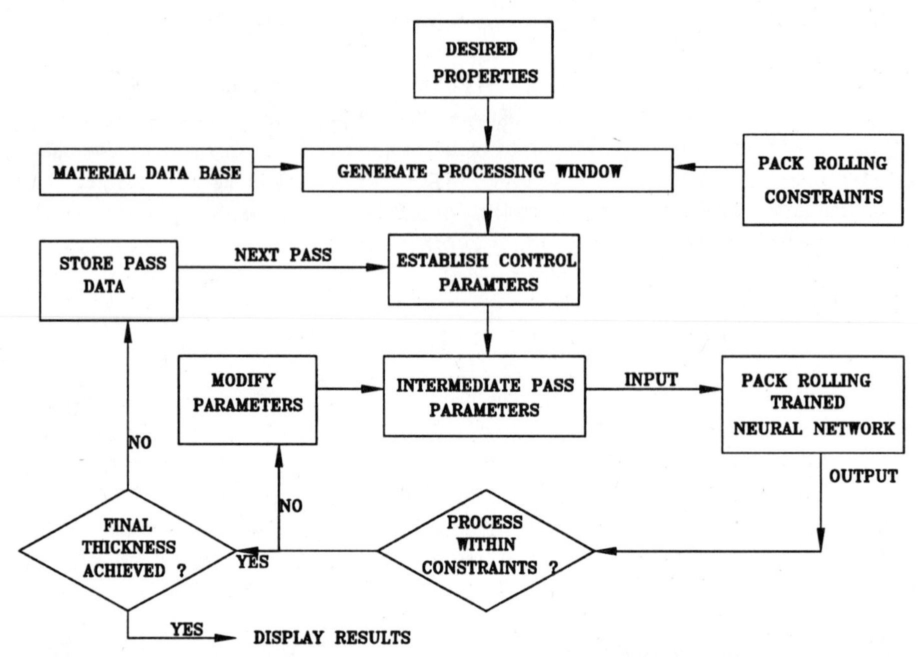

Figure 3-5: Intelligent Hybrid System Design and Control Methodology for Pack Rolling

- Reduction ratio,
- Torque, and
- Strip width.

The selection of the force and torque calculations are based on an estimated initial calculation. The force and torque values are refined in an iterative manner by sampling the output of the neural network model. The relative thickness of both the cover plate and the workpiece is employed as a criteria for proper force and torque selection. The trained pack rolling neural network model is invoked with the aforementioned input values and it yields a set of corresponding design information. The output from the neural network is passed back to the expert system which tests the stability of the pack rolling process in its entirety. The expert system also determines if all the design parameters lie within the pack rolling constraints. The outputs from the neural network to the expert system are:

- Output thickness for the pass,
- Insulation,
- Roll speed,
- Roll diameter,
- Initial billet temperature, and
- Starting roll temperature.

The thickness supplied by the last pack rolling pass serves as an input for the next passes. In a similar manner, the expert system generates a set of input conditions for the neural network. This process is repeated until the final desired thickness of the workpiece material is reached.

Each parameter from the neural network is checked by the expert system for stability. Specifically, the expert system checks for any violations of the roll speed based on the maximum and minimum roll mill speed, the initial billet temperature based on the furnace limitation, and the roll temperature based on the maximum allowable temperature of the rolls. If any of the above parameter constraints are not met, the expert system modifies the appropriate input parameters, by creating a new set of input data for the neural network.

3.6 VALIDATION

Experimental validation is an important and integral part of any software development process. The objective is to validate the hybrid system solutions, along with all the developed tools for the pack rolling process. During a typical pack rolling sequence a number of thermal and deformation steps can take place as shown in Figure 3-6. The pack rolling operation sequence comprises of a heating or annealing process, a rolling operation, and a dwelling operation. A dwelling operation is primarily the heat lost during the transfer time from the rolling mill to the furnace. The process is again followed by another annealing operation in preparation for another pack rolling process as shown in the figure. The validation of this approach will assist in acquiring a better insight of the process as well as in establishing a confidence in the hybrid system.

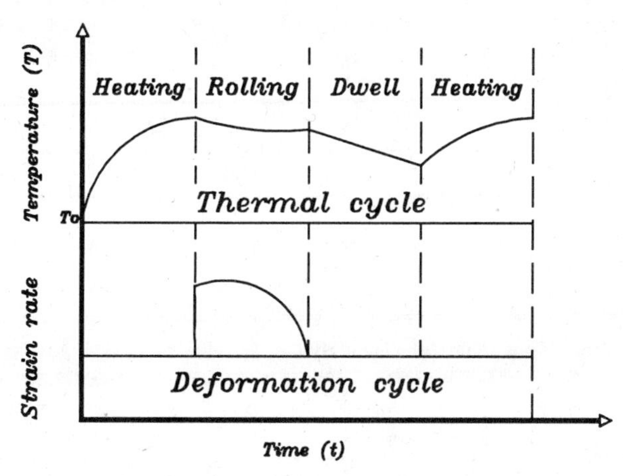

Figure 3-6: A Typical Thermo-Mechanical History During Multi-Pass Pack Rolling

The hybrid system was validated with the pack rolling model developed at Ohio University. The pack rolling model and the intelligent hybrid system model are different in that the pack rolling model is an analytical model, while the intelligent system is a design model. In other words, the input parameters for the pack rolling model are the outputs of the intelligent system and vice versa as shown in Figures 3-1 and 3-3. Tables 3-4 and 3-5 list the input and output parameters respectively, for a three pass, 10% pack rolling reduction using the pack rolling model. Table 3-5 was generated by furnishing the values listed in table 3-4 to the pack rolling model. The trained neural network for pack rolling is validated with the pack rolling experiment presented in tables 3-4 and 3-5. Tables 3-6 and 3-7 provide the input and output parameters for a corresponding pack rolling design using the intelligent hybrid system. The pack rolling design is accomplished in three 10% passes to match the pack rolling model. The hybrid system is provided with the output data from the pack rolling model which serves as input to the intelligent system. The output thickness of the cover plate and the workpiece, the initial billet temperature, and the roll speed match within 1% to 5% of the pack rolling model results.

Table 3-4: Input Data: Pack Rolling Model (3 passes at 10 % each)

Pass #	Workpiece input (in.)	Cover Plate input (in.)	Roll Speed (rpm)	Starting Pack temp. (°F)	Roll temp.	Reduction (%)	Insulation
1	.450	.172	6.0	2350	100	10	YES
2	.411	.151	6.0	2350	100	10	YES
3	.371	.136	6.0	2350	100	10	YES

Table 3-5: Output Data: Pack Rolling Model (3 passes at 10 % each)

Pass #	Workpiece output (in.)	Cover Plate output (in.)	Strain rate (1/sec)	Workpiece temp.	Roll Force (lbs)	Roll Torque (lb-in)
1	.411	.151	.7112	2330	11957	5740
2	.371	.136	.8353	2328	12707	6255
3	.331	.124	.9825	2322	13386	6397

Table 3-6: Input Data: Intelligent Hybrid System (3 passes at 10% each)

Pass #	Workpiece input (in.)	Cover Plate input (in.)	Strain rate (1/sec)	Workpiece temp. (°F)	Roll Force (lbs)	Roll Torque (lb-in)	Reduction (%)
1	.450	.172	.7112	2330	11957	5740	10
2	.404	.149	.8353	2328	12707	6255	10
3	.366	.131	.9825	2322	13386	6397	10

Table 3-7: Output Data: Intelligent Hybrid System (3 passes at 10 % each)

Pass #	Workpiece output (in.)	Cover Plate output (in.)	Roll Speed (rpm)	Starting Pack temp (°F)	Roll temp. (°F)	Insulation
1	.404	.149	6.11	2343	103	YES
2	.366	.131	6.18	2345	102	YES
3	.318	.112	6.36	2356	105	YES

Tables 3-8 and 3-9 show the input and output parameters, respectively, for a second pack rolling simulation using the pack rolling model. A three-pass operation with 20% reduction is compared with the results of the intelligent system. The input and output parameters from the intelligent system are shown in Tables 3-10 and 3-11, respectively. The hybrid system results for this case are within 5% of the pack rolling model results.

Table 3-8: Input Data: Pack Rolling Model (3 passes 20 % each)

Pass #	Workpiece input (in.)	Cover Plate input (in.)	Roll Speed (rpm)	Starting Pack temp. (°F)	Roll temp. (°F)	Reduction (%)	Insulation
1	.450	.172	10.0	2350	200	20	YES
2	.368	.133	10.0	2350	200	20	YES
3	.296	.106	10.0	2350	200	20	YES

Table 3-9: Output Data: Pack Rolling Model (3 passes 20 % each)

Pass #	Workpiece output (in.)	Cover Plate output (in.)	Strain rate (1/sec)	Workpiece temp. (°F)	Roll Force (lbs)	Roll Torque (lb-in)
1	.368	.133	1.598	2340	16228	12105
2	.296	.106	1.876	2332	16249	11495
3	.234	.086	2.216	2320	16050	10363

Table 3-10: Input data: Intelligent Hybrid System (3 passes 20 % each)

Pass #	Workpiece input (in.)	Cover Plate input (in.)	Strain rate (1/sec)	Workpiece temp. (°F)	Roll Force (lbs)	Roll Torque (lb-in)	Reduction (%)
1	.450	.172	1.598	2340	16228	12105	20
2	.367	.129	1.876	2332	16249	11495	20
3	.291	.102	2.216	2320	16050	10363	20

Table 3-11: Output data: Intelligent Hybrid System (3 passes 20% each)

Pass #	Workpiece output (in.)	Cover Plate output (.in)	Roll Speed (rpm)	Starting Pack temp. (°F)	Roll temp. (°F)	Insulation
1	.367	.129	9.89	2353	198	YES
2	.291	.102	9.91	2345	201	YES
3	.230	.087	10.03	2359	206	YES

In this section it is demonstarted that the trained neural network provides results which are in excellent agreement with the pack rolling model which has been tested and validated with actual experiments [12, 13]. The trained neural netwrok can thus be employed to design and control the pack rolling process and is described in the next section.

3.6.1 Sample Pack Rolling Design - A Case Study

The objective of the proposed design is to achieve a certain desired thickness, 0.3 in., of the workpiece material using the pack rolling process while satisfying all the material and machine constraints. The traditional pack rolling approach of a five-pass operation with a starting workpiece thickness of 0.5 in. is also presented for reasons of comparison. The salient parameters for this case study are as follows:

Desired Requirements
- Workpiece Thickness: 0.3 in., and
- User Specified Material.

Starting Material
- Workpiece Thickness: 0.5 in.,
- Cover Plate Material: 0.172 in., and
- Strip Width: 4 in.

Material Constraints
- Maximum Strain: 0.3,
- Strain Rate Limit: 0.05-3.0, and
- Minimum Temperature of the Workpiece: 1950 ^{0}F.

Machine Constraints
- Maximum Machine Force: 100,000 lbs,
- Maximum Machine Torque: 75,000 lb-in,
- Maximum and Minimum Roll Speed: 15 RPM and 3 RPM Respectively,
- Maximum Furnace Temperature: 2,400 ^{0}F, and
- Roll Radius: 4.095 in.

The pack rolling design using the intelligent hybrid system for this case study is presented in Tables 3-12 and 3-13. Based on the machine and material limitations, the pack rolling design for this case is accomplished in two pack rolling passes with one annealing operation between each pass. The design output for the case study is shown in Table 3-13. The two passes represent a 25% and an 18% reduction in achieving a workpiece thickness of 0.3005 in. from a starting workpiece thickness of 0.5 in. All process parameters satisfy the constraints of the process.

Table 3-12. Design Input: Intelligent Hybrid System - Pack Rolling Case Study

Pass #	Workpiece Input (in.)	Cover Plate Input (in.)	Roll Force (lbs)	Torque (lb-in)	Reduction (%)
1	.500	.172	22142	16895	25.9
2	.369	.139	17683	13357	18.9

Table 3-13: Design Output: Intelligent Hybrid System - Pack Rolling Case Study

Pass#	Workpiece output (in.)	Cover Plate output (in.)	Roll Speed (rpm)	Starting Pack temp.(°F)	Roll temp. (°F)	Insulation
1	.3699	.1399	8.21	2330	90	YES
2	.3005	.1064	8.18	2325	98	YES

A traditional five pass pack rolling simulation is shown in Tables 3-14 and 3-15. Table 3-14 and 3-15 show an actual case of pack rolling operation which is not optimized. As shown in the tables, the pack undergoes four annealing operations in between passes to relieve internal stresses. The traditional operation, in contrast to the design generated from the hybrid system, has three additional passes with three additional annealing cycles. Comparison of tables 3-13 and 3-15 show that the hybrid system is capable of optimizing the pack rolling process.

Table 3-14: Input Data: Traditional Pack Rolling Approach

Pass #	Workpiece input (in.)	Cover Plate input (in.)	Roll Speed (rpm)	Starting Pack temp. (°F)	Roll temp. (°F)	Reduction (%)	Insulation
1	.500	.172	6.0	2350	100	10	YES
2	.456	.152	6.0	2350	100	10	YES
3	.413	.136	6.0	2350	100	10	YES
4	.368	.124	6.0	2350	100	10	YES
5	.328	.113	6.0	2350	100	8	YES

Table 3-15: Output Data: Traditional Pack Rolling Approach

Pass #	Workpiece output (in.)	Cover Plate output (in.)	Strain rate (1/sec)	Workpiece temp.	Roll Force (lbs)	Roll Torque (lb-in)
1	.456	.151	.7009	2336	12583	6051
2	.413	.136	.8008	2330	13029	6646
3	.368	.124	.9510	2324	13674	6833
4	.328	.113	.9859	2320	12760	6052
5	.299	.105	.9712	2317	11166	4457

3.7 CONCLUSIONS AND RECOMMENDATIONS

The intelligent hybrid system developed here follows a methodology which may be applicable to any generic metal forming process and has been successfully demonstrated for the pack rolling processes. The case study demonstrates that a pack rolling process design was achieved by exploiting intelligent AI techniques which:

- Provides an optimum design,
- Provides a one-shot solution compared to iterative techniques,
- Does not violate any constraints,
- Provides a design process as opposed to an analytical tool, and
- Conserves resources.

The intelligent hybrid system has been successfully demonstrated and validated for the pack rolling process and provides results within a 5% of the pack rolling model results. The validations provide a great deal of confidence in this methodology, the intelligent system and their use in the metal forming industry. Some of the key benefits from this methodology are listed below:

- It provides a scientific approach to metal forming,
- It can be made property driven,
- It provides a design methodology instead of an iterative solution technique,
- It saves significant amount of computer time (as opposed to finite element models),
- It provides a better methodology to control manufacturing processes,
- It supports the identification of critical parameters,
- It yields the optimum operating conditions,
- It exploits the power of intelligent design tools,
- It harnesses the fruits of expert system and neural network synergism,
- It bridges the knowledge-base gap among material science, metal forming, and design, and
- It assimilates knowledge from AI, mechanical design, and metal forming to solve actual problems.

Proper design is critical for any product. The intelligent hybrid system and the design methodology presented in this paper can realize significant savings in resources, time and material, which are vital for a competitive metal forming industry. The proposed system also reduces the lead time for metal forming operations.

ACKNOWLEDGEMENTS

The authors would like to thank Mr. Daniel Baker and Mr. Sushil Jain of GM-Allison Gas Turbine Division, Indianapolis, Indiana, and Mr. William Kerr, Of WPAFB, Dayton, Ohio, for their support in the development of the pack rolling model. All proprietary information has been withheld.

REFERENCES

1. Dixon, J.R., *Design Research - Trends and Future Directions,* CIRP Working Seminar on Artificial Intelligence Based Product Design, 1990, pp. 89-98.
2. Suh, N.P., *Axiomatic Approach to Design of Products and Processes,* CIRP Working Seminar on Artificial Intelligence Based Product Design, 1990, pp. 1-12.
3. Ham, I., and Kumara, S., *AI Based Product Design - Review and Examples,* CIRP Working Seminar on Artificial Intelligence Based Product Design, 1990, pp. 21-53.
4. Yoshikawa, H., Tomiyama, T., *Design Philosophy and Intelligent Product Design,* CIRP Working Seminar on Artificial Intelligence Based Product Design, 1990, pp. 13-20.
5. Patterson, D.W., Introduction to Artificial Intelligence & Expert Systems, Prentice Hall, 1990.
6. Wright, P.K. and Bourne, D.A., Manufacturing Intelligence, Addison Wesley, 1988.
7. Knapp, G.M. and Hsu-Pin, W., *Neural Networks in Acquisition of Manufacturing Knowledge,* Intelligent Design and Manufacturing, Edited my Andrew Kusiak, John Wiley and Sons, 1992, pp. 723-744.
8. Zarefar, H. and Goulding, J.H., *Neural Networks in Design of Products: A Case Study,* Intelligent Design and Manufacturing, Edited by Andrew Kusiak, John Wiley and Sons, 1992, pp. 179-201.
9. Gunasekera, J.S., and Malas, J.C., *Conceptual Design of Control Strategies for Hot Rolling, Annals of CIRP,* 1987, Vol.40, No. 1, pp. 123-126.
10. Malas, J.C, *Methodology for Design & Control of Thermomechanical Processes,* Ph.D. Dissertation, Ohio University, 1991.
11. Suh, N.P., The Principles of Design, Oxford University Press, New York, 1990.
12. Ali, A.F., Rathinavel, A., Gunasekera, J.S., and Kerr, W.R., *Analytical Modeling of the Pack Rolling Process,* Transactions of the North American Manufacturing Research Institution of SME, May 1991, pp. 29-36.
13. Gunasekera, J.S., Ali, A.F., Rathinavel, A., and Dewasurendra, L., *Design, Analysis and Experimental Verification of the Pack Rolling Process,* Annals of the CIRP, June 1993.
14. Hillman, D.V., *Integrating Neural Nets and Expert Systems,* AI Expert, June 1990, pp. 54-59.
15. Kandel, A., and Langholz, G., *Intelligent Expert Networks in Communication and Control,* The World Congress on Expert Systems Proceedings, 1991, pp.239-245.

4

INTELLIGENT DIAGNOSTICS FOR INTEGRATED MANUFACTURING SYSTEMS

James H. Graham, University of Louisville

Jian Guan, University of Louisville

4.1 INTRODUCTION

The large scale automation and integration of manufacturing systems, which has become possible with the development of low cost digital computers and communications networks, permits more efficiency and flexibility in meeting production schedules and can potentially lead to lower cost and higher quality products. However, an integrated system is very dependent upon the trouble-free operation of all of its component parts; when a failure occurs it is critical to isolate the causes as rapidly as possible and to take appropriate corrective action.

This chapter describes a system for the monitoring and diagnosis of manufacturing systems which can best be described as a knowledge-based diagnosis system that uses a hybrid combination of symptom-based and functional reasoning. In this research we have used the functional hierarchy of the manufacturing systems, defined in terms of strata, as the backbone for the deep reasoning. Sets of production rules, classified in terms of echelons, are the basis for the shallow reasoning. An entropy-based cost function is used as the basis for selection of the next component to be examined. Two types of learning have been investigated to improve the performance of the shallow knowledge component: rule learning and probability updating.

Section 4.2 provides an overview of diagnostic reasoning in deep and shallow forms. Section 4.3 discusses the hybrid diagnostic model as applied to computer integrated manufacturing and Section 4.4 discusses the learning features of the model. Section V covers implementation issues in creation of the prototype diagnostic system and presents data from a case study. Section 4.5 summarizes the conclusions from this work and offers suggestions for future research directions.

4.2 OVERVIEW OF KNOWLEDGE-BASED DIAGNOSIS

Most knowledge-based diagnosis systems can be classified as either operating from a deep-knowledge (functional) basis or a shallow-knowledge (symptom recognition) basis. Shallow reasoning is typified by a rule-based system that attempts to match observed symptoms to fault causes by using a short series of inferences with the symptoms as inputs [1]. This is the traditional diagnostic expert system, which is easy to implement but which has essentially no knowledge of the structure of the domain. The advantages of this approach are that general purpose inference engines are readily available as commercial products, and the speed of reasoning is usually sufficiently fast to appear to act in real time for most small diagnostic problems.

By contrast, the deep-reasoning expert system attempts to construct a frame of inference that takes advantage of known functional relationships in arriving at a diagnosis [2-7]. When the observed behavior of the system does not match the correct operating behavior, then an inferential process is invoked to determine the system components which can explain the abnormal behavior. Several general approaches have been suggested, including the use of probabilistic reasoning [5], set covering [6], and theorem provers [7]. In dynamic systems, the deep, or structural, knowledge of the system has been represented by fault propagation digraphs [8-11], failure trees [12] and Petri nets [13-14].

Both the shallow reasoning approach and the deep reasoning approach have limitations which restrict their applicability. The shallow reasoning approach knows essentially nothing about the domain beyond the explicitly coded information in the rule base. It is generally incapable of dealing with situations which lie beyond these hard coded conditions. Shallow knowledge systems thus tend to be somewhat brittle unless exceptions are carefully, and exhaustively, programmed. In contrast, the deep reasoning approach can handle a wider range of input conditions, but typically requires extensive development time, and consumes excessive computational resources in creating a diagnosis.

Recently, hybrid diagnosis systems have been created for use in complex systems. These systems tend to combine both the ability to perform a rapid diagnosis and the ability to make use of functional information to avoid brittleness [15-19]. These systems can utilize either (1) a shallow reasoning phase, followed by a deep reasoning phase, (2) a deep reasoning phase followed by a shallow reasoning phase, or (3) some other combination of deep and shallow reasoning. Lee [18-19], for example, suggested a deep search based on system interconnections, leading to individual shallow knowledge bases for diagnosis of specific subcomponents.

One of the remaining shortcomings of knowledge-based diagnosis is the problem of acquiring the experiential knowledge for the shallow knowledge bases. Although machine learning has been investigated for some time in various areas of artificial intelligence, the literature on applications of learning techniques to diagnostic reasoning systems is relatively scarce. Pazzani [20] discusses an application of failure-driven learning to the construction of a diagnostic knowledge base. Koseki [21] describes a model-based diagnostic system which uses explanation based learning to generalize from single training samples. The AQ11 program of Michalski [22] demonstrated the ability to learn the rules for diagnosis of soybean disease from training samples. The balance of this paper will discuss a new approach to the application of knowledge-based learning in improving the performance of a fault diagnostic system for computer integrated manufacturing.

4.3 HYBRID DIAGNOSTIC ALGORITHM

The representation of a diagnostic model can be described from a variety of viewpoints, depending on the underlying system. This research has concentrated on a heirarchically intelligent manufacturing system, broken into three levels of physical control, coordination, and executive [24]. A block diagram on a typical system is shown in Figure 1.

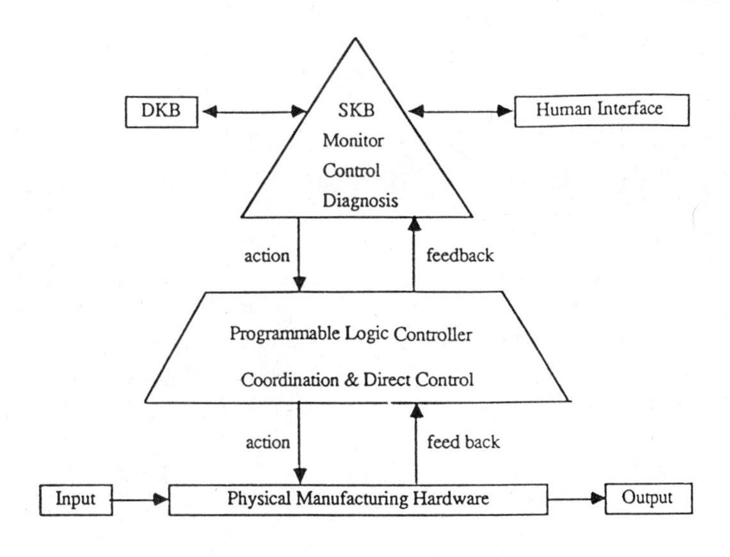

Figure 1 - CIM System with Monitoring and Diagnosis

The lowest level in the hierarchy is the physical hardware control level, which comprises a set of sensors and actuators which control and monitor the producion of the physical product. This level is controlled, in a typical manufacturing system, by a mid-level coordination and control computer, often implemented by a programmable logic controller (PLC), because of the simplicity of programming and relatively low cost of such units. Unfortunately, PLC's do not typically have good high level programming and user interface capabilites. Because of the limitations of the PLC, the task of monitoring and diagnosis is often given to an independent computer, which processes data from the PLC, and interfaces both with human operators of the manufaturing cell, and with plant-wide information systems [15].

Using the notation of Mesarovic [23], to describe the organization of the hybrid diagnostic system, the functional hierarchy of the manufacturing system is represented as a multistrata form, in which each level in the hierarchy represents a new level of functional decomposition of the system as shown in Figure 2. This information is typically available from the design specifications for the system. Each terminal node would be associated with a physical subsystem of the manufacturing process and would have an associated deep and shallow knowledge bases. The shallow knowledge component is represented in multiechelon form. Note that in this case, decision-making may potentially occur at each rule. Figure 3 is an example of the rules in a shallow knowledge base at one echelon level.

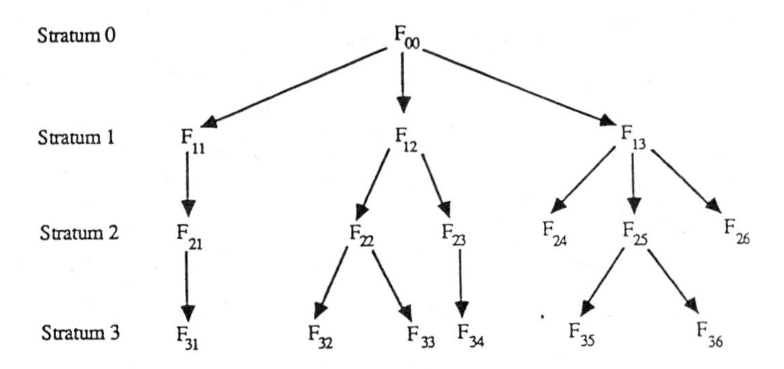

Figure 2 - Functional Hierarchy for CIM System

Within this hierarchical representation, the first task in diagnosis is to identify the subcomponent or subcomponents that need diagnostic activity. From a practical viewpoint, much of this initial decision making can done directly from the information obtained from the PLC. In a good system design, a number of the possible failure points will be instrumented with sensors, and the failure mode can be isolated to a specific subcomponent with a single rule in the top level shallow knowledge base. However, it is not usually possible to give complete coverage to failure points by sensors, and in some cases, one fault may trigger fault symptoms in other subcomponents through fault propagation. In these cases, methods based upon the use of fault propagation digraphs [8-11], can be used as a top level deep reasoning tool to isolate the proper subcomponent. The authors have developed an algorithm that produces a candidate set of possible faulty nodes using reachability analysis on the fault propagation digraph, then reduces the candidate set using sequential testing [26].

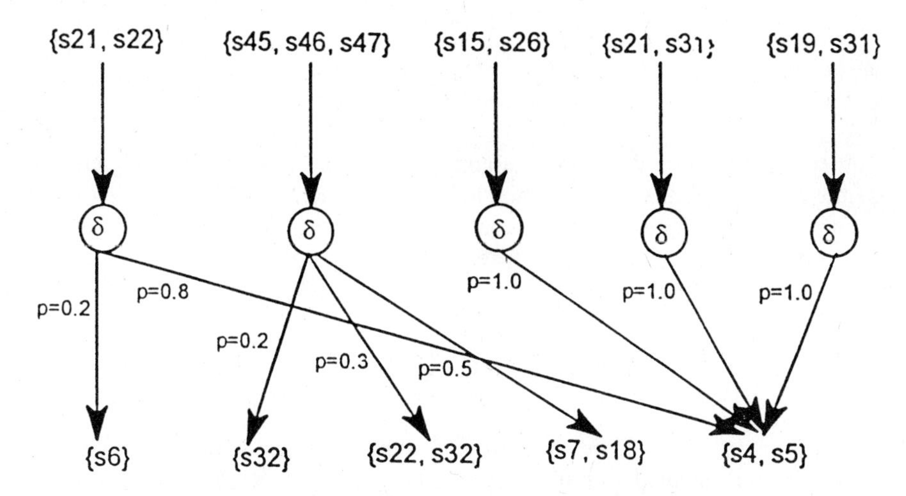

Figure 3 - Typical Rule-Base for CIM Subcomponent

At the subcomponent level, the previous procedure is largely repeated. The associated shallow (rule-based) knowledge base is checked first to see if there if the fault symtom is covered by an existing rule (or rules). If this fails, then the deep knowledge base for the subcomponent is activated, and attempts to find a candidate set of faulty nodes which will cover all observed symptoms. At the completion of this diagnostic activity, one or more nodes are physically tested, and appropriate repair action is taken by plant maintenance personnel. Based upon the success of the diagnostic process, the shallow and deep knowledge bases are updated based upon algorithms given in section 4.

The detailed steps in the diagnostic process are given as follows:

Step 1: Construct a functional hiearachy of the system and the associated deep knowledge bases. Incorporate any top level diagnostic information into a rule-based knowledge base. (Note - this is typically done off-line)

Step 2: When a symptom is observed, attempt to localize the fault to a specific subcomponent by checking first the top level shallow knowledge base, and then, if necessary, use top level deep knowledge to identify a faulty subcomponent.

Step 3: At the sucomponent level, check the shallow knowledge base for a failure mode. Use the cost weighted entropy criterion discussed below to make selection of test procedure, if multiple failure modes exist.

Step 4: If Step 3 fails, then use the deep reasoning fault propagation algorithm to determine a fault candidate set for the subcomponent.

Step 5: Physically test the indicated unit for the fault indicated.

Step 6: Based upon the results from step 5, update the diagnostic rules, using either:
a. the rule updating procedures of sections 4.1-4.2, or
b. the rule probability updating procedures of section 4.3

In step 3 of the hybrid diagnosis algorithm, the cost weighted entropy criterion is used to choose the next part of the shallow knowledge base to activate. This entropy criterion gives the maximum fault discernment per unit cost, and is defined by the following equations:

$$H_{ijk}(w,p) = -\sum_{x=1}^{t} w_{i+1,k,x}\, p_{i+1,k,x} \ln p_{i+1,k,x}\,, \; i \geq 2$$

$$\text{where } w_{i+1} \geq 0, \; \sum_{x=1}^{t} p_{i+1,k,x} = 1.$$

This entropy, a modification of the standard Shannon entropy equation, is calculated from the failure probabilities associated with the k-th group of the i-th echelon of strata j. The weighting factor, w, is a normalized cost, determined by dividing the actual cost of a measurement, by the maximum of the set of measurement costs for all components at the current rule level. Ties are broken, by choosing the lowest cost in absolute terms. Based upon the statistical properties of entropy, the following result can be established.

Proposition 1 - When a choice between different tests must be made, choosing the test with the lowest cost weighted entropy guarantees the greatest discrimination per unit test cost.

Proof - With unit weighting factors (i.e. all tests have equal cost), the formula is the standard Shannon entropy, which gives maximum information for the lowest entropy value. The weighting factors are all normalized to fall between 0 and 1.0, thus they tend only to scale the additive terms in the entropy results.

4.4 LEARNING CAPABILITIES

This section describes the learning capabilities that have been incorporated into the system to provide knowledge acquistion and knowledge updating during the on-line operation of the diagnostic system.

4.4.1 Conversion of a diagnostic case into rule form

A confirmed diagnostic case is represented as $D(S,C)$, where S = *{s|s is a subdevice sensed to be abnormal}* is the set of symptoms; and C = *{c|c is a subdevice whose failure has been confirmed}* is the set of causes or subdevices whose collective failure has caused the occurrence of the symptoms in S.

Two types of rules can be extracted from a diagnostic case $D(S, C)$. The first type of rule has the following form:

If {s1,s2,...,sn} then c

where $n = |S|$, $|C| = 1$, and $C = \{c\}$. In this case one failed subdevice c caused all the symptoms; i.e., there is only one failure source. The experience or heuristic expressed by the rule says that if the sensors or alarms attached to subdevices *s1* through *sn* go off, then possibly the subdevice *c* is the failure source.

In the second case, there are multiple failure sources or $|C| > 1$. The rule extracted takes the following form:

If {s1,s2,...,sn} then {c1,c2,...,cm}

where $n = |S|$ and $m = |C|$. This rule says that if subdevices *s1* through *sn* are sensed abnormal, then possibly subdevices *c1* through *cm* are the failure sources.

4.4.2 Integration of a new rule into the knowledge base

This is the most important step in the process of automatic diagnostic knowledge acquisition and rule updating. The following four cases are considered:

1) The new rule already exists in the current knowledge base. In this case the probability associated with the rule in the knowledge base is adjusted to reflect the fact that the rule has been reinforced so that more weight should be given to it in future diagnosis. This has a bearing on interpreting results of shallow reasoning when several rules fire at the same time, or the same set of symptoms leads to several different sets of failure sources. This is discussed in more detail in section 4.4.3.

2) The antecedent of the new rule *S1* is the same as the antecedent of an existing rule; and the consequent *C1* of the new rule is not the same as the consequent of the existing rule. Let the new rule be :

If S1 then C1

and the existing rule be :

If S2 then C2

Then

$S1 = S2$ *and* $C1 \neq C2$,

In this case the heuristic contained in the new rule is integrated with the existing knowledge as follows:

a. $B = B \cup C1$ *if* $C1 \notin B$.

b. $\delta(S1) = \delta(S1) \cup C1$.

c. *Update* P_i, *for* $i = 1, n + 1$, *where n is the number of existing consequent of S1.*

3) The consequent of the new rule *C1* is the same as the consequent of an existing rule; and the antecedent *S1* of the new rule is different from the antecedent of the existing rule. Let the new rule be :

If S1 then C1

and the existing rule be :

If S2 then C2

Then

$C1 = C2$ *and* $S1 \notin \{\alpha \mid$ *If* α *Then C1 is an existing rule*$\}$

In this case the heuristic contained in the new rule is integrated with the existing knowledge as follows:

a. $A = A \cup S1$ *if* $S1 \notin A$

b. $\delta(S1) = \{\{C1\}\}$
c. Let p = 1.0 for the newly established rule.

4) Neither the antecedent nor the consequent of the new rule can be found in the existing rule base. Let the new rule be

If S1 then C1

Then

$S1 \notin A$ *and* $C1 \notin B$
In this case the rule
If S1 then C1

is added to the rule base as follows:

a. $A = A \cup S1$
a. $B = B \cup C1$
c. $\delta(S1) = \{C1\}$
d. Let p = 1.0 for the newly established rule.

4.4.3 Updating Rule Probabilities

Whenever the rule leads to a correct diagnosis, the probability of applying that rule should be reinforced; otherwise the probabilities of other applicable rules should be reinforced. The reinforcement equations are:

$$p_i(k+1) = p_i(k) + X(k+1)\,[Y_i(k) - p_i(k)] \text{ for all } i$$

where Y(k) is one if the component is faulty and zero otherwise.

These equations have the format of classical stochastic approximations, which leads to the following result.

Proposition 2 - If the series X obeys the Dvoretsky conditions [25],

1. $\lim_{n\to\infty} X(n) = 0$
2. $\lim_{n\to\infty} \sum_{k=1}^{n} X(n) = \infty$
3. $\lim_{n\to\infty} \sum_{k=1}^{n} X^2(k) < \infty$

the p's will converge to the true probability values with probability one.

Proof - The probabilities, p, may be regarded as unknown, but fixed, parameters of the manufacturing system. The update equation, using a series that meets conditions 1-3 above, then meets the requirements for the contraction mapping specified by Dvoretsky's theorem for general stochastic approximations [25].

There are several choices for the correction coefficient series to satisfy the requirements of Proposition 2. The harmonic series is widely used in stochastic approximation applications, and was used for the initial case studies in this research, but it is by no means the only option. As the system runs, the diagnosis efficiency will increase, because the probabilities and entropy cost values become more accurate in directing the diagnostic system toward the component most likely to be faulty. Also, if there is a change in the system, such as an update of one component to make it more reliable, the system will automatically begin to update that component's failure probabilties to reflect this change.

4.5 IMPLEMENTATION

The object-oriented programming approach [27] was chosen for the implementation of the diagnosis system for several reasons. The hierarchical nature of the deep-reasoning diagnosis network was a natural for the class inheritance features of the object-oriented approach. The failure probabilities and associated cost values were easily stored as slot values in the units representing the associated components. The computation of the entropy-based cost functions were performed by attached procedures (methods) as required. Finally, the modular structure of the resulting software is highly flexible and adaptable to other diagnostic tasks. Although the initial implementation was done in Intellicorp's Knowledge Engineering Environment (KEE) system, other object-oriented languages such as C++ could be used for commercial implementation of this diagnostic approach.

Figure 4 shows the structure of the base node F00 in the top level deep knowledge base. Member slots Q and NO-OF-CHILDREN store values which are used to select the

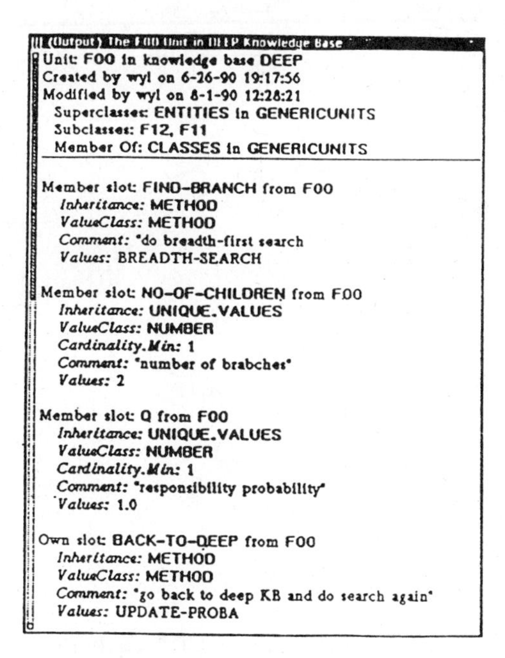

Figure 4 - F00 Unit in Structural Knowledge Base

next functional branch to search. These slots will be inherited by all descendents of F00. The branch selection process is performed by the method BREADTH-SEARCH which resides in the FIND-BRANCH slot, and is activated by sending a message to the appropriate unit (initially F00). Method UPDATE-PROBA is used during the backtracking operation, and the method ENTROPY-CALC which resides in the slot COMPUTE-ENTROPY, computes the cost weighted entropy associated with the next rule selection.

The system output produced during the search by the KEE implementation of the hybrid diagnostic procedure provides a very detailed description of the search process; much more detailed, in fact, than what is really needed by plant operating personnel. Eventually the output from the diagnostic expert system will be routed to a user friendly CIM monitoring display system that has been specifically designed for use on the manufacturing floor [15]. This display system separates the user screen into two areas representing global system information and local station information. A highlighted status bar allows the operator or maintenance worker to easily determine the status of any malfunctioning station. Diagnostic results are presented in a special diagnostics window.

4.5.1 CIM System Case Study

The CIM testbed at the Factory Automation Laboratory at the University of Louisville, shown in block diagram form in Figure 5, includes a Bridgeport Machining Center, a Cincinnati Milacron lathe, a GCA SCARA robot, a General Electric P50 process

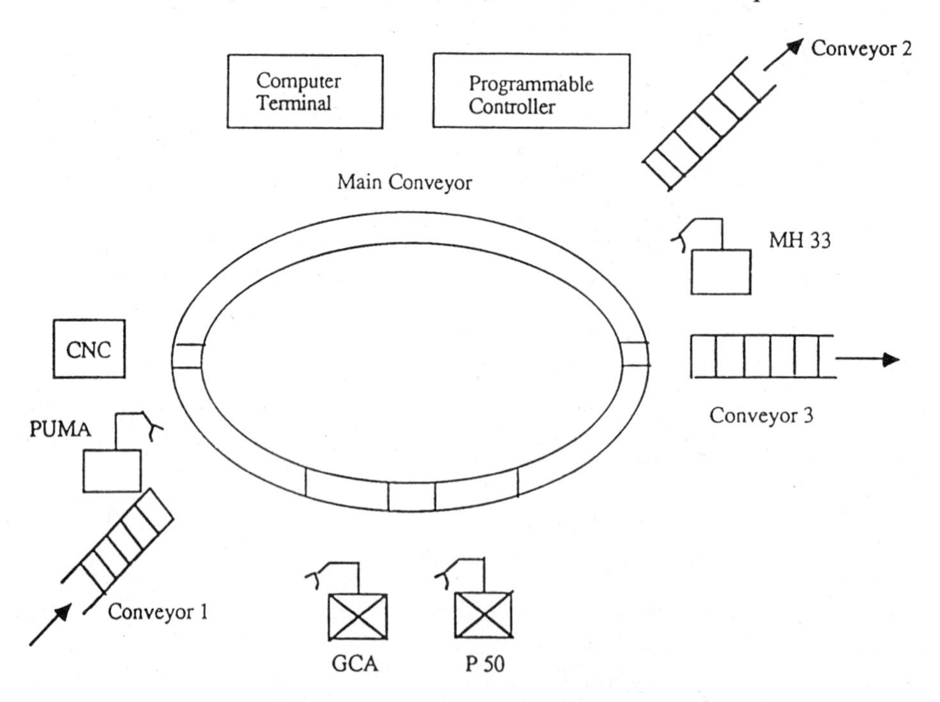

Figure 5 - Experimental CIM Layout

robot, and a Scantek conveyor system. The conveyor system includes a number of specialized stations for indexing, inspection, and barcode reading. This testbed allows the investigation of a wide range of manufacturing cell configurations for metalworking and/or assembly operations.

Several simulations were conducted of a typical knowledge-based diagnostic search for a problem with the material handling system. A search through the deep knowledge base, shown in figure 6, leads to the F32 strata which is a terminal deep knowledge node. Shallow knowledge base R111 is activated at this point, and the cost weighted entropies are computed for the R211 and R212. In this particular case, the GCA robot is not responding due to a perceived fault in the palleting system. Further investigation indicates that the problem may be either with the pallet system, or with a failure of the barcode reader at the station. The operator is advised to check both, and take corrective action. After the diagnosis and repair is completed, the addition of new rules and the update of the probabilities can be performed using the methods presented in section 4.4.

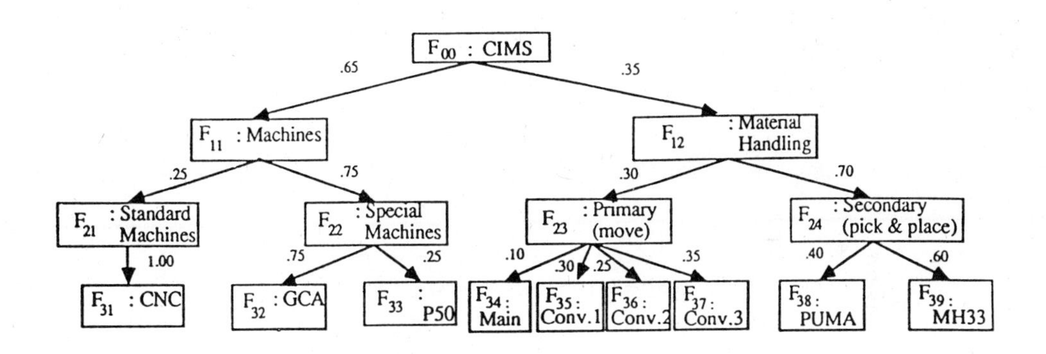

Figure 6 - Knowledge-based Diagnostic Search

Figure 7a shows the integration of the modified rule

If {s21, s22} Then {s11, s13}

into the rule base of Figure 3, using the procedures from section 4.4.2. This illustrates the augmentation of an exisitng rule with an expanded consequent. Figure 7b shows the integration of the new rule

If {s49, s48} Then {s7, s18}

into the rule base, demonstrating the case of finding a new set of symptoms for a previously known failure mode. Finally, Figure 7c shows the integration of a completely new failure mode given by the rule

If {s43, s49} Then {s17}

into the rule base. Repeated application of these procedures, and the updating of rule probabilities, will thus capture the experiential knowledge acquired through the diagnostic process.

(a)

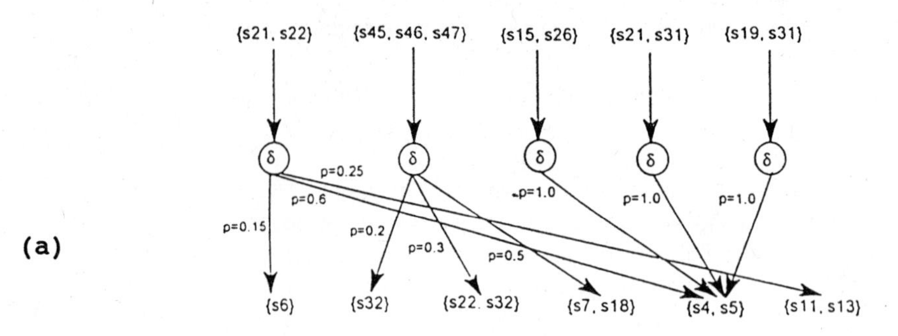

(b)

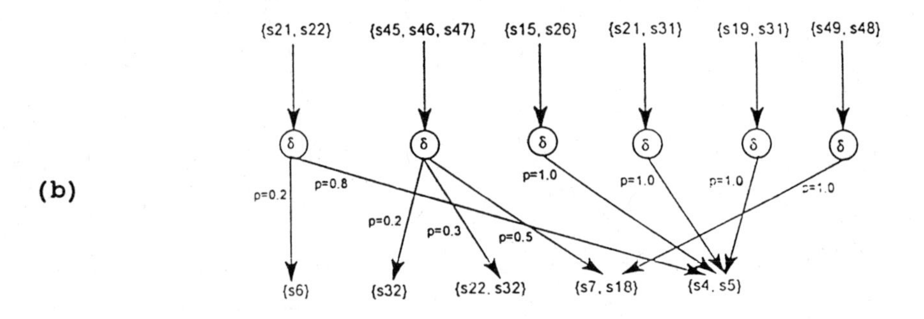

(c)

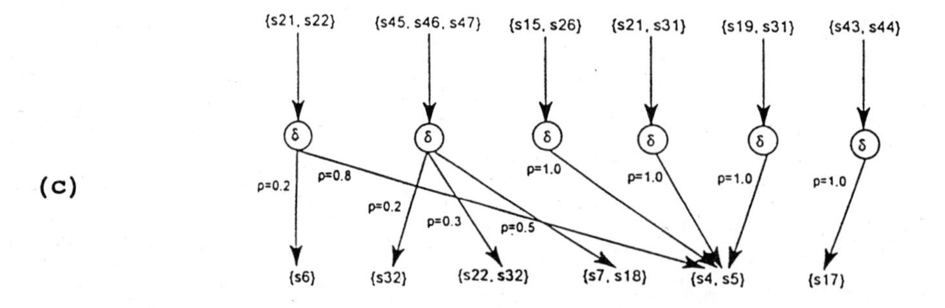

Figure 7 - Updated Rule Base for (a) Expanded Consequent (b) New Failure Symptom, (c) New Failure Mode.

4.6 CONCLUSIONS AND FUTURE DIRECTIONS

This paper has described a methodology for hybrid diagnostic search in computer integrated manufacturing systems with a two phase learning system to update the rule-based diagnostic knowldedge component. Initial testing with the KEE prototype indicates that the approach is both effective and efficient for determining faults in a medium sized manufacturing workcell. Future work will involve additional refinement and testing of the learning features, as there appear to be several additional learning possibilies within the hybrid framework that can be investigated.

ACKNOWLEDGEMENTS

The authors would like to acknowledge the financial support for this research which was provided by the Kentucky Center for Robotics and Manufacturing Systems, the General Electric Company, and the Henry Vogt Machine Company of Louisville, Kentucky. They would also like to acknowledge the use of the facilities of the Factory Automation Laboratory of the Speed Scientific School of the University of Louisville.

RERERENCES

[1] L. Brownston, et al, Programming Expert Systems in OPS5: An Introduction to Rule-Based Programming, Reading, MA: Addison-Wesley, 1985.

[2] R. Davis, "Diagnostic Reasoning Based Upon Structure and Behavior," Artificial Intelligence, 24(1-3), 1984, pp. 347-410.

[3] R. Reiter, " A Theory of Diagnosis from First Principle," Artificial Intelligence, 32(1), 1987, pp. 57-95.

[4] R. Milne, "Strategies for Diagnosis," IEEE Transactions on Systems, Man and Cybernetics, 17(3), 1987, pp. 333-339.

[5] Y. Peng, J. Reggia, "A Probabilistic Causal Model for Diagnostic Problem-Solving," IEEE Transactions System, Man and Cybernetics, V 17, 1987, pp. 146-162

[6] J. Reggia, D. Nau, Y. Wang, "Diagnostic Expert Systems Based on a Set Covering Model," Intl. Journal Man-Machine Systems, V 19, 1983, pp. 437-460

[7] M. R. Genesereth, "The Use of Design Descriptions in Automated Diagnosis," J Artificial Intelligence, V 24, 1984, pp. 411-436

[8] N.H. Narayanan, N. Viswanadham, "A Methodology for Knowledge Acquisition and Reasoning in Failure Analysis of Systems," IEEE Transactions on Systems, Man and

Cybernetics, V 17, March/April 1987, pp. 274-288.

[9] S.V. Rao, N. Viswanadham, "Fault Diagnosis in Dynamical Systems: A Graph Theoretical Approach," Automatica, Vol. 18, No. 4, 1987, pp. 687-695.

[10] S. Padalkar, "Graph Based Real-Time Fault Diagnosis," Ph.D. Thesis, Vanderbilt University, May 1990.

[11] S. Padalkar, et.al., "Real-Time Fault Diagnosis with Multiple Aspect Models," Proc. IEEE Intl. Conf. Robotics and Automation, Sacramento, CA, April 1991, pp. 803-808.

[12] S.J. Chang, F. DiCesare, G. Goldbogen, "Failure Propagation Trees for Diagnosis in Manufacturing Systems," IEEE Transactions on Systems, Man and Cybernetics, Vol. 21, No. 4, July/August 1991, pp. 767-776.

[13] J. F. Watson, A.A. Desrochers, "Applying Generalized Stochastic Petri Nets to Manufacturing Systems Containing Non-exponential Transition Functions," IEEE Transactions on Systems, Man, and Cybernetics, Vol. 6, Sept./Oct. 1991, pp. 1008-1017.

[14] M.C. Zhou, F. DiCesare, A.A. Desrochers, "A Hybrid Methodology for Synthesis of Petri Net Models for Manufacturing Systems," IEEE Trans. Robotics and Automation, Vol. 8, No. 3, June 1992, pp. 350-361.

[15] J. Graham, S. Alexander, L. Passafiume, "Enhanced Diagnosis for Integrated Manufacturing," Third Symposium on Advanced Manufacturing, Lexington, KY, 1989, pp. 29-32.

[16] J. Graham, J. Guan, S. M. Alexander, "A Hybrid CIM Diagnostic System with Learning Capabilities," Proc. Third Intl. Conf. Computer Integrated Manufacturing, Troy, NY, May 1992, pp. 308-314.

[17] G. H. Khaksari, "Expert Diagnostic System," First Intl. Conf. Industrial and Eng. Applications of AI and Expert Systems Tullahoma, TN, June 1988, pp. 43-53

[18] W. Lee, "A Hybrid Approach to a Generic Diagnosis Model for a Computer Integrated Manufacturing System," Ph.D. Thesis, Dept. of Industrial Engineering, Univ. of Louisville, Dec. 1990.

[19] W. Lee, S.M. Alexander, J.H. Graham, "A Diagnostic Expert System Prototype for CIM," Intl. J. Computers in Industrial Engineering, Vol. 22, No. 3, 1992, pp. 337-352.

[20] M.J. Pazzani, "Failure Driven Learning of Fault Diagnosis Heuristics," IEEE Trans. Systems, Man and Cybernetics, 17(3), 1987, pp. 380-394.

[21] Y. Koseki, "Experience Learning in Model-Based Diagnostic Systems," Proc. IJCAI, Detroit, MI, Aug. 1989.

[22] R.S. Michalski, R.L. Chilausky, "Learning by Being Told and Learning from Examples, J. Policy Analysis and Information Systems, 4, 1980.

[23] M. D. Mesarovic, D. Macko, Y. Takahara, Theory of Hiearchical Multilevel Systems, New York: Academic Press, 1972.

[24] G.N. Saridis, "Toward the Realization of Intelligent Controls," Proc. of the IEEE, Vol. 67, No. 8, August 1979, pp. 1115-1133.

[25] A. Dvoretsky, "On Stochastic Approximation," Proc. Third Berkeley Symposium on Mathematical Statistics and Probability, 1956, pp. 39-55.

[26] J. Guan, A Hybrid Model for Diagnostic Reasoning with Learning Capabilities, Ph.D. Dissertation, University of Louisville, December 1992.

[27] P. Coad, E. Yourdon, Object-Oriented Analysis, Englewood, NJ: Prentice Hall, 1990.

5

Fuzzy Logic Controller For Part Routing

David Ben-Arieh and E. Stanley Lee
Department of Industrial Engineering
Manhattan, Kansas, 66506

Modern computerized manufacturing systems increasingly accentuate small batch production with high variability of products. This tendency makes routing and scheduling combinatorially intractable problems. In addition to the increased complexity of the problem, the time frame in which the solution is needed keeps decreasing.

The approach presented in this paper remedies these difficulties by introducing an efficient real time control method for part routing. This control approach uses fuzzy parameters in order to decide to which machine to route the parts. The fuzzy control approach converts known and crisp system attributes into linguistic variables with corresponding membership functions. These membership functions are designed based on discrete event simulation of the system.

A rule base is then constructed based on all the applicable input parameters and for each control decision several rules are fired. These rules result in an aggregate fuzzy set that represents a particular routing decision. This fuzzy set is then converted into a crisp number that represents the degree to which this routing decision is attractive to the user. The process is repeated for every routing alternative, and the option that gets the highest score is selected.

The fuzzy logic controller is compared to three other routing heuristics in a simulation study, and proved to be a superior routing method. The advantage of this approach is especially evident when the system is congested or machines break down frequently, so formal and precise planning is impractical.

5.1 BACKGROUND

During the past recent years fuzzy control has emerged as an attractive application area of Fuzzy Set Theory [33]. Fuzzy control is founded on fuzzy logic which is attractive due to its similarity to the human way of thinking.

The purpose of a Fuzzy Logic Controller (FLC) is to automatically achieve and maintain some desired state of a process or a system. This is performed by monitoring several process variables and modifying selected control parameters. Usually this control task requires a precise mathematical model of the process. However, for many complex systems the construction of a realistic model is not practical. In these cases a human operator usually offers superior control to that of a conventional automatic controller, apparently through the use of an imprecise and robust control methodology. The FLC attempts to emulate such a human control capability, and appears very useful when the process is too complex for traditional automatic control. Fuzzy logic control can be viewed as a bridge between precise mathematical control and a human like decision making [3].

During the past several years, fuzzy logic has found numerous applications in fields ranging from finance to earthquake detection [9]. The pioneering research of Mamdani on fuzzy control [10, 11] was motivated by Zadeh's discussion on a linguistic approach and system analysis based on the theory of fuzzy sets [34, 35]. Notable applications of FLC include control of a heat exchange [15], activated sludge process [5], traffic control [16], cement kiln [8, 28], aircraft flight control

[7], turning processes [14], robotic control [18, 24], model car parking [21, 22], automobile speed control [12], power systems and nuclear reactor control [1], water quality control [29], train operation [32], and many others. In order to modify the modern computer to the task of fuzzy control, dedicated hardware systems have been developed [31], as well as fuzzy memory devices [25, 30].

Applications of fuzzy control to the area of production planning are discussed in [26, 27]. A different approach that uses fuzzy linear programming and approximate reasoning towards production planning and control is reported in [4].

5.2 INTRODUCTION

The basic FLC is composed of four parts, as shown in Figure 5-1. This structure is generic and suitable for a large variety of control tasks.

The system consists of four principal components: the condition interface, a knowledge base that consists of the fuzzy control rules and fuzzy set definitions database, the fuzzy controller itself, and an action interface.

The condition interface (also denoted as "fuzzification interface") is responsible for three main activities:

- Measurement of input variables.
- Scaling the input values into a corresponding universe of discourse.
- Mapping the input variables into suitable linguistic values. This activity is referred to as "fuzzification".

The knowledge base contains the knowledge of the application domain and the control goals. It consists of a set of control rules and a set of definitions. These definitions represent the descriptions of the fuzzy sets that correspond to the various input parameters.

The Fuzzy Logic Controller (FLC) is the unit that combines all the elements of the system into a decision making unit that operates in real time, in timely intervals, or based on interrupt signals.

The action interface is the unit responsible to communicate the decision made by the FLC back into the controlled system. This unit translates the decision variable into a corresponding universe of discourse and converts the fuzzy values into a crisp control action or decision (this activity is termed "defuzzification").

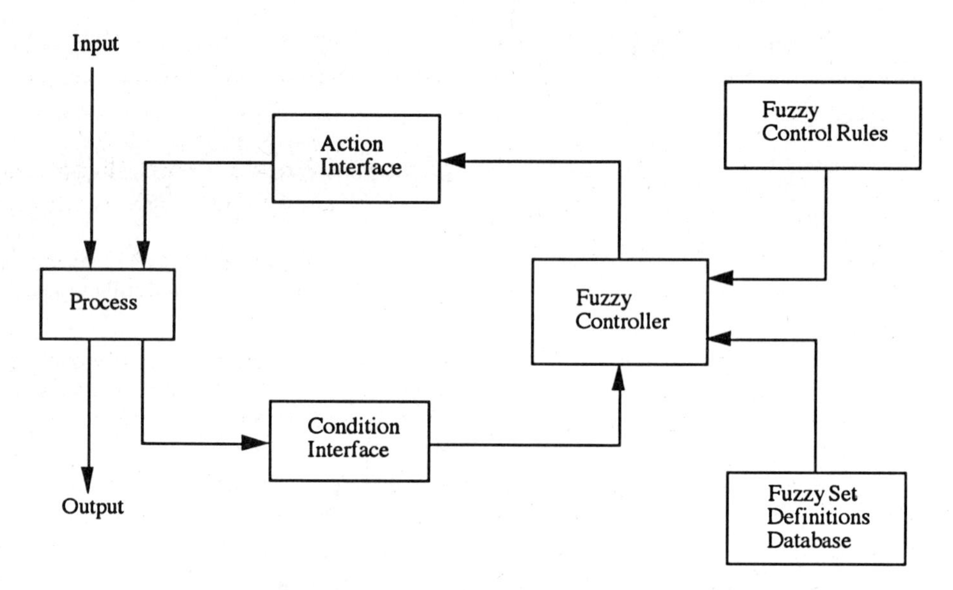

Figure 5-1 Fuzzy Control System Structure

The operation of the FLC consists of four steps. In the first one, the input variables are measured. Next, these variables are converted into fuzzy linguistic terms such a "long", "very short", etc. These linguistic terms are then used to evaluate the fuzzy control rules. The result of this evaluation is itself a fuzzy control decision. This fuzzy control decision is then converted into a crisp control action which is executed.

All the components and procedures that constitute the FLC are presented in detail in the following sections.

The Routing Problem Description

In our example, the FLC is required to route parts in a flexible manufacturing environment which consists of 6 similar machines. There are four types of parts to be routed; each one requires between 3 to 6 operations. Each operation can be performed on most machines with a different degree of efficiency. Each part has a computer generated due date which can be missed if an inefficient routing decision is made.

It is the task of the FLC to route the parts to the machines in a way that maximizes the production throughput of all part types combined. This means that when an operation is completed, the FLC chooses the machine for the next operation.

The processing times of the various operations of each part are stored in a database. For simulation purposes the processing time is assumed to be normally distributed with a standard deviation that is also a part of the process description.

The parts arrive continuously to the manufacturing environment with a given inter-arrival time interval. Also, the machines are subject to breakdowns, where each machine has a different probability for such an event (time between machine failure is exponentially distributed, while repair time is normally distributed).

5.3 THE LINGUISTIC VARIABLES: THE BASIS FOR THE FLC

This section presents the parameters (input variables) used by the *condition interface* for the control purposes. These parameters are also represented as fuzzy numbers in the *Fuzzy Set Definitions Database* component.

The FLC operates based on four input variables from the system. The input variables are:

- Job processing time.
- Queue length in each machine's buffer.
- Slack time available on each machine. This is the difference between the due date and the remaining operation time (the detailed definition is provided below).
- The machine breakdown rate.

Due to the fact that several machines can perform the same operations, the slack time for part *i* is calculated as follows:

$$\text{Slack Time}_i = \text{Due Date} - t_0 - P - \sum_{j \in J} Min(P_j + W_j) \qquad (1)$$

Where J is the set of all remaining operations, t_0 is the current time, P is the processing time of the current operation, and P_j and W_j are the processing and waiting times of operation j. In this case, the slack time on machine i is the time remaining after the current operation is performed on that machine, and the remaining operations are performed on the machines which require the least amount of time (processing and waiting in queue times).

The four input variables (also termed parameters) are dependent on the machine selected for the next operation. Since each machine has a different queue and a different processing time, the first three parameters have a different value for each routing alternative evaluated.

The four input variables are associated with linguistic values which are defined as follows:

Processing time: very short (VS), short (S), medium (M), long (L), and very long (VL).

Slack time: negative long (NL), negative medium (NM), negative small (NS), zero (ZO), positive small (PS), positive medium (PM), and positive long (PL).

Queue length: very short (VS), short (S), medium (M), long (L), and very long (VL).

Machine breakdown rate: very small (VS), small (S), medium (M), large (L), and very large (VL).

The linguistic values are represented using fuzzy triangular numbers of the form (a, b, c). The linguistic variable has a membership value of 1 (maximal membership value) if it carries a value of b. These values are defined based on a discrete event simulation of the system, or chosen heuristically. The simulation computes the various ranges of values that these parameters can have in the system to be controlled. Based on these ranges the values can be assigned to linguistic terms such as *Small* or *Very Large* with a membership value for each such term. For example, the maximal processing time is 35 min. and the minimal one is 12 min. Given the processing time distribution, 47 which is qualified as *Very Large,* is 3 σ above the maximal processing time. Simulation analysis shows that the average queue length is 23 while the maximal queue length that happened during the run is 40, and the smallest queue length is 4. These values are reflected by the definition of the fuzzy qualifiers, and values in between these values are generated considering the queue length distribution or even arbitrarily.

Once these triangular fuzzy numbers are determined, they are stored in the Fuzzy Definitions Database. The fuzzy numbers used in the current FLC are presented in Table 5-1, and in Figure 5-2.

Table 5-1 Fuzzy Values of Input Variables

Input Variable	Linguistic Value	Fuzzy Value
Processing Time	VS	(1, 1, 12)
	S	(1, 12, 24)
	M	(12, 24, 35)
	L	(24, 35, 47)
	VL	(35, 47, 47)
Slack Time	NL	(-1130, -1130, -700)
	NM	(-1130, -700, -174),
	NS	(-700, -174, 0)
	ZO	(-174, 0, 64)
	PS	(0, 64, 80)
	PM	(64, 80, 150)
	PL	(80, 150, 150)
Queue Length	VS	(0, 0, 4)
	S	(0, 4, 10)
	M	(4, 10, 23)
	L	(10, 23, 44)
	VL	(23, 40, 44)
Machine Breakdown	VS	(2%, 2%, 5%)
	S	(2%, 5%, 25%)
	M	(5%, 25%, 40%)
	L	(25%, 40%, 50%)
	VL	(40%, 50%. 50%)

Several other approaches for fuzzification are reported in the literature. One such approach represents fuzzy data as isosceles triangles [13]. The vertex of the triangle corresponds to the mean value of the data set, and the base is twice its standard deviation. Triangular fuzzy numbers are frequently used for representing data sets due to their ease of use [6]. A different approach uses a histogram of a measured data set in order to estimate the membership function [2].

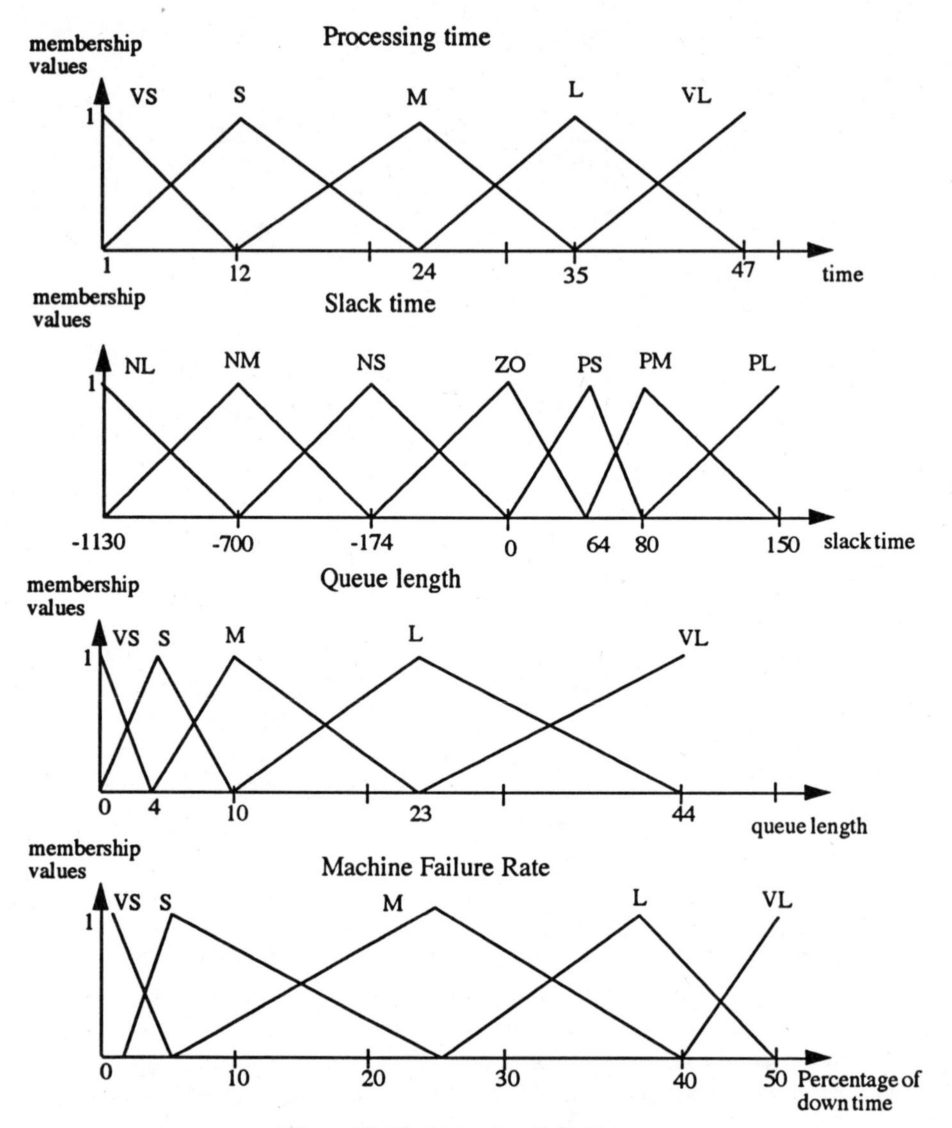

Figure 5-2 The Fuzzy Sets Definitions

5.4 FUZZY CONTROL RULES

The derivation of the control rules is a step of major significance to the performance of the fuzzy logic controller. A number of methods for rules derivation have been considered [20], which can be grouped into four approaches:

1. Based on a human operator's experience. Many experts have found that fuzzy control rules provide a convenient way to represent their domain knowledge [7, 8, 10, 15, 29].
2. Based on a fuzzy model of a process [23].
3. Based on observed operator's control actions [21, 22].
4. Based on learning [17, 18, 19, 24]. In this approach the controller improves its control capabilities by gradually developing better control rules.

The approach adopted in this implementation is a combination of methods 1 and 3, where the knowledgeable user sets the rules. The rules can later be tuned up in order to improve the system's performance (using simulation), thus reflecting somewhat a learning based approach.

The fuzzy control rules examine all the possible combinations of the fuzzy linguistic input variables. Hence, the knowledge base contains 5 x 7 x 5 x 5 = 875 rules.

An example of such a rule is given below:

IF Processing Time is *Low*
AND Queue Length is *Long*
AND Slack Time is *Zero*
AND Machine Breakdown Rate is *Very Small*

THEN Selectibility Factor is *Medium*

The example demonstrates that the purpose of the rules is to define a factor called Selectibility Factor. This factor is used to define the degree to which the candidate machines are suitable to execute the process in question.

In order to calculate the selectibility factor, a weight W_{ij} is associated with each input parameter *i*. In this case the parameter *j* represents the linguistic value of the input variable. The weights used in this research are heuristically decided and presented in Table 5-2 below.

The intention of the weights is to create preference of the machine with the shortest queue, fastest processing time, least break down rate and highest slack time. In all these cases the weight is the highest. Conversely, long queue, long processing time, high break down rate and negative slack time (reflecting lateness) carry small weight. Furthermore, the weights are decreasing exponentially rather than linearly when the parameters become less accommodating for a particular operation.

Table 5-2 The Value of the Parameters Weight

Parameter	Values					
	VS	S	M	L	VL		
Processing Time	1	$e^{-1/3}$	e^{-1}	e^{-2}	$e^{-7/3}$		
Queue Length	1	$e^{-1/2}$	e^{-1}	$e^{-3/2}$	e^{-2}		
Machine Breakdown Rate	1	$e^{-1/3}$	e^{-1}	e^{-2}	$e^{-7/3}$		
Slack Time	PL	PM	PS	ZO	NS	NM	NL
	1	$e^{-1/6}$	$e^{-3/6}$	e^{-1}	$e^{-10/6}$	$e^{-15/6}$	$e^{-21/6}$

Initially, the Selectibility Factor (SF) is calculated as:

$$SF = \frac{\sum_{i}^{n} W_{ij}}{n} \tag{2}$$

For example, the rule demonstrated above results in SF = 0.4316. This value is the result of the following calculation: $\dfrac{e^{-2} + e^{-\frac{3}{2}} + e^{-1} + 1}{4}$.

Once the SF in all 875 rules is assigned a numerical value, it is converted into a fuzzy term. This is done in order to reflect the degree of fuzziness or ambiguity in deciding on the routing based on the given input parameters.

The SF is assumed to have one of the following possible values: {Very Large, Large, Medium, Small, Very Small}. The maximal possible value of the SF is 1, and the minimal possible value is 0.08986. Thus, the membership value of VL is 1 for SF of 1, and similarly, the membership value for VS is 1 at the point where SF is 0.08986. The median SF is 0.3679; it is achieved at the point where all the input parameters are at their median value. This value carries a membership value of 1 for the linguistic value of M. The remaining values of *Small* and *Large* are calculated by averaging the SF for two possible alternative rules. For example, in the case of SF with the value of *Small* (S), the possible rules are: Processing Time is S, Queue Length is S, Machine Breakdown Rate is S and Slack Time is PS. The second rule is the same, except that Slack Time is PM. The fuzzy description of the SF is presented in Figure 5-3.

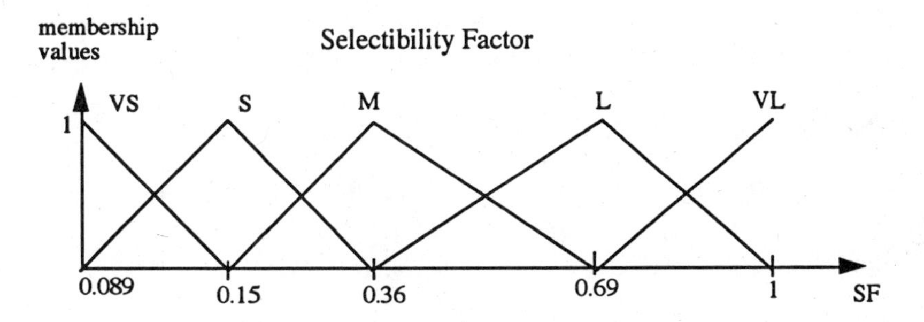

Figure 5-3 Fuzzy Representation of the Selectibility Factor

Therefore, as a fuzzy variable, the SF is defined using a triangular fuzzy value as follows: Very Large (0.6915, 1.0, 1.0), Large (0.3679, 0.6915, 1.0), Medium (0.1573, 0.3679, 0.6915), Small (0.08986, 0.1573, 0.3679), and Very Small (0.08986, 0.08986, 0.1573).

To summarize this step, in order to create fuzzy control rules, the input parameters that are all crisp are converted into an aggregate fuzzy value termed Selectibility Factor (SF) which reflects the degree to which a particular machine is useful. The SF is calculated separately for each routing alternative due to the different characteristics of each candidate machine (e.g. queue length, processing time, etc.)

5.5 SELECTING THE CONTROL ACTION

Once the fuzzy control rules and definitions are determined, the FLC converts the input parameters into a control action using the SF parameter. It is important to realize that due to the partial matching between the input parameters and their fuzzy descriptors, more than one rule can fire at the same time. In order to accommodate this property, the control action that is recommended (i.e. which machine to choose) is determined by using an aggregation procedure. This way, all the possible rules are fired and their various recommendations are aggregated into a unified solution. This is done in order to consider all the various possible solutions, which have various degrees of contribution to a good routing solution..

The FLC allows each input parameter to belong to two fuzzy sets at most. For example, Queue Length of 20 units belongs to both the *Medium* and *Long* sets. In order to accommodate all the possible rules that are applicable (with n input parameters there are 2^n applicable rules), a method termed Minimum Operation Rule is applied.

This approach can be demonstrated using the following example:

For simplicity, assume that only two fuzzy control rules are currently viable:

R_1: If PT is A_1, SL is B_1 and QL is C_1, THEN S_1.

R_2: If PT is A_2, SL is B_2 and QL is C_2, THEN S_2.

The rules consider only three input parameters, and can be represented graphically as shown in Figure 4. The input parameters are defined on the universal sets X, Y, and Z (e.g. Processing Time, Slack Time and Queue Length) and belong to the fuzzy sets of A_i, B_i and C_i (e.g. *Medium* and *Short* for A_1 and A2 respectively). In this example, the actual values of the input parameters are represented as *a*, *b*, and *c*. Using the Minimum Operation Rule, the minimal membership value of the rule parameters is projected to the consequent side (the THEN part) which represents the SF fuzzy number. The result of firing these rules are the resultant triangular fuzzy numbers that represent the various SF truncated by the minimal membership values.

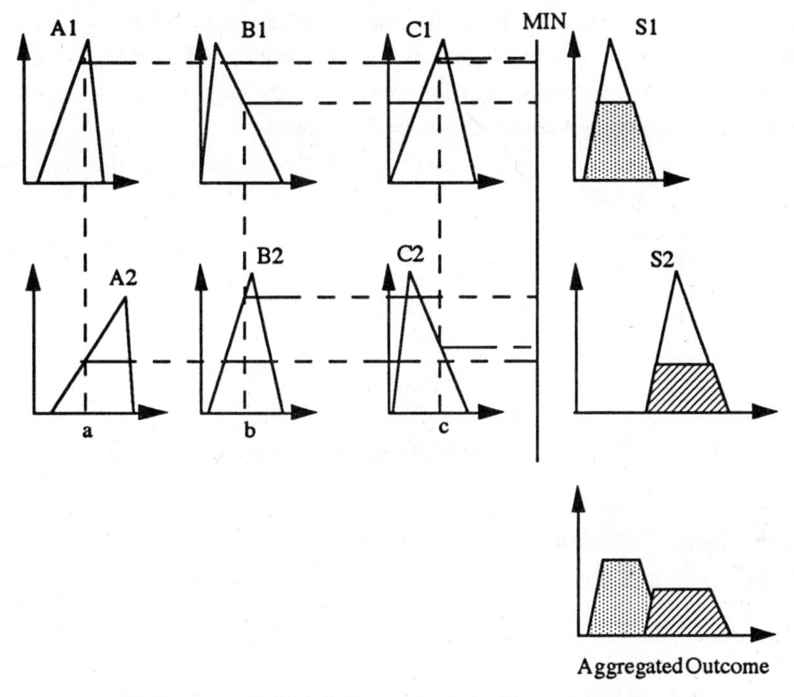

Figure 5-4 A Graphical Representation of Rules Aggregation

Once the viable rules result in a combined fuzzy set as shown in Figure 5-4, this result is converted into a crisp numerical value using the Center of Area method. This crisp number is calculated for every alternative decision (possible machine to select), and the alternative with the highest value is then selected. The Center of Area (COA) method calculates the center of gravity of the resultant fuzzy set as follows:

For a discrete universe of discourse:

$$Z = \frac{\sum_{i=n}^{m} Z_i \cdot \mu_c(Z_i)}{\sum_{i=n}^{m} \mu_c(Z_i)} \tag{3}$$

In this case, there are n-m+1 discrete quantification levels with Zi as the parameter value at level i, and $\mu_c(Zi)$ as the membership value of the parameter at this level. Since the membership functions used by the FLC are all linear, this approach is modified to a continuous case as follows:

$$S = \frac{\int_{t_1}^{t_2} S_t \cdot \mu_s(S_t)}{\int_{t_1}^{t_2} \mu_s(S_t)} \tag{4}$$

Example

A particular job has finished an operation and needs to be routed to the next operation. One of the candidate machines has the following parameters:

Processing Time of 8 minutes
Slack Time of 30 minutes
Queue Length of 6 units.

For brevity, the machine breakdown rate is not considered in the example.

Assume that the input parameters have the following membership values:

Processing Time: $\mu_{1,VS}(8) = 0.266$
$\mu_{1,S}(8) = 0.733$
$\mu_{1,M}(8)=\mu_{1,L}(8)=\mu_{1,VL}(8)=0.0$

Slack Time: $\mu_{1,NL}(30)=\mu_{1,NM}(30)=\mu_{1,NS}(30)=0.0$
$\mu_{1,ZO}(30) = 0.2188$
$\mu_{1,PS}(30) = 0.7813$
$\mu_{1,PM}(30)=\mu_{1,PL}(30) = 0.0$

Queue Length: $\mu_{1,S}(6) = 0.80$
$\mu_{1,M}(6) = 0.20$
$\mu_{1,VS}(6)=\mu_{1,L}(6)=\mu_{1,VL}(6)=0.0$

Due to the combination of membership values, 8 possible rules are in effect. The rules represent the combinations of parameters with membership values that are greater than zero.

For example one such rule is:

IF Processing Times is "Very Small"
AND Slack Time is "Zero"
AND Queue Length is "Short"
THEN Selectibility Factor is "Large".

From these 8 rules, 7 rules lead to Selectibility of value "Large" and one rule recommends value of "Medium". The minimum membership values are calculated as follows:

$\alpha_1 = \text{Min}\ \{\mu_{1,vs}(8), \mu_{1,zo}(30), \mu_{1,s}(6)\} = \text{Min}\ (0.2667, 0.2188, 0.8) = 0.2188$
$\alpha_2 = \text{Min}\ \{\mu_{1,vs}(8), \mu_{1,zo}(30), \mu_{1,m}(6)\} = \text{Min}\ (0.2667, 0.2188, 0.2) = 0.2$
$\alpha_3 = \text{Min}\ \{\mu_{1,vs}(8), \mu_{1,ps}(30), \mu_{1,s}(6)\} = \text{Min}\ (0.2667, 0.7812, 0.8) = 0.2667$
$\alpha_4 = \text{Min}\ \{\mu_{1,vs}(8), \mu_{1,ps}(30), \mu_{1,m}(6)\} = \text{Min}\ (0.2667, 0.7812, 0.2) = 0.2$
$\alpha_5 = \text{Min}\ \{\mu_{1,s}(8), \mu_{1,zo}(30), \mu_{1,s}(6)\} = \text{Min}\ (0.7333, 0.2188, 0.8) = 0.2188$
$\alpha_6 = \text{Min}\ \{\mu_{1,s}(8), \mu_{1,zo}(30), \mu_{1,m}(6)\} = \text{Min}\ (0.7333, 0.2188, 0.2) = 0.2^*$
$\alpha_7 = \text{Min}\ \{\mu_{1,s}(8), \mu_{1,ps}(30), \mu_{1,s}(6)\} = \text{Min}\ (0.7333, 0.7812, 0.8) = 0.7333$
$\alpha_8 = \text{Min}\ \{\mu_{1,s}(8), \mu_{1,ps}(30), \mu_{1,m}(6)\} = \text{Min}\ (0.7333, 0.7812, 0.2) = 0.2$

* This rule results in the value of "Medium" to the SF.

These rules can be represented graphically with an aggregate SF as shown in Figure 5-5.

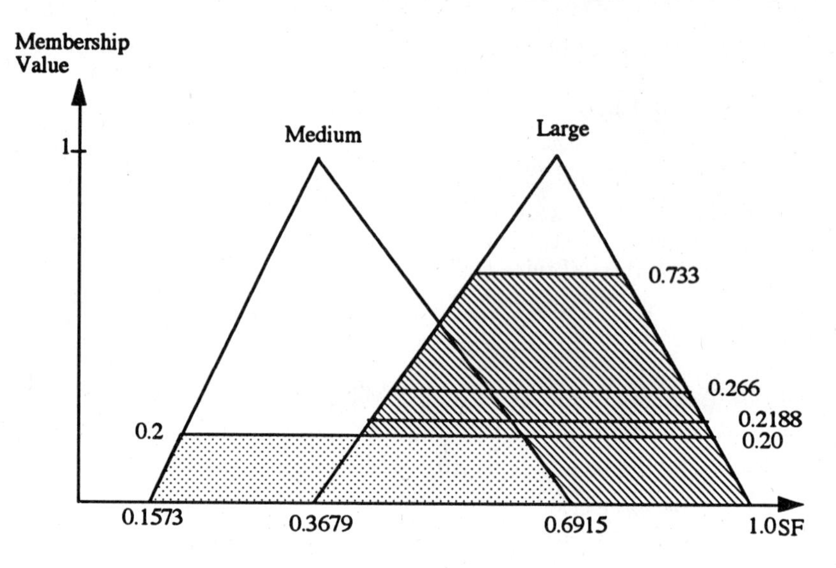

Figure 5-5 The Aggregated Outcome of Eight Rules

The example rules result in several consequent parts that show SF with values of *Large* and *Medium* and the different minimal membership values. Some of the minimal values are repeated on both the Large and Medium fuzzy sets.

Using the COA (Center of Area) method, the composite value of the rules is found to be 0.6045. This combined Selectibility Factor allows us to evaluate this routing alternative and compare it to the other optional machines available.

Now that all the components of the FLC are described, a more detailed view of the fuzzy logic controller is presented in Figure 5-6.

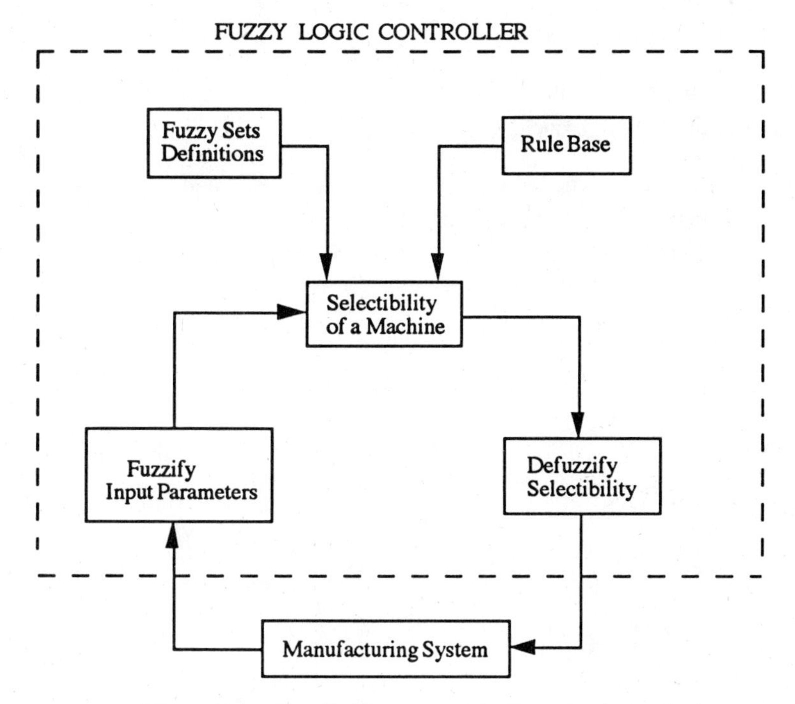

Figure 5-6 A Detailed Structure of the Fuzzy Logic Controller

5.6 EXPERIMENTAL RESULTS

In order to assess the decision quality of the FLC, a simulation study was conducted. In the simulation experiments, three other reference routing heuristics were used:

- Random routing (denoted as RAN).
- Short Processing Time (denoted as SPT).
- Shortest Queue Length (SQL).

The FLC and the three reference routing heuristics were compared based on four criteria:

- Mean flow time of the jobs.
- Percentage of tardy jobs.
- Mean job tardiness.

- Number of jobs completed.

The four criteria are not completely independent, but help to highlight the capabilities of the FLC approach.

The simulation experiments were run for 2400 simulated minutes under a variety of input conditions. For example, the inter-arrival rate of the jobs varied from 10 to 28 min. (exponentially distributed). More than 360 simulation experiments were conducted in order to evaluate the FLC under a broad range of conditions. Some of the results are presented graphically in Figures 5-7, 5-8, 5-9 and 5-10.

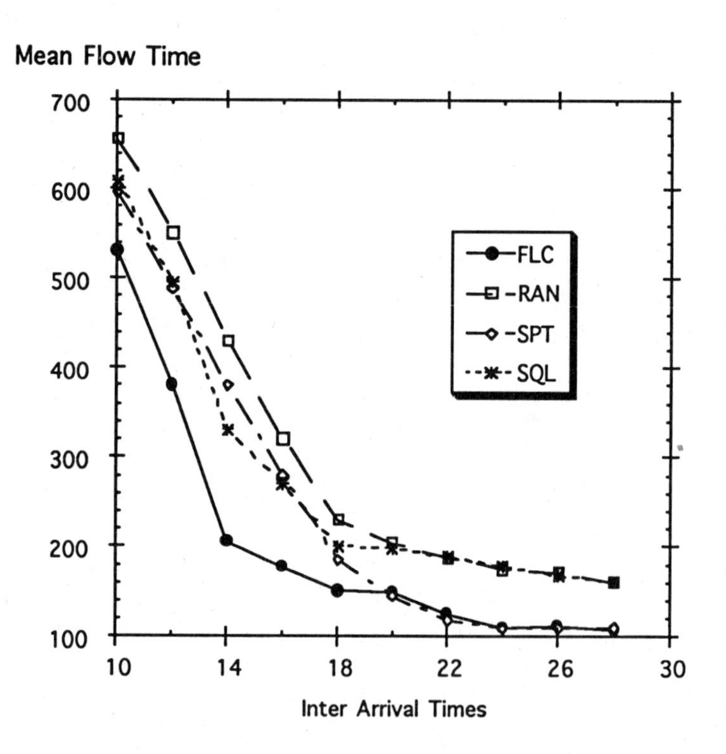

Figure 5-7 Mean Flow Time

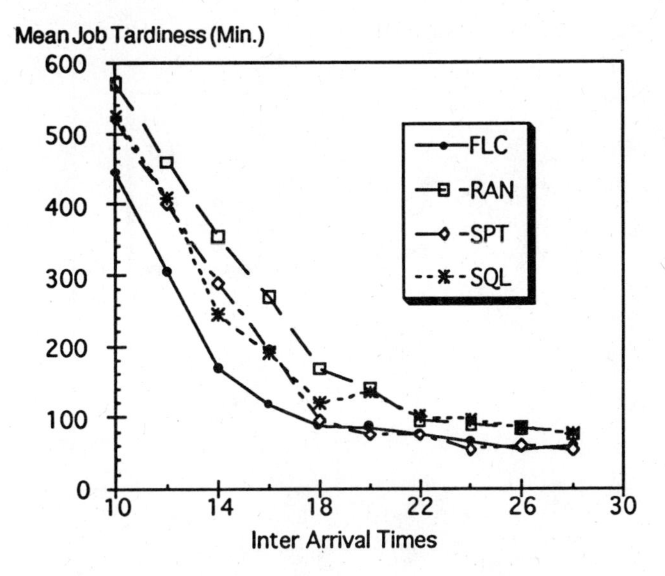

Figure 5-8 Mean Job Tardiness

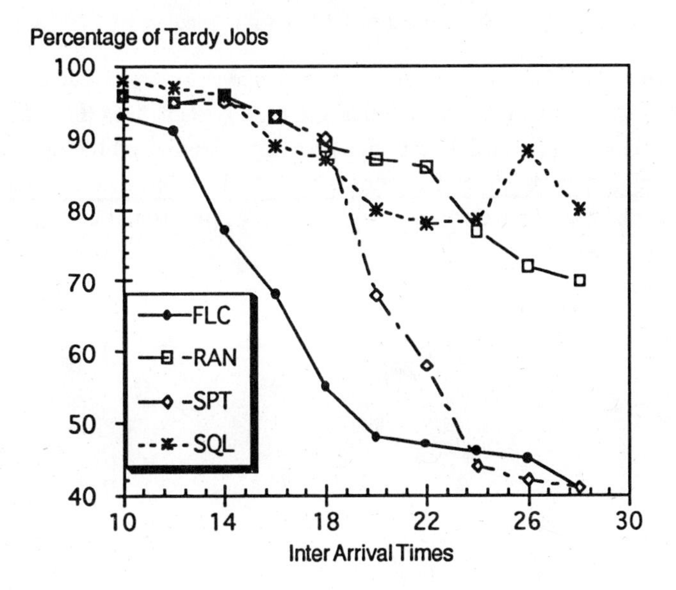

Figure 5-9 Percentage of Tardy Jobs

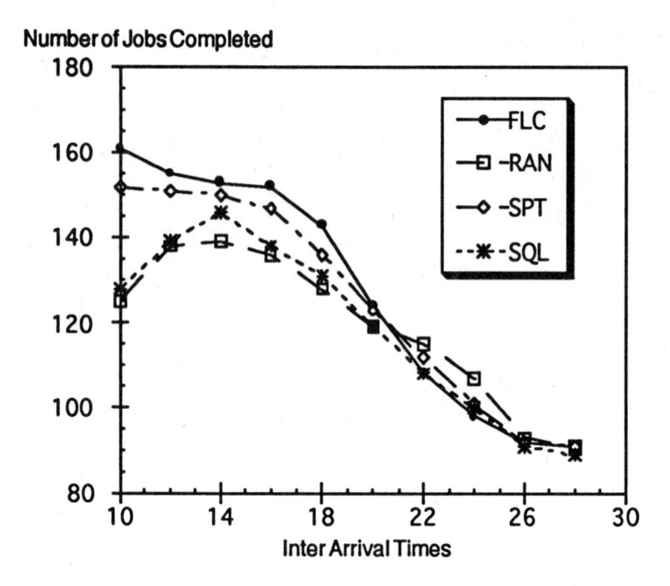

Figure 5-10 Number of Jobs Completed

The above four figures were all generated under a very small machine break down rate. The results demonstrate that the FLC outperforms the other reference heuristic methods. The FLC is especially advantageous when the inter-arrival rate of parts is high and the system is congested, or when unpredicted events such as machine breakdowns happen frequently. This is demonstrated in Figures 5-11 and 5-12 below.

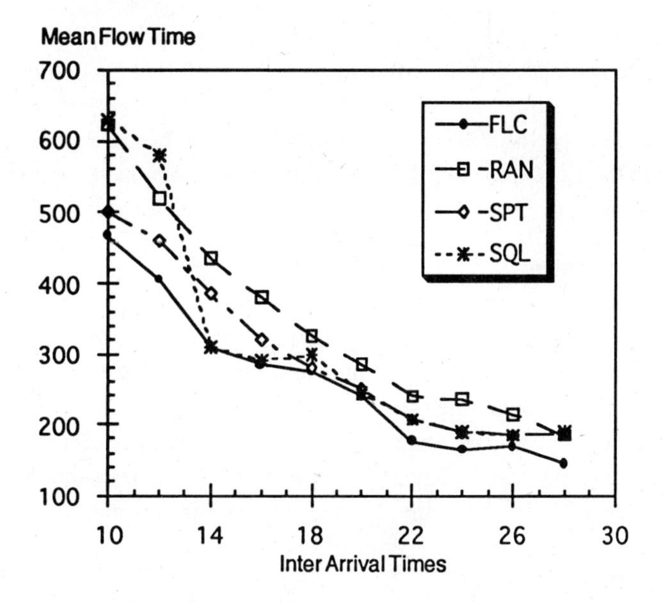

Figure 5-11 Mean Flow Time Under High Breakdown Rate

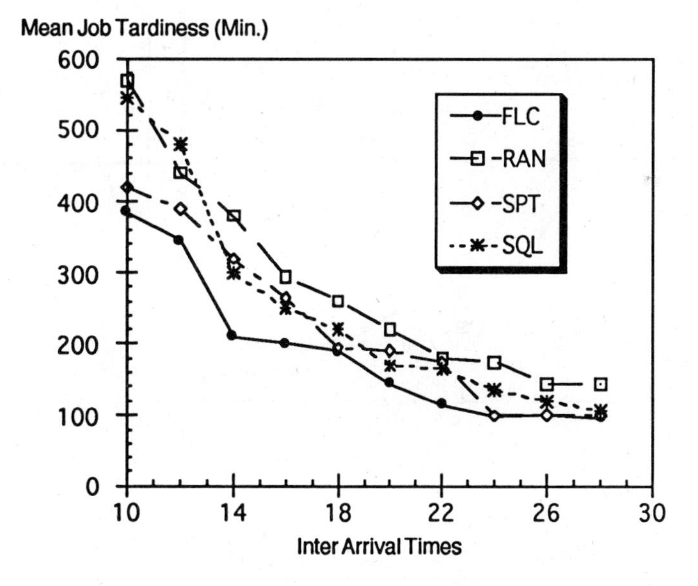

Figure 5-12 Mean Job Tardiness under High Breakdown Rate

In these Figures the failure rate of the machines is accelerated to failures every 400 min. (exponentially distributed) with repair times that are unique to each machine. The repair times are normally distributed with mean ranging from 8 minutes (machine 1) to 200 minutes (machine 6). These Figures show the advantage that the FLC has over the other routing methods in the entire range of operational parameters.

Fine Tuning the System

So far, the input parameters are weighted equally. It seems logical that the system can be improved by fine tuning the weight of the parameters according to the circumstances. The following Figure describes the mean flow time under heavy machine breakdown rate. In this case, changing the importance of the Machine Breakdown Rate parameter improves the performance of the system (the other input parameters remain equally weighted). The experiments have shown that a weight of 15% is about the optimal weight for the machine breakdown parameter.

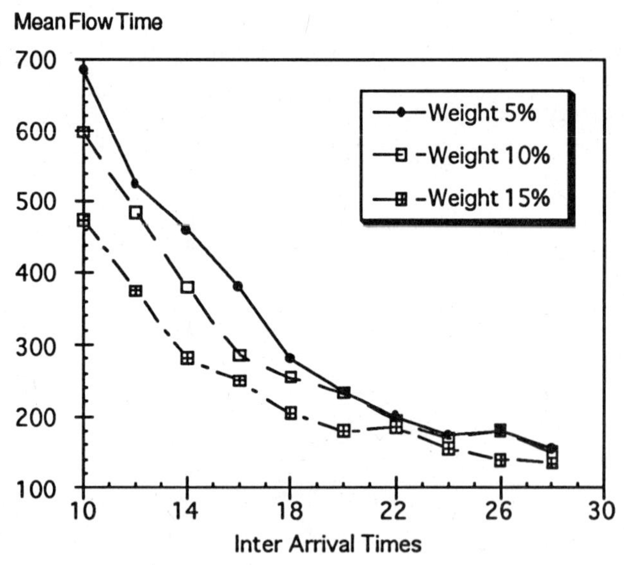

Figure 5-13 Comparison of Weights of Machine Breakdown Parameter

7. SUMMARY

This article presents a new approach to intelligently controlling a manufacturing system. The Fuzzy Logic Controller (FLC) is used to solve the routing problem which is a non polynomial problem. The FLC completes a control loop by reading several input parameters from the manufacturing system. These input parameters, which have crisp and precise values, are assigned a fuzzy value based on observations of the real system. All the possible combinations of these values are then used to constitute a rule base. This rule base advises a Selectibility Factor (SF) that shows the degree to which a routing alternative is attractive to the user. The SF is itself a triangular fuzzy number. Due to the fuzzy properties of the input parameters, several rules are applicable for every routing decision. In order to accommodate all the viable rules, an aggregation method is being used to consider the outcome of all these rules. The control loop is closed when the rules' outcomes are converted into a numerical value. This value represents the viability of each routing decision, therefore, the alternative routing with the largest value is selected and implemented in the system.

The performance of this approach is demonstrated using a simulation study. This study compares the FLC with three other routing heuristics which are random routing, routing based on the shortest processing time, and one based on the shortest queue length. The FLC clearly demonstrates an improvement over these heuristics.

REFERENCES

1. Bernard, J.A., 1988, Use of rule based system for process control, *IEEE. Contr. Syst. Mag.*, vol. 8, no. 5, 3-13.

2. Devi, B.B., and Sarma V.V.S., 1985, Estimation of Fuzzy memberships from histograms, *Information Science*, vol. 53, 43-59.

3. Gupta, M.M., and Tsukamoto, Y, 1980, Fuzzy Logic Controllers - A Perspective, *Proc. Joint Automatic Control Conf.*, San Francisco, pp. FA10 - C.

4. Hintz G.W., 1989, A method to control flexible manufacturing systems, *European J. of Operational Research*, vol. 41, 321-334.

5. Itoh, O., and Gotoh K., 1987, Applications of Fuzzy Control to Activated Sludge Process, *Proc. 2nd IFSA Conf.*, Tokyo, Japan July 1987, 282-285.

6. Kaufmann A., and Gupta M.M., 1985, *Introduction to Fuzzy Arithmetic Theory and Applications*, Van Nostrand Reinhold Inc.

7. Larkin, L.I., 1985, A Fuzzy Logic Controller for Aircraft Flight Control, in *Industrial Applications of Fuzzy Control*, Sugeno M. Ed., North Holland, Amsterdam, pp. 87-104.

8. Larsen, P.M., 1980, Industrial Applications of Fuzzy Logic Control, *Int. J. Man Machine Studies*, vol. 12, no. 1, 3-10.

9. Maiers, J. and Sherif Y.S., 1985, Applications of Fuzzy Set Theory, *IEEE Trans. on Systems, Man and Cybernetics*, vol. SMC-15, no. 1, 175-189.

10. Mamdani E.H., and Assilian S., 1975, An Experiment in Linguistic Synthesis With a Fuzzy Logic Controller, *Int. J. Man Machine Studies*, vol. 7, no. 1, 1-13.

11. Mamdani E.H., 1976, Advances in the Linguistic Synthesis of Fuzzy Controllers, *Int. J. Man Machine Studies*, vol. 8, no. 6, 669-678.

12. Murakami, S., and Maeda M., 1985, Application of fuzzy controller to automobile speed control system, in *Industrial Applications of Fuzzy Control*, Sugeno M. Ed., North Holland, Amsterdam, pp. 105-124.

13. Murayama Y. and Terano, T., 1985, optimizing control of a diesel engine, in *Industrial Applications of Fuzzy Control*, Sugeno M. Ed., North Holland, Amsterdam, 63-72.

14. Ollero A., and Garcia-Cerezo A.J., 1989, Direct Digital Control, Auto turning and Supervising Using Fuzzy Logic, *Fuzzy Sets and Systems*, vol. 33, 135-153.

15. Ostergaad, J.J., 1977, Fuzzy Logic Control of a Heat Exchange Process, *Fuzzy Automata and Decision Processes*, Gupta M.M., Ed., North Holland Amsterdam, pp. 285-320.

16. Pappis C.P., and Mamdani E.H., 1977, A Fuzzy Logic Controller for a Traffic Junction, *IEEE Trans. on Systems, Man and Cybernetics*, SME-7, no. 10.

17. Procyk, T.J., and Mamdani E.H., 1979, A linguistic self organizing process controller, *Automat*. vol. 15, no. 1, 15-30.

18. Scharf, E.M., Mandic N.J., 1985, The application of a fuzzy controller to the control of a multi-degree freedom robot arm, in *Industrial Applications of Fuzzy Control*, Sugeno M. Ed., North Holland, Amsterdam, pp. 41-62.

19. Shao, S., 1988, Fuzzy self organizing controller and its applications for dynamic processes, *Fuzzy Sets and Systems*, vol. 26, 151-164.

20. Sugeno, M., 1985, An introductory Survey of fuzzy control, *Information Science*, vol. 36, 59-83.

21. Sugeno, M., and Murakami K., 1985, An experimental study of a fuzzy parking control using a model car, in *Industrial Applications of Fuzzy Control*, Sugeno M. Ed., North Holland, Amsterdam, pp. 125-138.

22. Sugeno, M. and Nishida M., 1985, Fuzzy control of a model car, *Fuzzy Sets and Systems*, vol. 16, 103-113.

23. Sugeno, M., and Takagi, T., 1982, A new approach to design of fuzzy controller, *Advances in Fuzzy Set Theory and Applications*, Wang P.P. Ed., Plenum Press.

24. Tanscheit R., and Scharf E., 1988, Experiments with the use of a rule based self organizing controller for robotic applications, *Fuzzy Sets and Systems*, vol. 16, 195-214.

25. Togai, M., and Watanabe, H., 1986, Expert system on a chip: an engine real time approximate reasoning, *IEEE Expert Systems Mag.*, vol. 1, 55-62.

26. Turksen, I.B., 1985, Fuzzy sets and their applications in production research, *Proc. ICPR Towards the Factory of the Future*, 649-656.

27. Turksen, I.B., 1988, Approximate reasoning for production planning, Fuzzy Sets and Systems, vol. 26, 1-15.

28. Umbers, I.G., and King P.J., 1980, An Analysis of Human Decision Making in Cement Kiln Control and the Implications for Automation, *Int. J. Man Mach. Studies*, vol. 12, no. 1, 11-23.

29. Yagishita, O., Itoh, O., and Sugeno, M., 1985, Applications of fuzzy reasoning to the water purification process, in *Industrial Applications of Fuzzy Control*, Sugeno M. Ed., North Holland, Amsterdam, pp. 19-40.

30. Yamakawa, T., and Miki, T., 1986, The current mode fuzzy logic integrated circuits fabricated by the standard CMOS process, *IEEE. Trans. Comp.*, C-35, no. 2, 161-167.

31. Yanakawa, T., 1986, High speed fuzzy controller hardware system, Proc. 2nd. Fuzzy Systems Conf., Japan, pp. 122-130.

32. Yasunobu S., and Miyamoto, S., 1985, Automatic train operation by predictive fuzzy control, in *Industrial Applications of Fuzzy Control*, Sugeno M. Ed., North Holland, Amsterdam, pp. 1-18.

33. Zadeh, L.A., 1965, Fuzzy Sets, *Information and Control*, 8, 338-353.

34. Zadeh, L.A., 1972, A Rationale for Fuzzy Control, *Trans. ASME J. Dynam. Syst. Measur. Control*, vol. 94, 3-4.

35. Zadeh, L.A., 1973, Outline of a New Approach to the Analysis of Complex Systems and Decision Processes, *IEEE Trans. Syst, Man and Cybernetics*, vol. SMC-3, 28-44.

6

Fuzzy-Logic Control as an Industrial Control Language for Embedded Controllers

Sencer Yeralan, University of Florida

Baris Tan, Koc University

The industrial engineering profession has always been concerned with efficiency, effectiveness, and productivity, which translate into industrial competitiveness. Over the last few decades, the single most important factor that favorably affected efficiency, effectiveness, and productivity is arguably the computerization of almost all aspects of industrial activity. Effectiveness and efficiency is not only limited to the end products of industrial engineering studies but also to the methods that are used in creating the end products. We evaluate fuzzy-logic control as a general-purpose programming language for embedded control. This work goes further than analysis: a fuzzy-logic control code generator is designed and developed to increase the efficiency and effectiveness of the programmer many-fold. Moreover, we illustrate how automated tools and methods may be used to create a fuzzy-logic code generator itself.

6.1 Introduction

The embedded microcontroller is a technology which is expected to have profound effects on automation and control. Such microcontrollers offer a many-fold enhancement of efficiency in automating, monitoring, and controlling manufacturing tools as well as in end products, with orders of magnitude reduction in cost. The effective use of this technology requires the concurrent consideration of hardware, software, and specific application areas, all guided by the need to improve cost-effectiveness. Although the microcontroller chip manufacturing technology originated in the United States, foreign competitors, notably the Japanese industries, have been very successful in incorporating this technology into their end products, and thus gaining business advantages. It is argued that the implementation of this technology has been slow, in part due to a lack of applications development tools. There exists a need and an opportunity for industrial engineers to develop industrial control languages that take advantage of the powerful facilities offered by modern microcontrollers [40]. An industrial control language is a powerful applications development tool that can be expected to contribute to the implementation of microcontroller technologies. This paper investigates the advantages and the shortcomings of fuzzy-logic control as a general purpose programming environment for embedded microcontrollers. A prototype industrial control language is designed and developed. The code generator takes linguistic control statements and produces machine code for the microcontrollers.

An embedded controller is a computer system that is physically placed inside the application hardware. The increasing power and reducing size of microprocessors led to the proliferation of embedded controllers. Embedded controller systems cover a wide range of sophistication and complexity. Although embedded controllers are responsible of producing the high-technology appliances and products such as automatic cameras and facsimile machines, they have many powerful applications in industrial automation and control. With the unprecedented price-to-performance ratio, embedded controllers are prime candidates for distributed industrial control and for the implementation of "swarm intelligence." Components of a workcell, such as pick-and-place machines, conveyors, inspection machines, and parts orienting and presentation machines, may all be governed by embedded controllers. Similarly, the use of embedded microcontrollers simplifies and strengthens many control applications traditionally using programmable logic controllers (PLC), analog control such as proportional-integral-differential (PID) controllers, or hybrid digital-analog control. The original and current industry-wide standard control language is relay logic, implemented by programmable controllers. Early programmable controllers were electro-mechanical devices rather than electronic devices. Thus, relay logic does not take advantage of the facilities offered by new generation microcontrollers. Although there are expanded versions of relay logic languages offered by programmable controller manufacturers, it is the premise of this paper that there is a need for a new language for industrial control.

Most embedded controllers are built from microcontrollers. A microcontroller is an enhanced microprocessor. In addition to the microprocessor, a microcontroller typically contains several parallel and serial ports, system clock generators, data and program memory, timers, counters, interrupt logic, analog-to-digital converters, digital-to-analog converters, and even digital signal processing subsystems on the same silicon chip. Thus, a single-chip microcontroller may be placed in an application to perform as an embedded controller with no other support chips. Specialized single-chip microcomputers, or microcontrollers, that are used in embedded control applications have been around since the late 1970's, (for example the Intel 8031 family). Microcontrollers, compared to microprocessors, are required to operate in real-time, to track and execute several simultaneous or quasi-simultaneous tasks, and to be tolerant of external disturbances and hardware or software degradation. The new generation of single-chip microcontrollers possesses the necessary computing power, on-chip facilities, and speed to implement highly sophisticated dedicated controllers for industrial use. The 16-bit 80C166 series controllers from Siemens, for example, combine 10 channel, 10-bit analog-to-digital converters, up to 32 interrupt sources with 300 nsec typical (500 nsec maximum) latency, 76 input/output bits, two full-duplex serial ports with a dedicated baud rate generator, 16 channel capture/compare registers using 5 timers with 200 and 400 nsec resolution, hardware 16 bit by 16 bit multiply in 500 nsecs and 32 bit by 16 bit divide instructions in 1 microsec, 256K of total addressable memory with internal 8K of program and 1K of data memory on a single chip. The devices run at 40MHz with no wait states. The prices are in the $20.00 to $40.00 range.

Although the embedded controller is becoming a ubiquitous component in industry, it has not received the attention that is given to its counterpart, the general-purpose microcomputer. Specifically, operating systems and high-level programming languages that would improve applications software development are needed for embedded controllers.

This study focuses on the high-level control language for embedded microcontrollers applications. Specifically, it investigates the suitability of fuzzy-logic control as a general-purpose high-level language for programming embedded microcontroller applications. The syntax and grammar are developed for a prototype language. The language is implemented as a code generator. The input to the code generator is a high-level description of the control task. The output is a subroutine written in the native assembly language of a microcontroller. The PIC family of microcontrollers manufactured by Microchip Technology, Inc. are used in this project. These processors have a RISC (reduced instruction set computer) architecture. They are extremely cost-effective, with internal program and data memory, timers, a system clock generator, and input/output ports for under $1.00 per MIPS (million instructions per second). Complementary research into hardware which is capable of implementing control applications in an architecture similar to human perception of control is an ongoing project.

6.2 Industrial Control Languages

The most popular industrial control language is still ladder logic. It is possible to implement automation using relays with multiple contacts. Some of these contacts in turn energize other relays, thus allowing logic statements to be executed. In its simplest interpretation, ladder logic is a circuit diagram. Figure 1 shows a simple ladder logic program.

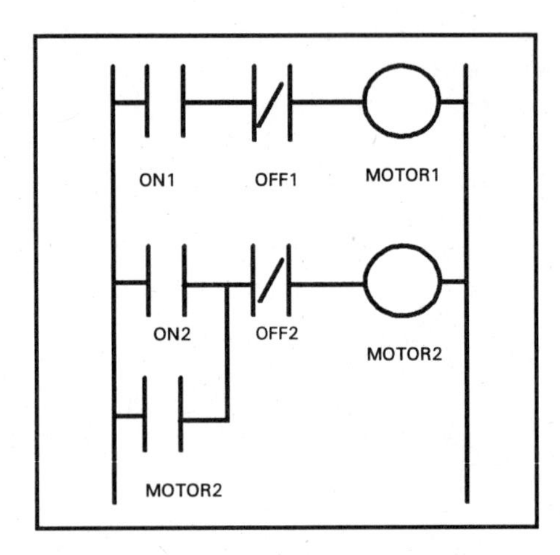

Figure 1. A PLC Ladder-Logic Diagram: Simple and Latched Outputs.

Each rung of the ladder is a Boolean (logic) statement. The first rung consists of a normally open switch, ON1, a normally closed switch, OFF1, and a contact or output MOTOR1. Provided that OFF1 is not pressed (is FALSE), pressing ON1 (making it TRUE) activates the output (MOTOR1). Rung 2 illustrates a latching circuit. It also demonstrates how the state of the output can be used as an input. MOTOR2 is on if OFF2 is not pressed and if ON2 is pressed or if MOTOR2 is on. That is,

MOTOR2 = (NOT OFF2) AND (ON2 OR MOTOR2).

Similarly, the logic implemented by the first rung can be written as a Boolean equation as,

MOTOR1 = ON1 AND (NOT OFF1).

Ladder logic has the advantage of being pictorial, and thus, easy to follow. Each rung is independent. If the program is implemented by relays, current will flow through each rung simultaneously. Thus, the Boolean equations implemented should be considered as a set which as evaluated simultaneously and instantly.

There are alternative pictorial media to illustrate logic statements. Figure 2 gives the same two Boolean equations as circuit built from logic gates.

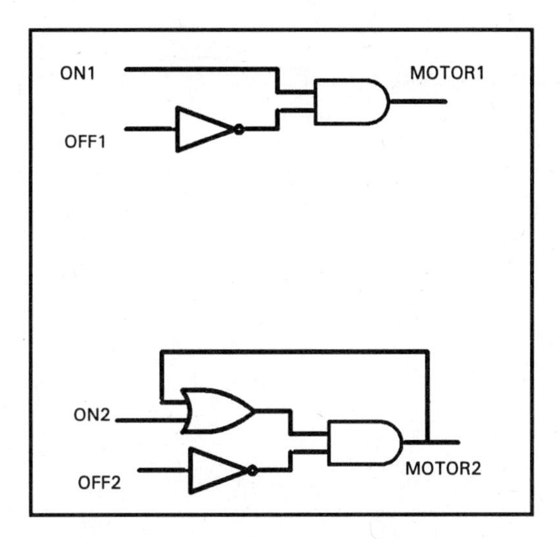

Figure 2. Logic Diagrams with AND, OR, and INVERT Gates.

Microcontrollers are often programmed in their own machine language. Assemblers, linkers, and cross-assemblers are available not only from their manufacturer, but also from third-party software developers. There are also several microcontrollers with on-chip (in ROM) interpreters. Versions of the 8073, 8052, and Z-8 have interpreted BASIC. Some manufacturers offer FORTH and PL/1 as additional languages. Languages such as BASIC and FORTH were developed as all-purpose programming languages, not for control applications. Many current control applications programming environments force the programmer to think in the format of the particular language that is used. In the past, most programs that have been developed to automate control systems have resulted in the requirement of more training of the users. Owing to its simplicity and popularity, BASIC has become the default language for many industrial applications ranging from robotics to industrial vision systems in spite of its inability to fully utilize the power and flexibility of microcontrollers. ADA, a general purpose language developed by the Department of Defense, was designed with practical programming in mind [6]. However, because ADA has a large overhead when handling concurrent modules, its performance in real-time applications is severely limited [12]. Also, ADA does not allow the addition of new software modules to an up and running system, thus making expanding distributed control systems very cumbersome. There are some third-party compilers which are becoming available for microcontroller code development [22], [23]. Almost all such compilers are C compilers, since C allows access to the low-level utilities of the microcontroller. In addition to their inability to fully utilize the power and flexibility of microcontrollers, currently popular high-level languages for industrial control, such as the Boolean-decision-based programmable controller languages, are still sequential in nature. Humans do not perceive control tasks as purely sequential activities. In fact, engineers are trained to think in a flow-chart-oriented manner to translate control problems to common computer languages such as FORTRAN. The perception of control tasks involves rules and exceptions, besides simple procedural routines.

Consider the two Boolean equations given in Figures 1 and 2. In structured assembly language one may be inclined to implement each output as a subroutine. Let ComputeMotor1 and ComputeMotor2 be the two subroutines. Let ON1, ON2, OFF1, OFF2, MOTOR1 and MOTOR2 be addressable bits of some input/output port of a microcontroller. A typical microcontroller assembly language program subroutine for the first output may be as follows.

```
GetMotor1:
      jb     OFF1, NoOff1          ; if OFF1=1 jump to NoOff1
NoOn1:
      clr    MOTOR1                ; else clear output bit MOTOR1
      ret
NoOff1:
      jnb    ON1, NoOn1            ; jump if ON1=0 to NoOn1
      setb   MOTOR1                ; else set output bit MOTOR1
      ret
```

Similarly, for the second output,

```
GetMotor2:
      jb     OFF2, NoOff2          ; if OFF2=1 jump to NoOff2
      clr    MOTOR2                ; else clear output bit MOTOR2
      ret
NoOff2:
      jnb    ON2, NoOn2            ; jump if ON2=0 to NoOn2
      setb   MOTOR2                ; else set output bit MOTOR2
NoOn2:
      ret
```

It is observed that the second equation is actually easier to implement, since if neither button is pressed, no action is taken. In the first equation, if neither button is pressed, the motor needs to be turned off. Both subroutines require labels to implement conditional jumps. The first subroutine requires a backward reference (jump) to clear the motor. As the control tasks become more involved, the assembly language becomes cumbersome. Attention needs to be paid to bit assignments, labels, subroutine entry and exit points, along with the application logic. Compared to the assembly language implementation, the ladder diagram and the circuit diagram hide these lower-level concerns from the programmer. The ladder logic and the circuit diagrams only convey the application logic, or the underlying Boolean expressions. Furthermore, assembly language is microcontroller specific and difficult to learn each time a new microcontroller is to be used.

Observations of graduate student projects in the Industrial Research Laboratory at the University of Florida suggest that writing and debugging automation programs with ladder logic is at least 5 times less time consuming that writing programs in assembly language. The advantage seems to increase further as the automation tasks grow in size.

A parallel development in the last decade concerns the general adoption of Fuzzy-Logic Control (FLC) in many applications [18], [30], [33], [34], [35], [36]. FLC is intended to bring more intelligence to the application, resulting in a smoother, more human-like quality to control. It is commonly mentioned in advertisements that fuzzy-logic control mimics the human decision making process [16]. This may be a statement more of a commercial nature than one of rigorous artificial intelligence. Nonetheless, fuzzy-logic control is a well-accepted ingredient of commercial applications. Some of the early applications were single-lens-reflex camera exposure and focus control systems and the all-electric automobile transmissions. It is noteworthy that most of the fuzzy-logic controlled appliances are Japanese in origin. A review of trade magazines reveals that the Japanese have invested heavily in the early to mid '80s in developing fuzzy-logic control systems but that there is a very large interest and a flurry of activity among American manufacturers. There are a few commercial FLC development systems which assist the user in constructing and evaluating a rule base. Also, there are macros and other utilities for specific microcontrollers to assist in FLC applications development [29].

The potential of FLC as a general-purpose industrial language is favorably influenced by the fact that, compared to assembly language, it is a higher-level language, implementing Boolean equations as a set of rules. In this respect it is comparable to ladder logic. Moreover, since the rules are given in linguistic terms, no specialized training is necessary for the operators to program FLC. Debugging, and revising the rule base is equally straightforward. In fact, by its nature, FLC is tolerant to contradictory or redundant rules. Modifications may be accomplished by simply changing the membership functions or adding new rules. Perhaps the most important feature of FLC is the fact that multi-input-multi-output nonlinear control paradigm may readily be implemented. Compared to PID control, for example, one may incorporate nonlinear behavior such as thresholds, response saturation, and hystheresis. The fuzziness provides a general-purpose smoothing mechanism. That is, when the rules are evaluated, switching from one rule action to the other may be accomplished in a gradual fashion as a weighted sum of rules is computed.

It must be mentioned that any controller design must start with a good understanding of the system to be controlled. Specifically, the response time, the precision and resolution, and the ranges of linearity or nonlinearity of the system must be understood. Controller design is not an exact science with established recipes applicable to all control problems. Fuzzy-logic control and its supporting utility tools should be viewed as engineering aids to the designer.

6.3 A Code Generator for FLC

Traditionally, computer hardware/software systems are developed from the top-down. This approach starts with chip design and produces a computer system with a programming and interfacing environment in which the end users must be trained [5].

Each subsequent phase is limited by the assumptions and design decisions of the previous phases. The operator inherits the most restrictions. The top-down approach is reasonable when hardware and software development efforts are considerably demanding. A most important result of rapid advancements in computer technologies is the proliferation of high-powered system development tools, ranging from computer-aided design (CAD) systems for chip-level and board-level computer system design utilities to compiler development tools. These tools bring hardware and software design within the realm of serious users who would like to implement specialized computer environments. Specifically, with the currently available compiler development tools, language design development is a feasible task for the users. It would be desirable that the applications programming language for microcontrollers closely follow the way humans perceive control tasks [27]. High-level languages which translate the human-described control tasks to sequential code suitable to run on microcontrollers is needed. Ideally, the nature of control tasks should drive the specific form of the high-level applications programming language, which in turn determines the operating system, the board-level controller architecture, and finally the chip design. This requires the users to be involved starting with language development . The bottom-up development requires a systems approach, combining issues of human factors, computer hardware architectures, industrial control, and high-level languages [13]. From the viewpoint of the end user, e.g. the operator, the control language must fulfill the following criteria.

1. Be programmable at a level higher than assembly language and be easily understood by users,
2. Be easily modifiable and maintainable,
3. Be capable of implementing both Boolean and analog control,
4. Be robust, stable, and responsive, with a performance comparable to PID/PLC control,
5. Generate code for speedy execution necessitated by real-time applications,
6. Generate compact code for memory efficiency,
7. Allow full access to on-chip facilities such as interrupts, ports, and timers,
8. Allow code development for applications with multiple microcontrollers.

PICFPL is a small fuzzy logic control code generator for PIC microcontrollers. It compiles files written with higher level language (fuzzy control language) and generates an output file written with the PIC microcontroller assembly language. In this first version of PICFPL.exe we show the feasibility of the project. For an extensive review of PICFPL, reader is referred to [1].

PICFPL requires that the source code file has the following structure:

- input declaration block (initiated by the keyword %input),
- output declaration block (initiated by the keyword %output),
- term declaration block (initiated by the keyword %terms),
- rules specification block (initiated by the keyword %rules),
- options block (initiated by the keyword %options).

A typical input to PICFPL is given below.

```
; *************** input variables ***********************************
;
;name  type  address            remark
;
;-------------------------------------------------------------------------
% input
 var_1  *8      8         ; var_1
 var_2  *8      9         ; var_2

; ************** output variables **********************************
;name    type  address            remark
;-------------------------------------------------------------------------
% output
  out_1 *16   10

; ************** variable terms ***********************************
; !!!  the same terms may be used with several input or variables  !!!
; !!!                  this saves memory                           !!!
;name              shape                          remark
;-------------------------------------------------------------------------
% terms
ZR_SB        (1@0, 1@20, 1@70, 0@90)              ; zero
PS_SB        (0@80, 1@100, 1@170, 0@190)          ; positive small
PL_SB        (0@180, 1@200, 1@235, 0@255)         ; positive large
LOW          (1@75)                               ; Singleton for output
MED          (1@128)                              ; Singleton for output
HIGH         (1@225)                              ; Singleton for output
; ************** rules *******************************************
; rule                                                          remark
;-------------------------------------------------------------------------
% rules
if (var_1 is PL_SB)  then out_1 = LOW                     ; comments here...
if (var_1 is PS_SB) and (var_2 = ZR_SB)  then out_1 = MED

; ************** options *****************************************
; options                                       remark
;-------------------------------------------------------------------------
% options                            ; "%options" directive is required
                                     ; ... but no options need be listed
                                     ; ... if the defaults suffice

  centroid
  tvfi
  mamdani  corlprod
  corlmin


; --- end of file --------------------------------------------------------
```

Figure 3. Sample FLCL Input File

Each block except the options block consists of one or more declarations. For example, the input declaration block consists of one or more input declarations.

Each input declaration (input linguistic variables) must have 3 elements: identifier, type, and register address. The identifier (name of the input variable) must not exceed 16 characters. The type should one of the following types: *8 (unsigned byte), *8s (signed byte), *16 (unsigned word) or *16s (signed word). The register address specifies where the value of the input variable is found.

The output declaration block is similar to the input declaration block. Each output declaration (output linguistic variable) also consists of the identifier, type, and register address. All the intermediate calculations are made in double precision and then converted to the desired format.

Each term declaration (term set) must have two elements: identifier and shape. The identifier(term name) must not exceed 16 characters. The shape is a collection of points defining the membership function associated with its term. For simplicity we allow only trapezoidal shapes with 6 points. The term sets are not tied to linguistic variables. This way the same term set may be used for more than one linguistic variable, which allows more flexible programs and especially saves memory on the microcontroller internal RAM. The points defining the shape of the term set should be unsigned bytes for this version (between 0 and 255).

Each rule starts with an IF keyword and should have conditions and actions. A condition is a statement between two parentheses that ties an input linguistic variable with a term set, i.e., meaning: input linguistic variable equal or is term set. the conditions are separated by a (fuzzy) logical AND operator. The conditions are separated from the actions by means of the keyword THEN. The actions are similar to conditions but they apply to the output linguistic variables. The following example shows a rule specification:

```
IF(temperature is high) AND (humidity is high) THEN (AC = HIGH);
```

The options block specifies which fuzzy inference methods and which defuzzification methods are to be used.

The prototype language is implemented using compiler support tools LEX and YACC, both standard UNIX utilities. LEX is a code generator for a lexicographic analyzer [10], [24]. The input is a specification of the syntax of the language elements. The output is a C language subroutine that reads the input stream and returns tokens. YACC (Yet Another Compiler Compiler) is a parser generator. The input is the language grammar in the Backus-Naur Form (BNF). The output is a C language subroutine which calls the lexicographic analyzer to extract tokens from the input stream and takes user-defined actions when a language construct is detected.

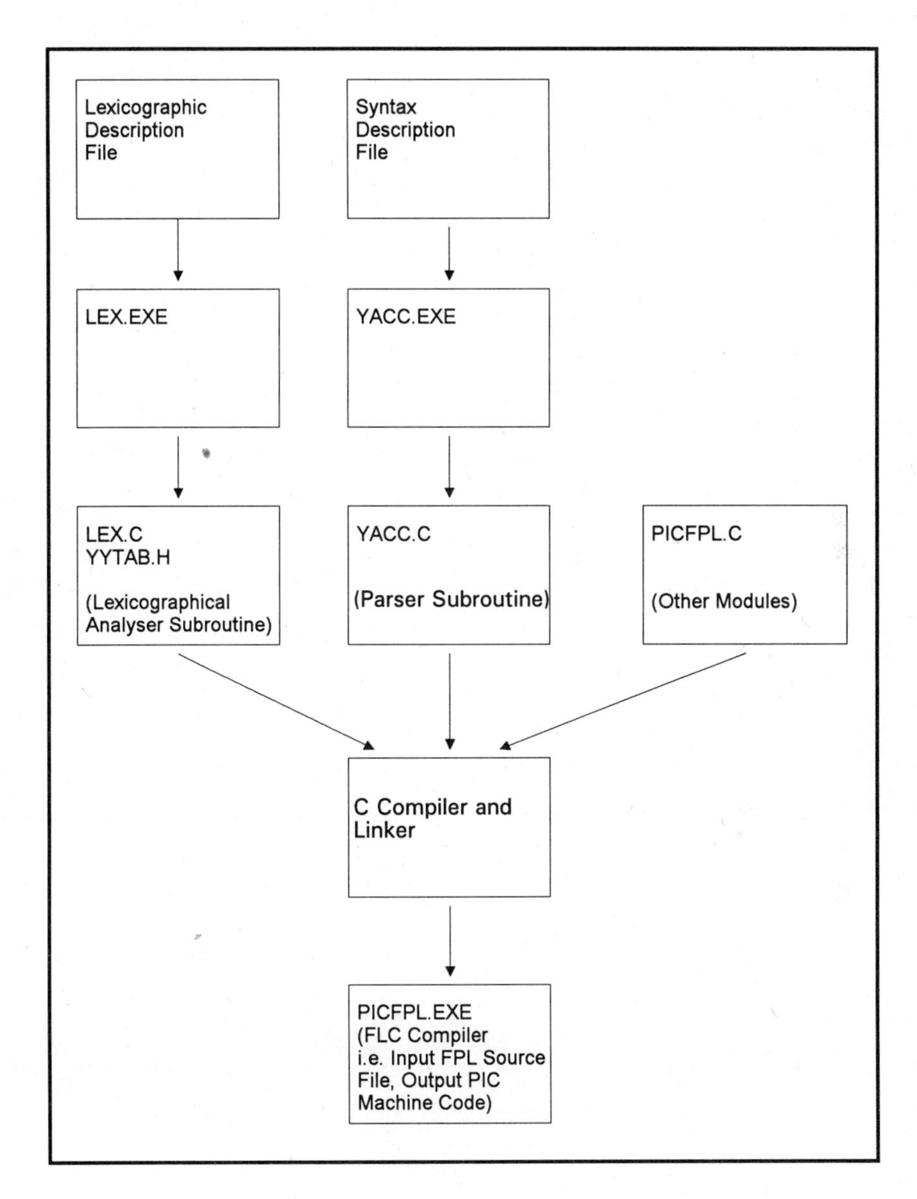

Figure 4. Use of Code Generators LEX and YACC in Creating the Fuzzy-Logic Code Generator

When language statements are detected, the information is stored in data structures. In a skeletal implementation, the main program simply prints out the input and output terms, the conditions, and the rules. An error checker may be placed after this step to report any inconsistencies. For example, if an identifier used in a condition is not previously defined in an input declaration, the parser does not report an error. It is more convenient for the error checker to verify that all identifiers are previously defined.

```
%{
  #include <stdio.h>
  #include "global.h"
%}
%union
{
  int   i;
  char *s;
}
/*----------------------- constants --------------------------------*/
%token      INTEGER_CONSTANT
/*----------------------- identifiers ------------------------------*/
%token      IDENTIFIER
/*----------------------- key words --------------------------------*/
%token UBYTE SBYTE UWORD SWORD
%token INPUT    OUTPUT TERMS   RULES    OPTIONS
%token CENTROID TVFI   MAMDANI CORLPROD CORLMIN
%token IF       IS     THEN
/*---------------------- Fuzzy logic operators ---------------------*/
%token AND OR NOT XOR
/*----------------------- start symbol -----------------------------*/
%start program
%%
program :   InputBlock OutputBlock TermsBlock RulesBlock OptionsBlock;
InputBlock    : '%' INPUT   inputs;
OutputBlock   : '%' OUTPUT  outputs;
TermsBlock    : '%' TERMS   terms;
RulesBlock    : '%' RULES   rules;
OptionsBlock  : '%' OPTIONS options;
/* ----------- inputs and outputs blocks -------------------------- */
inputs        :   variable_declaration | inputs variable_declaration;
outputs       :   variable_declaration | outputs variable_declaration;
variable_declaration :   IDENTIFIER type address;
type          :   UBYTE | SBYTE | UWORD | SWORD;
address       :  INTEGER_CONSTANT;
/* --------------- terms block ------------------------------------ */
terms         :   term | terms term;
term          : IDENTIFIER shape;
shape         : '(' point_list ')';
point_list    :   point | point_list ',' point;
point         : INTEGER_CONSTANT '@' INTEGER_CONSTANT;
/* --------------- rules block ------------------------------------ */
rules         :   rule | rules rule;
rule          : IF conditions THEN IDENTIFIER '=' IDENTIFIER;
conditions    :   condition | conditions logic_sym condition;
condition     : '(' IDENTIFIER IS IDENTIFIER ')';
logic_sym     : AND | OR | XOR;
/* --------------- options block ---------------------------------- */
options       : | options_list;
options_list  :   option | options_list option;
option        :   CENTROID | TVFI | MAMDANI | CORLPROD | CORLMIN;
%%
```

Figure 5. Syntax Description File (Input to the Parser Generator YACC)

The generation of microcontroller code is achieved by retrieving the information from the data structures and emitting the microcontroller code. This step is specific to the target microcontroller. The language implemented for the PIC family of microcontrollers is fully discussed in [1], along with complete source code.

The PICFLP is further enhanced by off-line productivity and utility tools. These are software subsystems that assist the programmer in generating fuzzy-logic control applications. Before these tools are discussed, we first summarize the elements and basic principles of fuzzy-logic control. An extended example is given in the Appendix.

6.4 Overview of Fuzzy-Logic Control Methods

Fuzzy-logic control algorithms implemented on microcontrollers consist of seven stages as given below. The different implementations of fuzzy-logic control algorithms are due to the alternative inferencing and defuzzification methods. This document describes alternative methods and illustrates their differences by a common example. For an extensive discussion of the subject, reader is referred to [3], [4], [8], [9], [11], [16], [17], [37], [41], [42], and [43]

Stages						
Crisp Inputs	Fuzzification	Rule Evaluation	Inference	Aggregation	Defuzzification	Crisp Output
Read inputs from files (registers)	Compute membership grade of each input in each term	Compute membership grade for each rule	Alternatives: (1) Inference output is a set of singletons (2) Inference output is a fuzzy set	Used only if the Inference output is a collection of fuzzy sets.	Alternatives: (1) Centroid (2) Leftmost Max (3) Rightmost Max (4) Average Max	Place output into files (registers)

6.4.1 Acquisition of crisp inputs

In microcontroller applications, the crisp inputs are bytes or words read from files (registers) or ports. The inputs may be signed or unsigned values. If signed integers are used, the input byte or word is interpreted to be in two's complement notation.

6.4.2 Fuzzification

Fuzzification refers to computing the membership grade of each input in each of the terms. In the above example, the terms Cool, Warm, and Hot are evaluated for the temperature read from the sensor. The three terms of humidity are also evaluated.

6.4.3 Rule Evaluation

Rule evaluation is performed one rule at a time, using the membership grade of each condition obtained in the fuzzification step. That is, the condition

LINGUISTIC VARIABLE is *term*

is the membership grade of the current input in the fuzzy set *term*. The membership grade of each rule is the minimum of the condition memberships, since the conditions are combined by AND operations.

6.4.4 Inference

The output of the fuzzy-logic controller depends on the rules, or more specifically, on the membership grades of the rules, depending on the current input values. Typically, there are two opposing requirements that need to be balanced in this step. First, the rule or rules with higher membership grades should be weighted more in constructing the output. Secondly, the effect of the rules with lower membership grades should nonetheless be taken into account to assure that the output switches smoothly from one value to another. This means that the weights given to the rules with higher membership grades need be allocated conservatively. Depending on the application and the number of possible outputs, there are alternative interfacing methods. For example, if the output is restricted to only a few selections, as in the example above (low, medium, high), then a simpler inferencing method may be used. If the output is a continuously variable quantity, such as in motor speed, then the smooth transients of the controller output become more important. In the latter case, a more involved inference method may be preferable.

The inferencing methods may be classified according to the nature of the action part of each rule. If the rules have actions that are singletons then the output of the inference operation is a set of singletons with their associated membership grades. If the output of each rule is a term, similar to the terms associated with the inputs, then the output of the inference operation is a fuzzy set. The first class is sometimes referred to as the TVFI (Truth Value Flow Inference) method. The Mamdani method is perhaps the best known method of the second class.

The TVFI method uses singleton outputs. Thus, the result of the inference operation is the set of membership grades for each of the singleton outputs. The singletons are constants determined by the rules. Note that this set of singletons and associated membership grades may be viewed as a fuzzy set. The crisp output of the fuzzy-logic controller is obtained from the set of singletons at the defuzzification stage.

The Mamdani method uses fuzzy terms for rule actions. The result is a collection of terms or fuzzy sets, as prescribed by the rules. In addition, each output term must be weighted by the rule membership, so that the output terms associated with the rules that have higher membership grades may have a greater effect on the controller output. The Mamdani method "chops" the fuzzy sets at the level of the membership grade of its rule. Suppose rule R_i is evaluated to have a membership grade of F_i. Let the output term of R_i be O_i. O_i is a fuzzy set, that is, for every element x, O_i assigns the membership grade $f_i(x)$. The term set is weighted by the rule membership grade F_i by assigning the membership grade

$$f_i'(x) = \max \{ f_i(x), F_i \} \qquad (1)$$

to element x. That is, the weighted fuzzy set O'_i assigns membership grade $f_i'(x)$ to element x. The output of the inference method is the collection of weighted fuzzy sets O'_i. This method of weighting the term sets is also referred to as the min-correlation method.

The min-product method is similar to the min-correlation method. It is also used to weight fuzzy sets. The term set is weighted by the rule membership grade F_i by assigning the membership grade

$$f_i'(x) = F_i \, f_i(x) \qquad (2)$$

to element x. That is, the weight F_i is multiplied by the membership grade $f_i(x)$.

If either the min-correlation or the min-product method is used, the output of the inference operation is the collection of weighted term sets. Before a crisp output is generated, the collection of weighted fuzzy sets O'_i are combined in the step known as aggregation.

6.4.5 Aggregation

When the output of the inference operation is a collection of weighted term sets, before a crisp output is obtained, these term sets are combined. Note that if the output of the inference operation is a collection of singletons, the aggregation operation is not performed.

The Mamdani method prescribes a specific aggregation method to combine the weighted fuzzy sets O'_i into a single fuzzy set, O. The output set O assigns membership grade

$$g(x) = \min \{ f_i'(x) \} \qquad (3)$$

to each x. The output O may be viewed as the union of all of the weighted terms O'_i. Again, the crisp output needs to be computed from this collection of fuzzy sets. Compared to the TVFI method, the Mamdani method requires more computations and data storage, since it generates a collection of fuzzy sets rather than a collection of singletons.

The min-correlation or the min-product inference methods also generate a collection of term sets. The aggregation of these sets is usually performed by computing the center of gravity of each fuzzy set O'_i using the following formula.

$$COG_i = M_i / F_i \tag{4}$$

where

$$M_i = \int x f'_i(x)dx \tag{5}$$

$$F_i = \int f'_i(x)dx \tag{6}$$

and where $f_i'(x)$ is computed as described above. The center of gravity of the fuzzy set may be viewed as an equivalent singleton, and F_i, as the equivalent weight (membership grade) of the singleton. (Note, however, that F_i may be greater than 1).

6.4.6 Defuzzification

The output of the inference stage for the TVFI (Truth Value Flow Inference) method and the output of the aggregation stage for the min-correlation or the min-product method is a set of center-of-gravities (equivalent singletons). The output of the inference stage of the Mamdani method is a fuzzy set. Defuzzification refers to evaluating a crisp output from a set of singletons, a set of center-of-gravities, or a fuzzy set. The defuzzification methods may be partitioned into two: the Centroid method and the Max (maximum) method. The Max method actually contains three variants, the Leftmost Max, the Rightmost Max, and the Average Max methods. All inference/aggregation methods may be used with any one of these defuzzification methods.

The Centroid method computes the center of gravity of the fuzzy set. In case of the TVFI method, the set of singletons may be viewed as a fuzzy set. The centroid is then computed as

$$\frac{\sum_i S_i F_i}{\sum_i F_i} \tag{7}$$

where S_i is the singleton action value for rule R_i and F_i is the membership grade for rule R_i. In case of the min-correlation or min-product methods, the centroid is computed by the above formula, except that the value S_i is replaced by COG_i, as,

$$\frac{\sum_i COG_i F_i}{\sum_i F_i} \tag{8}$$

or equivalently by,

$$\frac{\sum_i M_i}{\sum_i F_i} \tag{9}$$

where M_i and F_i are given by equations (5) and (6).

The Mamdani method generates a single fuzzy set at the end of the inference stage. The centroid of this fuzzy set is readily computed as,

$$\frac{M}{F} = \frac{\int xg(x)dx}{\int g(x)dx} \tag{10}$$

where g(x) is defined by equation (3).

The Max method simply selects the output value which has the highest membership grade. For the TVFI method, the output from the inference stage is a set of singletons with the associated membership grades. The Max method selects the controller output to be the singleton with the highest membership grade. In case there are several singletons with equally high membership grades, the Leftmost Max method selects the singleton with the lowest output value, and the Rightmost Max, the singleton with the highest output value. The Average Max method takes the arithmetic average of the singletons with the high membership grade.

The Max method may also be applied to the Mamdani method. When the output from the inference stage is a fuzzy set, the Max method selects the output value which has the highest membership grade. Again, in case there are more than one such output values, the Leftmost Max, the Rightmost Max, and the Average Max methods select the smallest, the largest, and the average output value, respectively.

6.4.7 Output Generation

The output of the fuzzy-logic controller is a crisp value. In case the output may be any value within the range, the output of the defuzzification stage is simply forwarded as the controller output. If the fuzzy-logic control system outputs are limited to a number of selections, then the output of the defuzzification stage needs to be rounded to the nearest selection. In the air conditioning example, if the AC_Speed may assume any value between 0 and 0FFh, for example, the output of the defuzzification stage is readily used as the system output. If AC_Speed needs to be one of the values Low, Medium, and High, denoted by the numerical values 1, 2, and 3, then the output of the defuzzification stage must be rounded to the nearest integer value.

6.4.8 Remarks

All the inference methods may be adapted to be used with any one of the aggregation methods. For example, the Mamdani method may be modified to first evaluate the min-product output terms. The modified method may then OR the output terms to obtain a single output fuzzy set.

The TVFI method is the computationally least demanding [9]. The Mamdani and the min-product / min-correlation methods require more computation and memory to store the intermediate results. In general, computations for the centroid method can exploit the trapezoidal structure of fuzzy sets to eliminate the need for numerical integration.

The min-product method results in fuzzy sets whose slopes are reduced. Compare, for example, Figures 11b and 14b. The min-correlation "chops" the fuzzy set, thus reducing its "spread". The min-product method may be desirable to assure that the controller output varies smoothly as the dominant set of rules change.

Alternatively, the smoothness of the controller output as the dominant subset of rules change may be improved by using many overlapping terms. This would assure that many rules fire for any given input combination, and thus, the relative effect of a changing rule is reduced.

It is observed from equations (7), (8), (9), and (10) that if all rules have membership grade 0, then the centroid method fails to produce a result. More precisely, the division-by-zero operation makes the result undetermined. It is important that at least one rule fires in the given range of input values.

6.5 Productivity and Utility Tools for Rule Generation

There are four novel approaches to assist in rule base generation with PICFPL. The Rule-Base Reviewer, The PID translator, The fuzzy rule learner, the adaptive fuzzy logic controller. The first is a utility tool to generate fuzzy-logic control rule bases which implements a control function as close to a given PID controller as possible. The second one is a knowledge-base optimizer which scans the rules and identifies the degree of contradiction and redundancy among pairs of rules. The third is a software environment, in the spirit of virtual reality, to generate an artificial environment in which a human performs the control function as the software system extracts the rules from the experiment. The fourth approach is in the spirit of adaptive control, where the rule base is kept in memory and is modified in order to improve the performance of the controller.

6.5.1 The PID Translator (PIDX)

PID control is perhaps the most commonly used industrial control paradigm. The performance of Fuzzy Logic Controller is often compared to the performance of PID controller [19], [20], [21], [28], [39]. The PID Translator is a useful program that generates a specified number of rules, each rule with a specified number of conditions, to implement single-input single-output fuzzy-logic controllers. More specifically, the input to the PID translator is a set of PID coefficients, and the output, a set of rules for the fuzzy-logic controller. PIDX will optimize the term sets and the rules to minimize the difference between the performance of the PID controller and the fuzzy-logic controller.

6.5.2 The Rule-Base Reviewer (RBR)

Once a set of rules are generated, the rule-base reviewer scans the rules and identifies the degree of contradiction and redundancy among pairs of rules. This information may be used to revise the rule base. For example, redundant or nearly-redundant rules may be removed to improve run-time efficiency. Similarly, contradicting rules need to be re-examined to assure that there are no misconceptions about the control task.

6.5.3 The Fuzzy Rule Learner (FRL)

The Fuzzy Rule Learner consists of a user-friendly software environment which allows the user to change the output signals as he observes the changes in the inputs. The inputs may be received from the environment, from a history file, or directly from the user. The experiment may run in real-time for moderately fast events, or in modified time, either accelerated or slowed down. This approach can be used to derive fuzzy control rules by modeling an operator's control actions [31].

First consider a single-input single-output system. The user screen may display slide bars may determine the input and the control signal. That is, the user may change the input during the experiment. Similarly, the control signal is set by the user by dragging the slide bar. The correction signal in turn determines the system response. The system response must be computed by a user-supplied subroutine or table. The system response may be displayed by a meter. The experiment consists of changing the input and the control signal while trying to keep the system under control. As the experiment progresses, the input-output patterns are learned by the software and written into rules. These rules are later reviewed by the user and modified if necessary.

6.5.4 The Adaptive Fuzzy-Logic Controller (AFLC)

The Adaptive Fuzzy-Logic Controller differs from the previous two tools in that it resides within the PIC. The rule base is kept in memory and is modified according to the performance of the controller. A necessary component to AFLC is a subsystem which evaluates the performance of the controller. This may be a subroutine which keeps track of the average absolute difference between the actual system output and the desired system output. The fuzzy-logic control routines in program memory will periodically vary the rule base and observe its effects on the performance of the controller. If the controller performance increases, the new rule base is adopted. If not the previous rule base is restored. AFLC is especially useful in controlling equipment in unknown, or in changing environments, such as when rapid aging is expected [2], [28], [32], [38].

6.6 A Detailed Example

The six stages are illustrated by the example of air conditioning control. The inputs are temperature and humidity. Associated with each input is a linguistic variable. Each linguistic variable may take a number of possible values. Values of the linguistic variables are called terms. Each value, or term, is represented by a fuzzy set. In our example, we have the terms Cool, Warm, and Hot associated with the linguistic variable Temperature; and the terms Dry, Moderate, and Damp, associated with the linguistic variable Humidity.

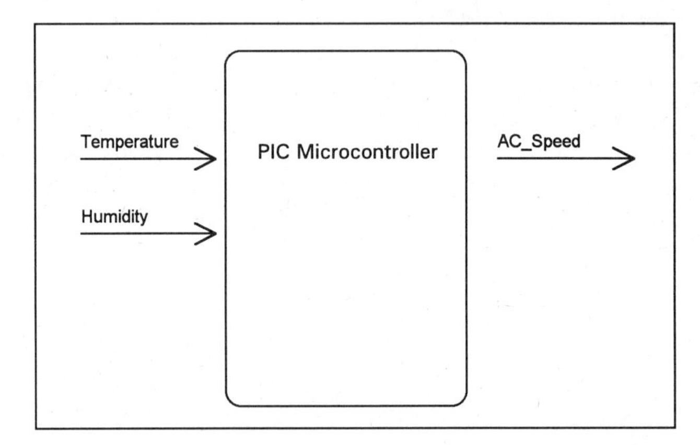

Figure 6. An Air Conditioner Controller.

The temperature and humidity inputs which may be bytes or words read from a sensor. The linguistic variables are also called **TEMPERATURE** and **HUMIDITY**. However, the linguistic variables are used in the abstract sense. That is, the linguistic variables are not directly involved in computations. Rather, given the inputs, the membership grades of the terms of these linguistic variables are evaluated.

The membership grade for each term set is a function of the corresponding input. Let the membership grade for term "Cool" be given as a function of temperature t as follows.

$$m_{Cool}(t) = \begin{cases} 1 & \text{for } (t < 60) \\ 1 - (t-60)/10 & \text{for } (60 \le t \le 70) \\ 0 & \text{for } (t > 70) \end{cases}$$

Note that the actual input from the sensor will be a byte (or perhaps a word) in the range of say, 0 to 0FFh which varies linearly with the temperature. The membership grade for the term "Cool" must internally be computed from the input value. Let the sensor return 0 for 45 degrees and 0FFh for 100 degrees. That is, the sensor will return 69 (or 46h) when the temperature is 60 degrees, and 116 (or 74h) when the temperature is 70 degrees. Equivalently, the membership grade, $m_{Cool}(b)$, for the term Cool is computed as a function of the input b as follows.

$$m_{Cool}(b) = \begin{cases} 1 & \text{for } (b < 46h) \\ 1 - (b\text{-}46h)/(74h\text{-}46h) & \text{for } (46h \leq b \leq 74h) \\ 0 & \text{for } (b > 74h) \end{cases}$$

The microcontroller code must use the sensor scale rather than the degree (^{O}F) scale when computing the membership grades. However, the user interface is more intuitive if ^{O}F scale is used. As far as the examples are concerned, the computations may be carried out using either of the two scales. The discussion below uses the application scales degrees (^{O}F) and relative humidity in percent.

It is convenient to represent membership functions graphically using the application scales as shown below. A single graph is used to show all three terms of the linguistic variable "**TEMPERATURE**".

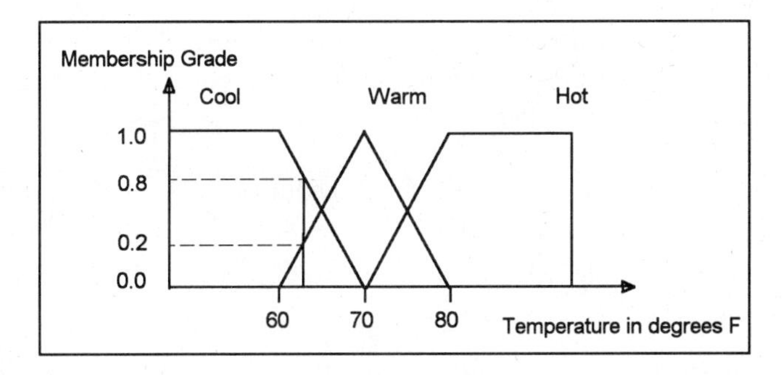

Figure 7. Terms of the Linguistic Variable "TEMPERATURE"

Similarly, the three terms of linguistic variable "**HUMIDITY**" are given below, using the relative percent humidity scale.

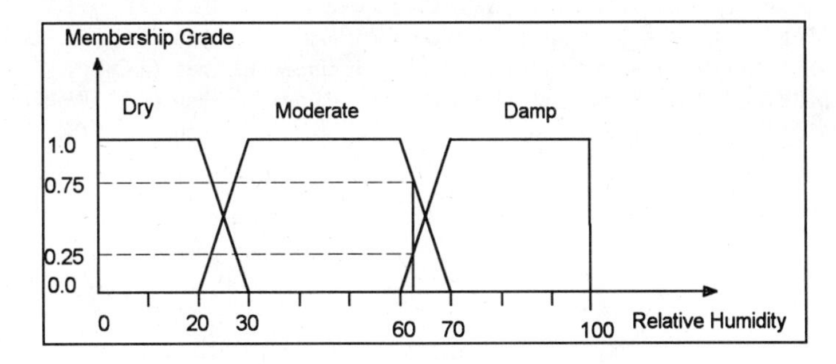

Figure 8. Terms of the Linguistic Variable "HUMIDITY"

The output is a level referred to as AC_Speed. Temperature and humidity is measured by sensors and placed into registers. Alternatively, they may be available on one of the microcontroller ports. The output AC_Speed is a value sent to the air conditioner. There are two possible alternatives:

1. The air conditioner has a limited number of settings. In manual air conditioner controls, this corresponds to the common "push button" type speed selection controls. Consider three speed settings: low, medium, and high. Usually, the settings are represented by numerical values such as,

   low = 20
   medium = 50
   high = 70.

   These values may represent the percent speed or rating of the compressor.

2. The air conditioner has a range of settings, for example, an analog input whose value may be described digitally be a byte or a word. In manual air conditioner controls, this corresponds to the dial-type speed controls which may be varied continuously by rotating a knob.

Analogous to the duality between the inputs and the linguistic variables is the relationship between outputs of the inference operation and the outputs to be sent to the environment. The output of the inference operation may be fuzzy sets, also called terms of the output, or singletons.

Once the input and outputs are established, and the terms of the linguistic variables are determined, the control task is written as a set of rules. For this example, consider the following seven rules.

R1: If (**TEMPERATURE** is *cool)* and (**HUMIDITY** is *dry)* then (**AC_SPEED** is *low)*
R2: If (**TEMPERATURE** is *cool)* and (**HUMIDITY** is *moderate)* then (**AC_SPEED** is *medium)*
R3: If (**TEMPERATURE** is *cool)* and (**HUMIDITY** is damp) then (**AC_SPEED** is *medium)*
R4: If (**TEMPERATURE** is *warm)* and (**HUMIDITY** is *dry)* then (**AC_SPEED** is *medium)*
R5: If (**TEMPERATURE** is *warm)* and (**HUMIDITY** is *moderate)* then (**AC_SPEED** is *high)*
R6: If (**TEMPERATURE** is *warm)* and (**HUMIDITY** is damp) then (**AC_SPEED** is *high)*
R7: If (**TEMPERATURE** is *hot)* then (**AC_SPEED** *is high)*

Each rule has an action in the form

... then **OUTPUT** is *term*

In the most general sense, the terms are fuzzy sets, similar to the terms associated with the input variables. In many applications, singletons are used as the output terms. Singletons are fuzzy sets with only one member with grade 1 and all other members with grade 0. The following examples illustrate the general fuzzy output terms and the singleton outputs.

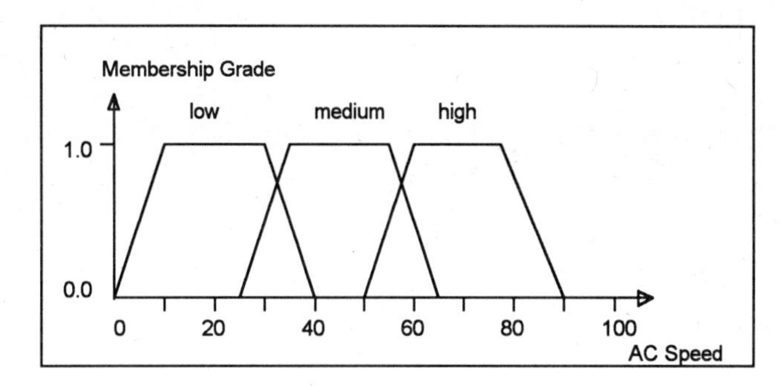

Figure 9. Terms of the Output

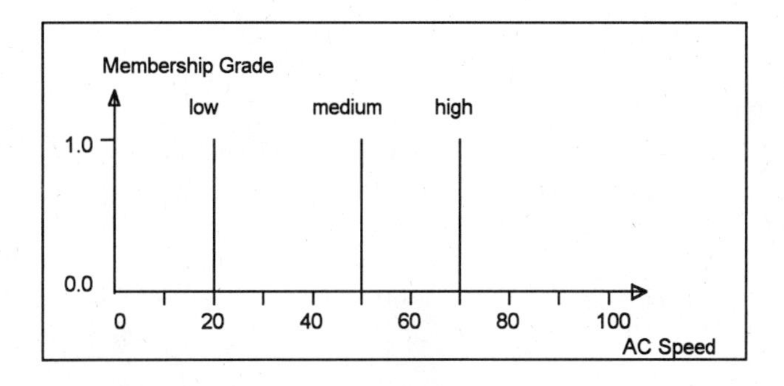

Figure 10. Singleton Terms of the Output

The various inference and defuzzification methods are illustrated by numerical examples. Consider the air condition controller discussed above. The acquisition of inputs, fuzzification, and rule evaluation steps are the same for all inference, aggregation, and defuzzification methods. The numerical examples are given for the TVFI, Mamdani, min-correlation, and product-correlation methods.

The output of the fuzzy-logic controller, AC-Speed, is set to *low (1)*, *medium (2)*, or *high (3)*, which are singleton terms as given in Figure 10.

6.6.1 Input Acquisition

Consider the air condition control example given above. Let the temperature and humidity be read from the sensors as

Temperature = 62' F
Humidity = 62.5%.

6.6.2 Fuzzification

Irrespective of the inference and defuzzification method used, the first stage simply evaluates the membership grade of each input in each of the terms. From Figure 7 we observe the following membership grades.

$\mu_{cool}(68) = 0.8$
$\mu_{warm}(68) = 0.2$
$\mu_{hot}(68) = 0.0$

$\mu_{dry}(62.5) = 0.0$
$\mu_{moderate}(62.5) = 0.75$
$\mu_{damp}(62.5) = 0.25$

6.6.3 Rule Evaluation

Each of the 7 rules of the rule base is evaluated next. Since rules combine conditions by AND operations, the membership grade of a rule is the minimum of the condition memberships which were computed in the previous stage.

R1: If (**Temperature** is *cool)* and (**Humidity** is *dry)* then (**AC_Speed** is *low)*
F1= min {0.8, 0.0) = 0.0

R2: If (**Temperature** is *cool)* and (**Humidity** is *moderate)* then
(**AC-speed** is *medium)*
F2= min {0.8, 0.75 } = 0.75

R3: If (**Temperature** is *cool)* and (**Humidity** is damp) then (**AC-speed** is *medium)*
F3= min {0.8, 0.25 } = 0.25

R4: If (**Temperature** is *warm)* and (**Humidity** is *dry*) then (**AC-speed** is *medium)*
F4= min {0.2, 0.0 } = 0.0

R5: If (**Temperature** is *warm)* and (**Humidity** is *moderate*) then (**AC-speed** is *high)*
F5= min {0.2, 0.75 } = 0.2

R6: If (**Temperature** is *warm)* and (**Humidity** is damp) then (**AC-speed** is *high)*
F6= min {0.2, 0.25 } = 0.2

R7: If (**Temperature** is *hot)* then (**AC-speed** *is high)*
F7= min {0.0 } = 0.0

A quick review of the rule memberships reveal that rule R2 has the highest membership grade, 0.75, and rules R1, R4, and R7 have the lowest membership grade, 0.0.

6.6.4 Inference, Aggregation, and Defuzzification

6.6.4.1 The TVFI (Truth Value Flow Inference) Method

The TVFI method uses the singleton outputs of the rules and the associated rule memberships. These are summarized for the given temperature and humidity inputs in the following table.

Rules	Singleton Outputs (S_i)	Rule Membership Grades (F_i)	S_iF_i
R1	20	0.0	0.0
R2	50	0.75	37.5
R3	50	0.25	12.5
R4	50	0.0	0.0
R5	70	0.20	14.0
R6	70	0.20	14.0
R7	70	0.0	0.0
Total		1.40	78.0

If the Centroid defuzzification method is used, the output is 78.0/1.4=55.71. Since the controller output must be 20, 50, or 70, the value 55.71 is rounded to the nearest output value, 50. Thus, the controller sets the AC-Speed to medium.

If the Max defuzzification method is used, the rule with the largest membership grade is identified. This is rule R2, with the action (**AC_Speed** is *medium*). Thus, the AC_Speed is set to medium. Since only one rule has the highest membership grade, all three Max methods would yield the same result. If there were more than one rule with membership grade 0.75, then either the leftmost or the rightmost, or the average of the rule actions would be selected.

6.6.4.2 The Mamdani Method

Consider next the rule actions being fuzzy sets as shown in Figure 9. The fuzzification and rule evaluation stages are the same as in the first case.

In this example, only rule R1 prescribes the *low* output. There are three rules, R2, R3, and R4, which prescribe the output to be *medium* and three rules R5, R6, and R7 which prescribe the output to be *high*. Since the rules are considered to be combined in the rule base by OR operations, the membership grade associated with each output term is obtained by identifying the rule which prescribes that output term with the maximum membership grade. The membership grade for each of the output terms is computed below.

Output Term	Rules, (R_i)	Membership Grade (F_i)
low	1	0.0
medium	2, 3, 4	max { 0.75, 0.25, 0.0 } = 0.75
high	5, 6, 7	max { 0.20, 0.20, 0.0 } = 0.20

The Mamdani method uses the min-correlation technique to weight the output terms, as given by equation (1). The weighted ("chopped") output terms are shown in Figure 11.

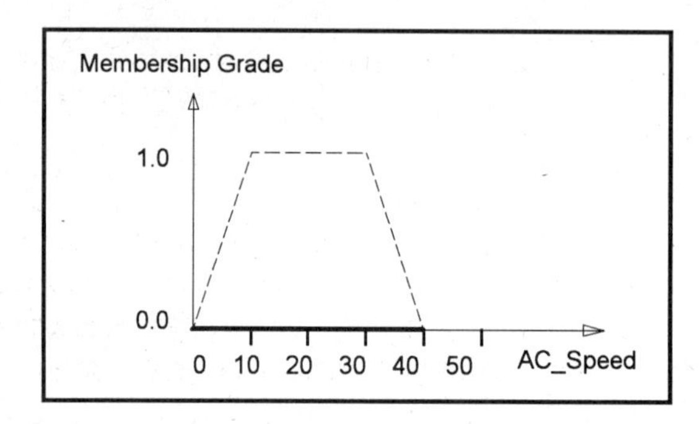

Figure 11a. Output Term *Low*

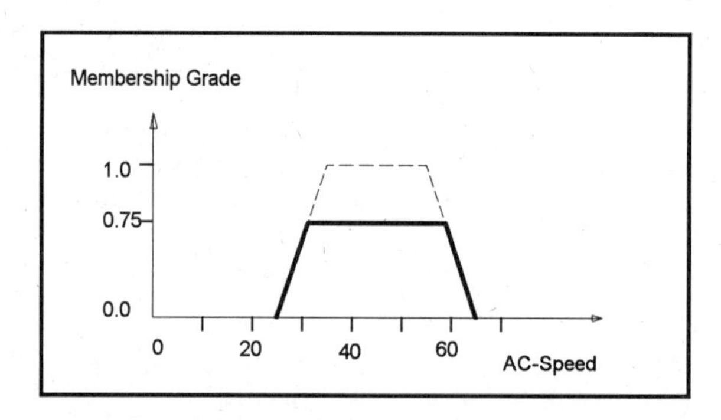

Figure 11b. Output Term *Medium*

$$m_{medium}(x) = \begin{cases} 0.1x-2.5 & \text{for } 25.0 \le x \le 32.5 \\ 0.75 & \text{for } 32.5 \le x \le 57.5 \\ -0.1x+6.5 & \text{for } 57.5 \le x \le 65 \end{cases}$$

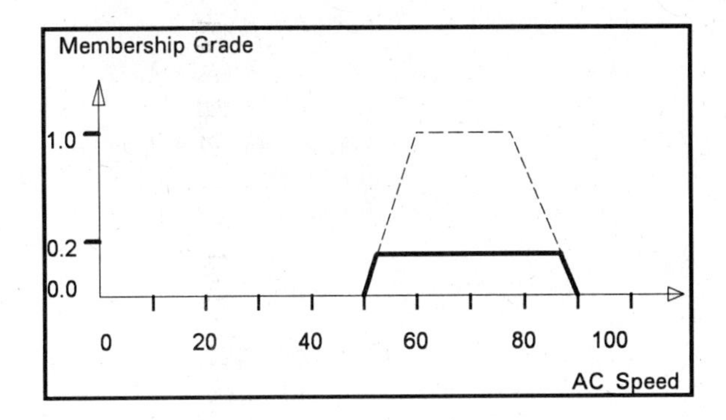

Figure 11c. Output Term *High*

$$m_{high}(x) = \begin{cases} 0.1x-5 & \text{for } 50 \le x \le 52 \\ 0.2 & \text{for } 52 \le x \le 88 \\ -0.1x+9 & \text{for } 88 \le x \le 90 \end{cases}$$

The aggregation step combines the weighted term sets as given by equation (3). The results of this aggregation is illustrated below.

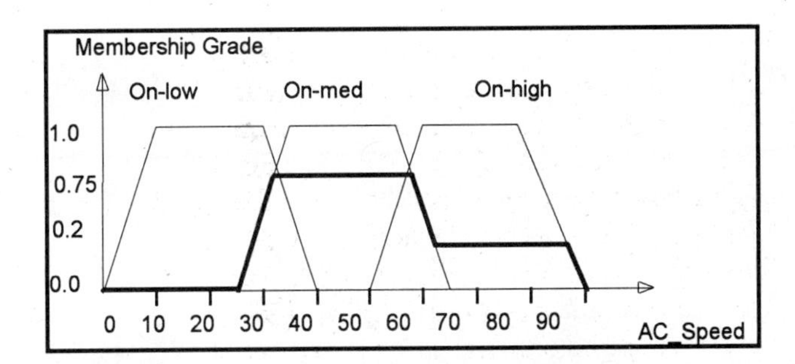

Figure 12. Output Fuzzy Set of the Mamdani Method

In the Mamdani method, the crisp output is obtained by computing the center of gravity of the output fuzzy set.

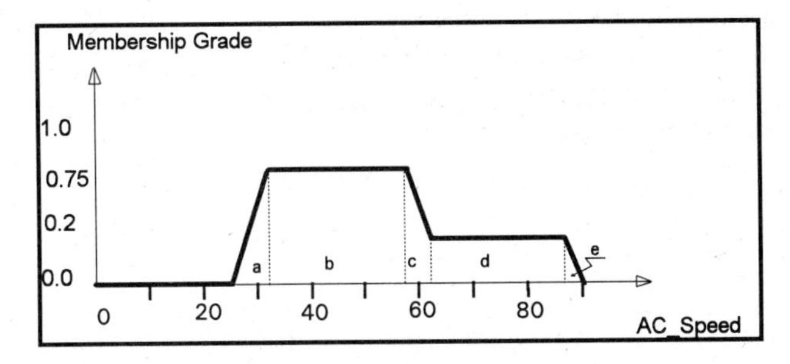

Figure 13. Partitioning of the Output Fuzzy Set

$$m_{output}(x) = \begin{cases} 0 & \text{for } 0.0 \le x \le 25.0 \\ 0.1x-2.5 & \text{for } 25.0 \le x \le 32.5 \\ -0.75 & \text{for } 32.5 \le x \le 57.5 \\ -0.1x+6.5 & \text{for } 57.5 \le x \le 63.0 \\ 0.2 & \text{for } 63.0 \le x \le 88.0 \\ -0.1x+9 & \text{for } 88.0 \le x \le 90.0 \end{cases}$$

The output is partitioned into five areas. The center of gravity is computed from equation (10) as follows.

$$\frac{\int_{25}^{32.5} x(0.1x-2.5)dx + \int_{32.5}^{57.5} x(0.75)dx + \int_{57.5}^{63} x(-0.1x+6.5)dx + \int_{63}^{88} x(0.2)dx + \int_{88}^{90} x(-0.1x+9)dx}{\int_{25}^{32.5} (0.1x-2.5)dx + \int_{32.5}^{57.5} (0.75)dx + \int_{57.5}^{63} (-0.1x+6.5)dx + \int_{63}^{88} (0.2)dx + \int_{88}^{90} (-0.1x+9)dx}$$

Alternatively, one may use the Max method. The crisp outputs for the Max methods are observable from Figure 12.

Rightmost maximizer = 57.5
Leftmost maximizer = 32.5
Average maximizer = 45

6.6.4.3 The Min-Product Method

The product-correlation method uses the product form, given by equation (2), of weighting the output terms. The weighting scheme is illustrated by Figure 14.

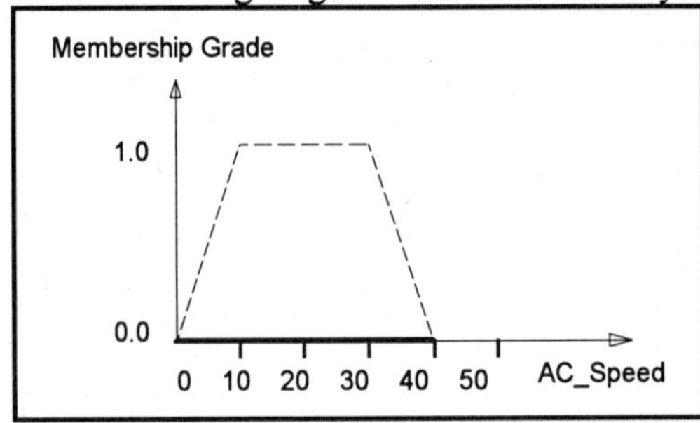

Figure 14a. Output Term *Low*

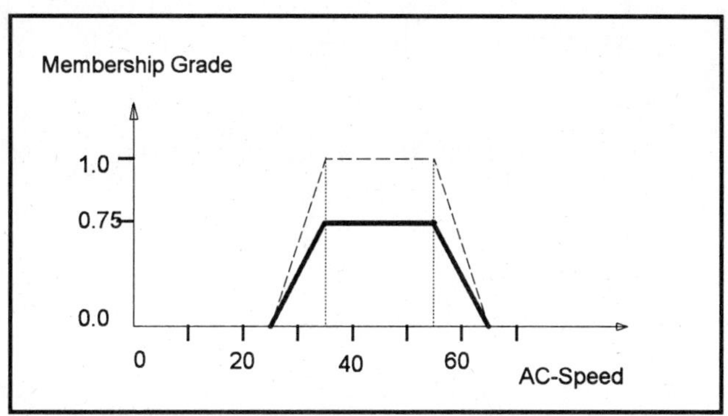

Figure 14b. Output Term *Medium*

$$m_{medium}(x) = \begin{cases} 0.075x-1.875 & \text{for } 25 \le x \le 35 \\ 0.75 & \text{for } 35 \le x \le 55 \\ -0.075x+4.875 & \text{for } 55 \le x \le 65 \end{cases}$$

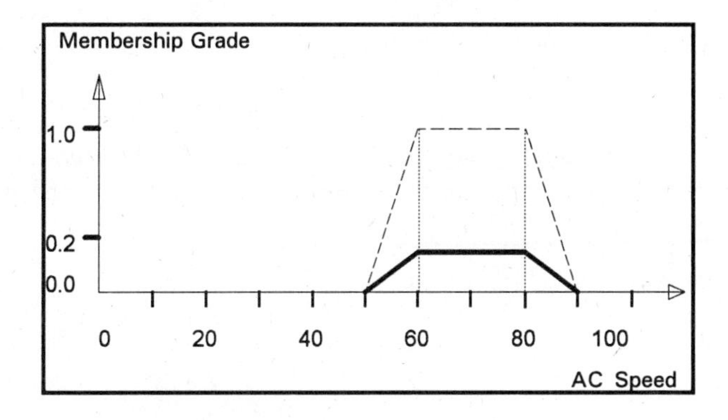

Figure 14c. Output Term *High*

$$m_{high}(x) = \begin{cases} 0.02x-1 & \text{for } 50 \le x \le 60 \\ 0.2 & \text{for } 60 \le x \le 80 \\ -0.02x+1.8 & \text{for } 80 \le x \le 90 \end{cases}$$

The centroid defuzzification operation is carried out as given by equations (5), (6), and (9).

$M_{low} = 0$
$F_{low} = 0$

$$M_{medium} = \int_{25}^{35} x(\frac{0.75}{10}x - 1.875)dx + \int_{35}^{55} x(0.75)dx + \int_{55}^{65} x(-\frac{0.75}{10}x + 4.875)dx = 1012.5$$

$$F_{medium} = \int_{25}^{35} (\frac{0.75}{10}x - 1.875)dx + \int_{35}^{35} 0.75dx + \int_{55}^{65} (-\frac{0.75}{10}x + 4.875)dx = 22.5$$

$$M_{high} = \int_{50}^{60} x(\frac{0.2}{10}x - 1.0)dx + \int_{60}^{80} x(0.2)dx + \int_{80}^{90} x(-\frac{0.2}{10}x + 1.8)dx = 420.0$$

$$F_{high} = \int_{50}^{60} (\frac{0.2}{10}x - 1.0)dx + \int_{60}^{80} 0.2dx + \int_{80}^{90} (-\frac{0.2}{10}x + 1.8)dx = 6.0$$

The centroid defuzzification method produces the output,

$$\frac{M_{low} + M_{medium} + M_{high}}{F_{low} + F_{medium} + F_{high}} = 50.26.$$

If the Max defuzzification method is to be used, it is observed from Figure 14 that the membership grade for medium is the highest (0.75). From Figure 14b, observed that the leftmost max method gives, 35, the rightmost max method gives 55, and the average max method gives 45. If the outputs need to be selected among the three singletons, 20, 50, 70, the leftmost method, with value 35, may be rounded to either 20 or 50, thus yields either a low or a medium output. The rightmost and the average max methods set AC_Speed to 50 (medium).

6.6.4.4 The Min-Correlation Method

It was mentioned that except for the aggregation method, the Mamdani method is the min-correlation method. The min-correlation method uses the centroid method or max method of aggregation. Consider the output fuzzy sets given in Figure 11. Applying the centroid method (equations 5, 6, and 9) yields,

$M_{low} = 0$
$F_{low} = 0$

$$M_{medium} = \int_{25}^{35} x(\frac{1.0}{10}x - 2.5)dx + \int_{35}^{55} x(0.75)dx + \int_{55}^{65} x(-\frac{1.0}{10}x + 6.5)dx = 1125.0$$

$$F_{medium} = \int_{25}^{35} (\frac{1.0}{10}x - 2.5)dx + \int_{35}^{55} 0.75dx + \int_{55}^{65} (-\frac{1.0}{10}x + 6.5)dx = 25.0$$

$$M_{high} = \int_{50}^{60} x(\frac{1.0}{10}x - 5.0)dx + \int_{60}^{80} x(0.2)dx + \int_{80}^{90} x(-\frac{1.0}{10}x + 9.0)dx = 980.0$$

$$F_{high} = \int_{50}^{60} (\frac{1.0}{10}x - 5.0)dx + \int_{60}^{80} 0.2dx + \int_{80}^{90} (-\frac{1.0}{10}x + 9.0)dx = 14.0$$

and the singleton output is computed as,

$$\frac{M_{low} + M_{medium} + M_{high}}{F_{low} + F_{medium} + F_{high}} = 53.92.$$

If the Max defuzzification method is to be used, it is observed from Figure 11 that the membership grade for medium is the highest (0.75). From Figure 11b, observed that the leftmost max method gives, 32.5 the rightmost max method gives 57.5 and the average max method gives 45. If the outputs need to be selected among the three singletons, 20, 50, 70, the leftmost method sets AC_Speed to 20 (low), while the rightmost and the average max methods set AC_Speed to 50 (medium).

6.7 Conclusions

Fuzzy-logic control is a strong candidate for a general-purpose industrial control language. It has several major advantages over assembly language, ladder logic, and other languages such as C and Basic.

1. FLC focuses on the logic pertaining to the application and hides the details of system bookkeeping.
2. FLC is rule-oriented. The order of rules is arbitrary. It is easily modified.
3. FLC uses linguistic terms in rules. No special training is required. Operators and technicians may easily write programs.
4. FLC tolerant to contradictory or redundant rules.
5. FLC is capable of generating non-linear multi-input-multi-output rules without the need of further special features.

There are also disadvantages as listed below.

1. The response of FLC is difficult to predict with analytical tools. Simulation is often the only course of evaluation.
2. It is difficult to program in machine language. A Code generator is almost always compulsory.

Perhaps the most significant conclusion based on experience gathered during this work is that it is the natural language rule-based nature of FLC which makes it an attractive choice for industrial embedded control. The fuzziness aspects are important in "smoothing" the response of the controller as one rule gradually looses weight and another gradually gains weight but does not intrinsically enhance the control task. A similar smoothing process may also be obtained by, say, a statistical Bayesian approach. The fuzzy theoretical approach is empirical, easy to understand and is sufficient to accomplish the task of response smoothing, and thus, should be preferable to statistical methods in most cases.

Acknowledgment

This research was partially supported by Microchip Technology, Inc., (grant number 222659-400).

References

1. Arrestam R., and J. Holmlund. *Fuzzy Logic Control Language (FLCL) for Embedded Controllers.* Research Report, Industrial Research Laboratory, University of Florida, 1991.

2. Bartolini, G., G. Casalino, F. Davoli, M. Mastretto, R. Minciardi, and E. Morten. *Development of Performance Adaptive Fuzzy Controllers with Application to Continuous Casting Plants.* Industrial Applications of Fuzzy Control, Edited by M. Sugeno, Elsevier Science Publishers B.V.(North~Holland), 1985, pp. 73-87.

3. Bellman, R., and L. Zadeh. *Decision Making in a Fuzzy Environment.* Management Science,(1970),17,B.141-b.164.

4. Bellman R., and M. Giertz. *On the Analytic Formalism of the Theory of Fuzzy Sets.* Inf. Sciences 5, 1973.

5. Boehm, H. W. *Improving Software Productivity*, IEEE COMPUTER, September 1987

6. Buhr, R.J.A. *System Design with ADA*, Prentice-Hall Inc., 1984

7. Chang, S. and L. A. Zadeh. *On Fuzzy Mapping of Control.* IEEE Transactions on Systems, Man, and Cybernetics, Vol. SCM-2, pp 30-42, 1972.

8. Dobois,D., and H. Parade. *Fuzzy Sets and Systems: Theory and Practice.* Academic Press, 1980.

9. Dubois, D., and , R. Martin-Clouaire, and H. Prade. *Practical Computing in Fuzzy Logic.* Fuzzy Computing Theory, Hardware, and Applications, Edited by M.M.Gupta and T.Yamakawa, Elsevier Science Publishers B.V.(North~Holland), 1988, pp. 11-34.

10. Earhart, S.V. *UNIX Programmer's Manual.* CBS College Publishing, 1986.

11. Evans, G. W., W. Karwowski, and M. R. Wilhelm. *An Introduction to Fuzzy Set Methodologies for Industrial And Systems Engineering.* Applications of Fuzzy Set Methodologies in Industrial Engineering, Edited by Evans, G. W., Karwowski, W., Wilhelm, M. R., Elsevier Science Publishers B.V.(North~Holland), 1989, pp. 3-11.

12. Goldsack, S.J. *ADA for Specification : Possibilities and Limitations*, Cambridge University Press, 1985

13. Hoare, C.A.R. *An Overview of Some Formal Methods for Program Design*, IEEE COMPUTER, September 1987

14. Isik, C. *Inference Hardware for Fuzzy Rule-Based Systems.* Fuzzy Computing Theory, Hardware, and Applications, Edited by M. M.Gupta and T.Yamakawa, Elsevier Science Publishers B.V.(North~Holland), 1988, pp. 185-199.

15. Johnson, D., T. Nishizeki, A. Nozaki, and H. S. Wilf. *Discrete Algorithms and Complexity.* Academic Press, 1987.

16. Kaufmann, A. *Logics for Expert-Systems: With Fuzzy Sets, Interval of Confidence, Probabilistics Sets and Expertons.* Fuzzy Computing Theory, Hardware, and Applications, Edited by M.M.Gupta and T.Yamakawa, Elsevier Science Publishers B.V.(North~Holland), 1988, pp. 225-241.

17. Kaufmann, A., and M. M. Gupta. *Introduction to Fuzzy Arithmetic.* Van Nostrand Reinhold, 1985.

18. King, R. E., and F. C. Karonis. *Multi-Level Expert Control of A Large-Scale Industrial Process.* Fuzzy Computing Theory, Hardware, and Applications, Edited by M.M.Gupta and T.Yamakawa, Elsevier Science Publishers B.V.(North~Holland), 1988, pp. 323-340.

19. Larsen, P. M. *Industrial Applications of Fuzzy Logic Control.* International Journal of Man-Machine Studies, 12, 1980, pp. 3-10.

20. Lembessis, E. *Dynamic Learning Behaviour of a Rule-Based Self-Organising Controller*, Ph.D. Thesis, Queen Mary College, University of London, 1984.

21. Mamdani, E. H. *Applications of Fuzzy Algorithms for Control of a Single Dynamic Plant.* Proceedings of IEEE, 121-12, 1974, pp. 1585-1588.

22. Mange, D. A. *A High-Level-Language Programmable Controller Part II-Microcompilation of the High-Level Language Micropascal*, IEEE MICRO, April 1986

23. Mange, D. A. *A High-Level-Language Programmable Controller Part I-A Controller for Structures Microprogramming*, IEEE MICRO, February 1986

24. Mason, T., and D. Brown, *Lex & Yacc*, O'Reilly & Associates, Inc., 1990.

25. Novák V. *Fuzzy Sets And Their Applications.* Adam Hilger 1989.

26. Pedrycz, W. *Methodological and Applicational Aspects of Fuzzy Models for System Engineering.* Applications of Fuzzy Set Methodologies in Industrial Engineerind, Edited by Evans, G. W., Karwowski, W., Wilhelm, M. R., Elsevier Science Publishers B.V.(North~Holland), 1989, pp. 13-26.

27. Pennington, N. *Stimulus Structures and Mental Representations in Expert Comprehension of Computer Programs*, Cognitive Psychology, 19, pp. 295-341, 1987

28. Scharf, E. M., and N. J. Mandic. *The Application of a Fuzzy Controller to the Control of a Multi-Degree-of-Freedom-Robot Arm.* Industrial Applications of Fuzzy Control, Edited by M. Sugeno, Elsevier Science Publishers B.V.(North~Holland), 1985, pp. 41-61.

29. Sibigtroth, J. M., *Creating Fuzzy Micros,* Embedded System Programming, December 1991.

30. Sugeno, M. (ed), *Industrial Applications of Fuzzy Control.* North-Holland, Armsterdam, 1985.

31. Sugeno, M., and K. Murakami. *An Experimental Study on Fuzzy Parking Control Using a Model Car.* Industrial Applications of Fuzzy Control, Edited by M. Sugeno, Elsevier Science Publishers B.V.(North~Holland), 1985, pp. 125-138.

32. Sugiyama, K. *Analysis and Synthesis of the Rule-Based Self-Organising Controller*", Ph.D. Thesis, Queen Mary College, University of London, 1986.

33. Takaki, T., M. and Sugeno. *Fuzzy Identification of Systems and its Applications to Modeling and Control.* IEEE Transactions on Systems, Man, and Cybernetics, Vol. SMC-15, pp 116-132, 1985.

34. Tong, H. R. *An Annotated Bibliography of Fuzzy Control.* Industrial Applications of Fuzzy Control, Edited by M. Sugeno, Elsevier Science Publishers B.V.(North~Holland), 1985, pp. 249-269.

35. Tong, R. M. *A Control Engineering Review of Fuzzy Systems.* Automatica, 13, 1977, pp. 559-569.

36. Willaeys, D., N. Malvache, and P. Hammad. *Utilization of Fuzzy Sets for System Modeling and Control.* Proceedings of 16th IEEE Conference on Decision and Control, New Orleans, December, 1977.

37. Wulf, W.A., M. Shaw, P. N. Hilfinger, and L. Flon. *The Concept of a Linguistic Variable and Its Application to Approximate Reasoning.* 1973

38. Yamakazi, T. *An Improved Algorithm for a Self-Organizing Controller and its Experimental Analysis.* Ph.D. Thesis, Queen Mary College, University of London, 1982.

39. Yasunobi, S., and S. Miyamoto. *Automatic Train Operation System by Predictive Fuzzy Control.* Industrial Applications of Fuzzy Control, Edited by M. Sugeno, Elsevier Science Publishers B.V.(North~Holland), 1985, pp. 1-19.

40. Yeralan, S., and D. J. Ramcharan. *A New Standard for Industrial Control Languages.* Proceedings of the 12th Computers and Industrial Engineering Conference, Orlando, Florida, March, 1990.

41. Zadeh, L. A. *Outline of a New Approach to the Analysis of Complex Systems and Decision Processes,*" IEEE Transactions on Systems, Man, and Cybernetics, Vol. 3, pp 28-44, 1973.

42. Zadeh L.A. *The concept of a Linguistic Variable and its applications to Approximate Reasoning*, 1973.

43. Zimmermann, O.J. *Fuzzy Set Theory and its Applications.* Kluwer-Nijhoff Publishing, 1985.

7

USING NEURAL NETWORKS FOR THE AUTOMATIC MONITORING AND RECOGNITION OF SIGNALS IN MANUFACTURING PROCESSES

T.-H. Hou, National Yunlin Institute of Technology, Taiwan

L. Lin, State University of New York at Buffalo

The design and implementation of effective control of manufacturing processes depends on the successful monitoring and recognition of process signals. In this chapter, we describe the application of neural network to the monitoring and recognition of both periodic and aperiodic process signals that may be encountered in manufacturing processes. In a developed prototype system a preprocessor first converts collected signals from time domain to the frequency domain using digital signal processing techniques. Then a signal recognition system with a back propagation neural network implemented in programming logic identifies the signals by examining their characteristic frequencies. The system's performance in identifying several types of signals is demonstrated through running examples. The promising results indicate the potential application of neural networks in automatic manufacturing process control.

7.1 INTRODUCTION

As the design of a manufacturing process advances, the benefits of developing and implementing an effective monitoring system become more significant. The direct benefits of having a properly designed monitoring system include improved product quality, better plans for maintenance, more effective control of operating machines, and better manufacturing decisions such as process plans and workload plans, etc. Therefore, process monitoring plays a vital role in an integrated manufacturing system rather than just as a maintenance support [1].

Process monitoring is to evaluate and detect changes or differences in a system during operation from sensed signals. Monitoring procedures usually involve signal collection, signal analysis, and fault identification. Signals in forms of vibration, sound, light, current, pressure, flow rate, etc. are collected by various types of sensors. In general, these signals can be categorized as periodic and aperiodic.

The raw sensory data acquired in a manufacturing process is usually random in nature; therefore, signal analysis is to simplify the signal representation and to extract meaningful characteristic components from the signal. Preprocessing techniques, such as filtering, smoothing, and windowing, are sometimes used to reduce noise and improve the signal-to-noise ratio. Commonly used feature extraction methods include statistical analysis, spectral analysis, prediction error analysis, autocorrelation analysis, short time signal analysis, and damping factors [2].

Prediction error analysis has been successfully implemeted for spindle monitoring [3]. Applications of autoregressive and moving average (ARMA) models with damping factor analysis have been used for tool wear [2]. Short time data processing, which processes a windowed waveform of sensed signals, has been used for monitoring bearing conditions [2]. Syntactic analysis has been claimed to be able to characterize a particular primitive vibration in a filtered signal [4].

In general, signal processing techniques are based on time and/or frequency domains. Frequency based analysis techniques include spectrum analysis, waterfall plot, cepstrum analysis, and matched filtering [5]. One of the most effective techniques in process monitoring is spectrum analysis due to its capability of recognizing each rotating element in a manufacturing process from their individually identifiable frequencies [6]. Changes at a given frequency can be related to damage of the corresponding component.

In the fault detection stage, heuristic rule based routines, statistical pattern recognition algorithms, Fuzzy set theory, and neural networks are usually used. Using rule based routines, attribute value of a signal is compared with a threshold and an IF-THEN production rule is fired if the specified condition is met. This method has been used in some preventive monitoring expert systems [7,8] and other manufacturing process applications such as machine faults diagnosis [2]. However, the detection accuracy depends on the threshold value, which is normally determined empirically and heuristically.

Statistical pattern recognition routines identify a current process status based on the nearest neighbor algorithm. The method assumes a Gaussian data distribution in a pattern set and is therefore only applicable to those patterns with a linearly separable decision space.

Fuzzy set based pattern recognition has been applied in monitoring wear of drilling bits [9]. Detection accuracy of this method depends on a membership function. However, it is difficult to determine a membership function.

Artificial neural networks, which simulate human information processing, are inherently a parallel processing technique and have a desirable learning ability. When a neural net is presented with a series of training data that represent a known causal relationship, it is able to learn the governing relationship and use it in subsequent pattern recognition tasks. They have been widely applied in visual and signal recognition and in control, such as robot control systems [10,11].

Back propagation neural net is one of the most useful nets. It has been implemented to detect tool wear states [12]. The results showed that the net could learn rapidly and was able to integrate various sources of information. Reported identification accuracie were over $90%. Another study used the net to detect sensor failure in a control system and compared the results to those of statistical nearest neighbor techniques [13]. The net has been found with significantly improved accuracy the statistical nearest neighbor algorithm. The main advantage of back propagation net over the nearest neighbor clssifier lies in its ability to capture nonlinear characteristics. This leads to a possibility of on-line training and implementation in manufacturing. An application of the net to sonar signal classifications also showed over 90% identification accuracy [14]. From the above examples, neural network appears to be a promising and viable technique in signal classification.

Next we wlll introduce the fundamental concept of back propagation neural network and discusses a prototype neural network based system for signal recognition. The objectives are to develop a general framework for both periodic and aperiodic signal monitoring for manufacturing processes and to implement the back propagation neural network in the monitoring system. Programming logic and computation results of the system will also be discussed.

7.2 THE BACK PROPAGATION NEURAL NETWORK

The following is a brief review of the neural network approach we used in this development. The back propagation neural network contains an input layer, a hidden layer, and an output layer [15,16]. The system structure is shown in Figure 7-1.

The system uses a set of examples in a training database as input, the generalized delta rule learning algorithm to adjust weights (w) and biases (θ), and a sigmoidal activation function to derive an output. After a few iterations, the system will learn the relationships between input and output and converge to a set of weights that represent the relationships and are then used to recognize a new pattern presented to the system. The computation logic is shown in Figure 7-2.

The computation procedure starts from initializing a set of system weights by randomly setting its value between 0 and 1. Then the attribute values and the corresponding known category of a training example are presented and fed into nodes in the input layer (o_i), and then transmitted to nodes (j) in the hidden layer through weighting factors (w_{ji}). The net input to node j is

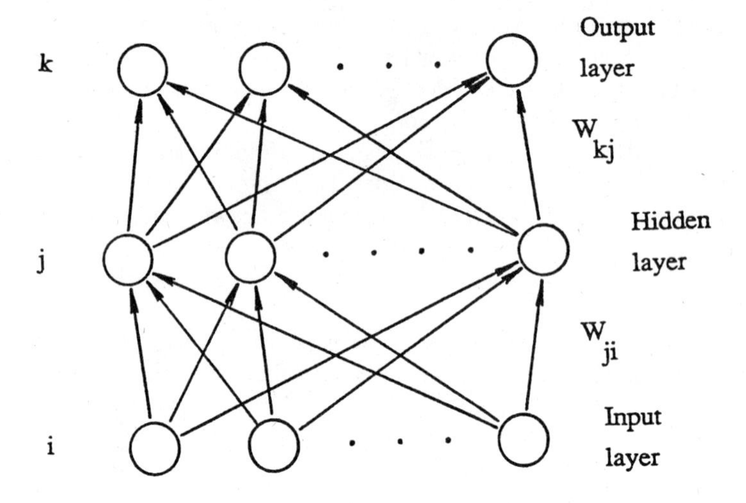

Figure 7-1: System diagram of the multi-layer neural network.

$$net_j = \sum_i w_{ji} o_i \tag{1}$$

Through the activation function f, the output is

$$o_j = f\,(net_j) \tag{2}$$

A sigmoidal activation function is defined as

$$o_j = \frac{1}{1 + \exp\left(\dfrac{net_j + \theta_j}{\theta_o}\right)} \tag{3}$$

where θ_j is the threshold or bias and θ_o is a scaling factor. Similarly, output from nodes in the hidden layer are fed into nodes in the output layer. This process is called the feed forward stage.

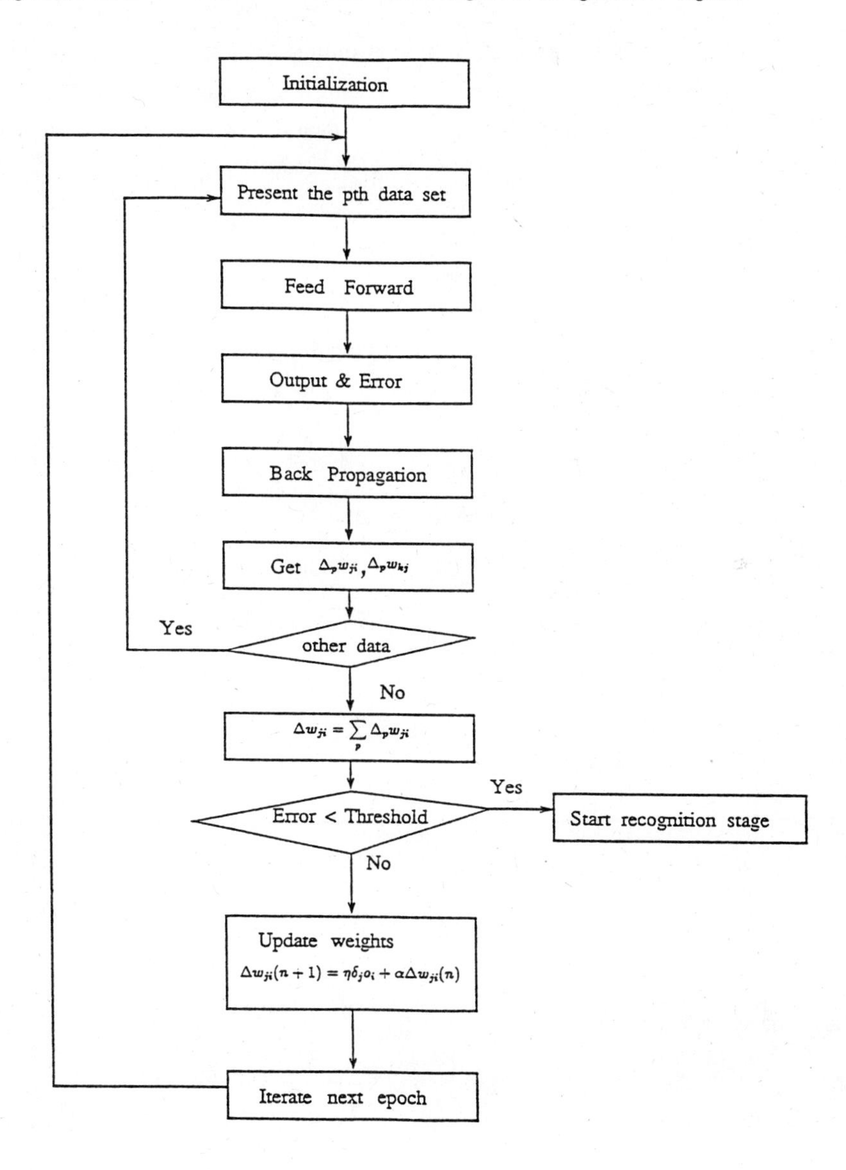

Figure 7-2: Computation logic of the generalized delta rule neural network.

After the feed forward process, values of output (o_{pk}) are calculated from nodes in the output layer. Since in general, the output o_{pk} may not match the desired known category t_{pk}, to determine whether further process is necessary the average system error is calculated as

$$E = \frac{1}{2P} \sum_p \sum_k (t_{pk} - o_{pk})^2 \tag{4}$$

where k is the output node number and p is the example number.

The error is then back propagated from nodes in the output layer to nodes in the hidden layer using a gradient search method [16]:

$$\Delta_p w_{kj} = -\eta \frac{\partial E}{\partial w_{kj}} \tag{5}$$

After a few chaining calculations, we obtain

$$\Delta_p w_{kj} = \eta \delta_k o_j \tag{6}$$

where η is the learning rate given by the user and

$$\delta_k = (t_k - o_k)\, f_k'(net_k) \tag{7}$$

Similarly the delta value in the hidden layer is

$$\delta_j = f_j'(net_j) \sum_k \delta_k w_{kj} \tag{8}$$

and

$$\Delta_p w_{ji} = \eta \delta_j o_i \tag{9}$$

This process is called the back propagation stage. If there are other examples in the training data set, the system will present the next example and continue the feed forward and back propagation stages. After all the examples are trained, the system will collect adjusted weights according to $\Delta w_{ji} = \sum_p \Delta_p w_{ji}$. If system errors in nodes of the output layer are less than a specified threshold value, the current weights are deemed desirable. The system will stop training and use the weights in the subsequent recognition stage. Otherwise, the system will update the weights according to $\Delta w_{ji}(n+1) = \eta \delta_j o_i + \alpha \Delta w_{ji}(n)$, where α is the momentum coefficient. Then the system will continue the next epoch (or iteration).

Different learning rate, momentum coefficient, and number of nodes in a hidden layer affect computation efficiency. Some general rules for their selection have been documented in [15,16,17].

7.3 A GENERALIZED MANUFACTURING PROCESS MONITORING SYSTEM

Since many machine faults generate vibratory signals with identifiable frequencies [6], spectrum analysis can be used to detect these characteristics. In a monitoring system, continuous signals are sampled and converted into discrete signals that can be periodic and aperiodic. A discrete periodic signal can be represented as a discrete Fourier series as follows [18]:

$$x[n] = \sum_{k=\langle N \rangle} a_k e^{jk(2\pi/N)n}, \quad k = 0,1,2,\ldots,N-1. \tag{10}$$

where a_k is the Fourier series coefficient and can be determined from the following equation:

$$a_k = \frac{1}{N} \sum_n x[n] e^{-jk(2\pi/N)n} \tag{11}$$

Equations 10 and 11 are called discrete-time Fourier series pairs.

For a general a periodic sequence $x[n]$ with finite duration N, the Fourier transform of $x[n]$ is defined as follows [17]:

$$X(k) = \frac{1}{N} \sum_{n=0}^{N-1} x[n] e^{-jk(2\pi/N)n}, \quad k = 0,1,2,\ldots,N-1 \tag{12}$$

It has been shown that $X(k) = a_k$ [18]. In other words, Fourier series coefficients of a periodic signal can be obtained from Fourier transform of one period of that sequence. Therefore, in order to express signals in the frequency domain it is only necessary to sample and collect one period of data for periodic signals and some finite sequence of data for aperiodic signals.

Based on the above techniques, a framework of the prototype system is shown in Figure 7-3. Sensors are first used to collect signals that are further sampled and converted into discrete signals by a sampling and A/D conversion circuit. These discrete signals are then transformed into a frequency representation by a Fast Fourier Transform (FFT) routine.

Frequency domain analysis is used because: (1) in a frequency analysis, there are only a limited number of specific characteristic elements that will facilitate the simplification of a signal representation, (2) changes in an elememt of a manufacturing process will be reflected in its frequency response, and (3) FFT is able to improve Fourier transform computation efficiency.

The next step after the FFT computation is to extract key features from the frequency response. Typical features include the average value, Root-Mean-Square, peak-to-peak value, peak location, peak magnitude, crest factor, Kurtosis, and spectral profile [2]. Features are used to simplify signal representation and should be able to represent the original signal without losing its primary characteristics. Note that the usefulness of these features are mainly for the recognition of complex aperiodic signals. For periodic signals, frequency components are normally adequate for a successful recognition of the signals. Finally, the back propagation neural net uses these feature values to learn the relationship between examples and then to identify an incoming signal. Before the learning stage, the user needs to prepare a training data set, which includes feature values for each example and its known category. The neural net then starts learning and converges to a set of system weights that is then used to classify an incoming sensed signal in the recognition stage. The recognition process is able to work in a real-time process because the process only needs a single step feed forward computation, resulting a rapid recognition.

The proposed system can be used for manufacturing processes that generate either periodic or aperiodic signals. With a frequency domain analysis, the system is especially useful for rotating machinery processes. Potential applications may include the detection of imbalance, resonance, looseness of mechanical components, misalignment, oil whirl, seal rub, bearing failure, and component wear.

7.4 APPLICATIONS

Applications in Periodic Signals

In order to demonstrate the feasibility of the system to periodic signals, a DYNASCAN 3010 function generator is used to generate sinusoidal, saw tooth, and square signals as test patterns. The incoming signals all have a known frequency 2.5 Hz. PCLAB, a computer I/O interface system [19], is used for sampling and A/D conversion. Sampling frequency is set at 25 Hz. This means that 10 data points are acquired in one sampling period.

FFT is then applied to the 10 data samples to transform time domain data into frequency domain and then obtain the power spectrum. Time domain data and power spectrums of one period data for a sinusoidal signal, a saw tooth signal, and a square signal are shown in Figure 7-4.

After the FFT operation there are only five values in the power spectrum that represents a simple frequency response. Therefore, the whole spectrum data is used in the pattern recognition routine without using a feature extraction process.

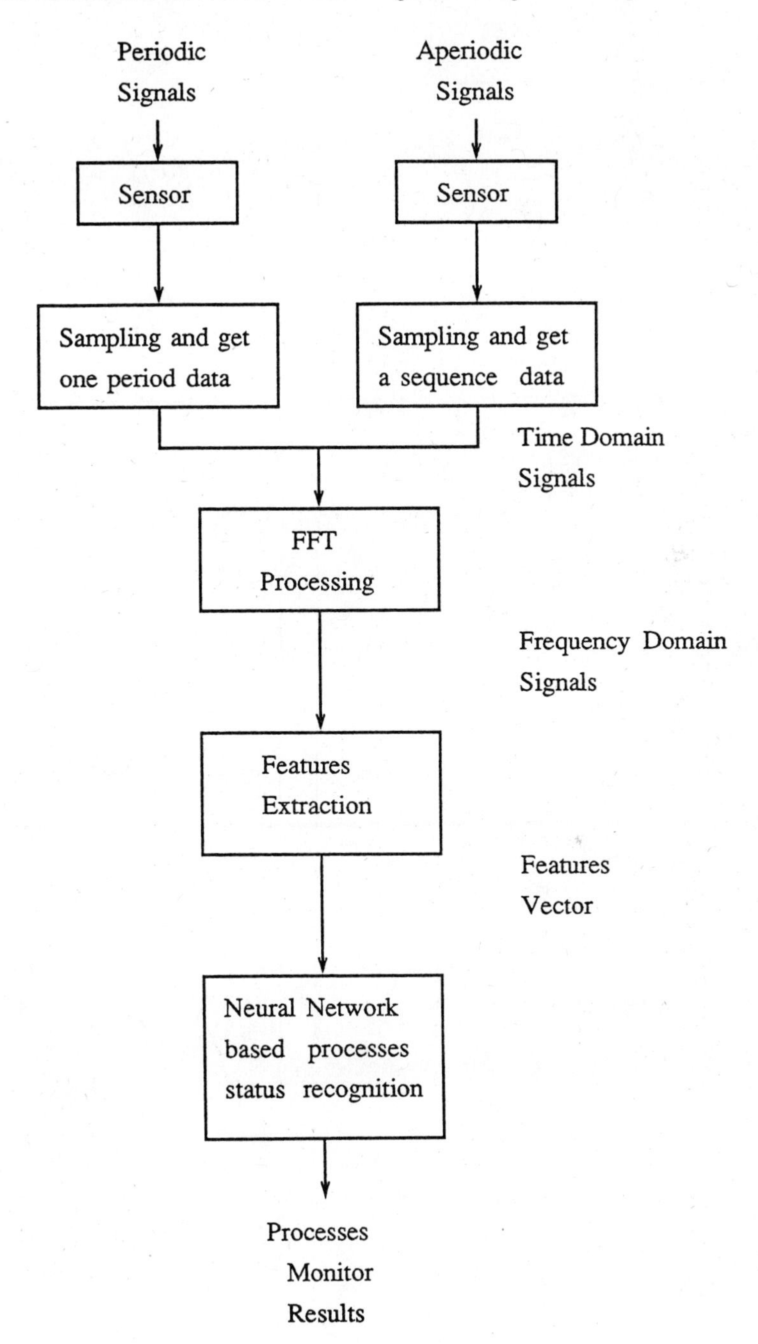

Figure 7-3: Framework of the monitoring system.

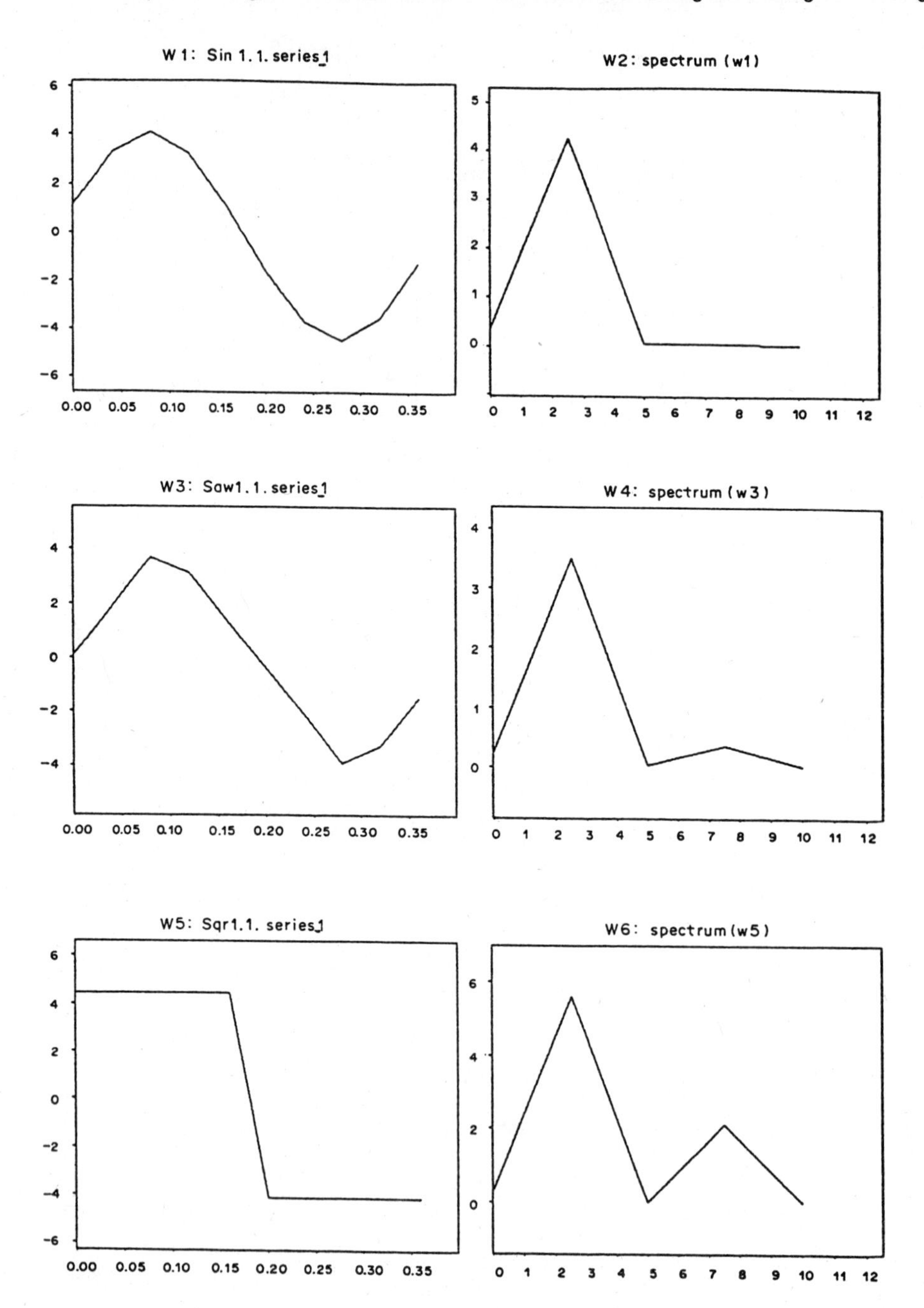

Figure 7-4: Time domain and power spectrum of periodic signals.

Figure 7-5 shows some examples of training data set (a); an input frequency values to the input nodes of the neural net (b); and calculated output values at the output nodes (c), which show that the output value at node 1 exceeds 0.9 (0.9542) therefore the pattern is classified as type 1 (sinusoidal). An actual computer run of the system is shown in Figure 7-6. In the net, there are five nodes in the input layer, eight nodes in the hidden layer, and three nodes in the output layer. The training data set has 10 examples and learning rate (η) 0.4, momentum (α) 0.7, scaling factor (θ_o) 1.0, and threshold (θ_o) 0.1 are used. After the user keys in the training data file, the neural net begins to learn the pattern relationships. The average system error calculated according to equation (4) is decreasing rapidly and becomes very small after 131 iterations. The system then converges to a set of weights and the learning stage completes. In the recognition stage, attribute values of an incoming signal are fed to the input layer. Through the feed forward process, calculated outputs are obtained. In this running example, an incoming signal is classified as category i if the calculated output in output node i is greater than 0.9. On the other hand, if the calculated output in all nodes are below 0.9, the signal is regarded as an unclassified pattern. The use of a tight detection criteria instead of the conventional 0.5 results in a more discriminating performance.

Input Value					**Output Category**		
0.3320	4.2514	0.0496	0.0487	0.0303	1.0	0.0	0.0
0.3545	4.2348	0.0388	0.0299	0.0306	1.0	0.0	0.0
0.3184	4.2321	0.0573	0.0313	0.0284	1.0	0.0	0.0
0.2275	3.5222	0.0367	0.4048	0.0149	0.0	1.0	0.0
0.2070	3.6472	0.0366	0.5517	0.0169	0.0	1.0	0.0
0.2051	3.5689	0.0349	0.4656	0.0171	0.0	1.0	0.0
0.2002	3.4776	0.0356	0.3353	0.0158	0.0	1.0	0.0
0.2891	5.5712	0.0019	2.1273	0.0028	0.0	0.0	1.0
0.2871	5.5707	0.0021	2.1278	0.0015	0.0	0.0	1.0
0.2900	5.5701	0.0017	2.1283	0.0031	0.0	0.0	1.0

(a) Training data set

0.332031
4.251389
0.049598
0.048737
0.030266

(b) Input values at the input node

0.9542 0.0892 0.0045

(c) Calculated output values at the output nodes

Figure 7-5: Data used in the neural net.

```
*****************************************
*                                       *
*               WELCOME                 *
*                 TO                    *
*             MULTI-LAYER               *
*        NEURAL NETWORK SYSTEM          *
*                                       *
*****************************************
Enter training filename  --> sig.dat

Enter the number of iteration per screen printout:
10
THE NUMBER OF TRAINING DATA SET = 10
NUMBER OF NODES IN THE INPUT LAYER =  5
NUMBER OF NODES IN THE HIDDEN LAYER =  8
NUMBER OF NODES IN THE OUTPUT LAYER =  3
LEARNING RATE =    .4  MOMENTUM =    .7  SCALING FACTOR =  1.0  THRESHOLD =    .1
START THE LEARNING STAGE

   EPOCH NO      RMS ERROR        ABSOLUTE ERROR
    11           .37956           12.44702
    21           .29874           12.52670
    31           .18796            9.46244
    41           .15118            7.80424
    51           .11305            6.64178
    61           .06633            5.10753
    71           .08066            5.12480
    81           .02091            2.87227
    91           .01287            2.31154
   101           .00938            1.99134
   111           .00731            1.76829
   121           .00595            1.60178
   131           .00499            1.47213
 LEARNING STAGE HAS FINISHED

START THE PATTERN RECOGNITION STAGE
Enter NEW pattern filename   --> sinp1
  9.253902E-001  8.644641E-002  1.559505E-002
the pattern belongs to category  1
Do another pattern(Y/N)? y
Enter NEW pattern filename   --> sawp1
  7.123075E-002  9.093804E-001  2.109412E-002
the pattern belongs to category  2
Do another pattern(Y/N)? y
Enter NEW pattern filename   --> sqrp1
  6.066082E-004  5.340857E-002  9.677480E-001
the pattern belongs to category  3
Do another pattern(Y/N)? n
Stop - Program terminated.
```

Figure 7-6: Running results of back propagation net.

Effects of varying number of nodes in the hidden layer on learning and detection accuracy have been tested. Nine examples in the training set and 40 test patterns are used. In the neural net, five nodes in the input layer, three nodes in the output layer, learning rate 0.4, and momentum 0.7 are used. Table 7-1 shows a summary of the results.

The first column represents number of nodes in the hidden layer, the second column number of iterations to converge, and the third column the number of correct detection over 40 test patterns. The result shows that the system with eight nodes in the hidden layer has the minimum number of iterations among the four configurations. More nodes in the hidden layer may slow down the learning process due to increased complexity of network structure. On the other hand, less nodes may also increase learning iterations because of decreased extent of nonlinearity.

Table 7-1: Performance of various networks.

No. of nodes	No. of iterations	No. of correct detection
12	184	35
8	159	35
6	201	35
3	208	35

The third column shows the same detection number for various numbers of nodes. It seems from this example that number of nodes in the hidden layer affects learning speed, but it does not necessarily impact detection accuracy.

Figure 7-7 shows that the 12-node and 8-node systems have peaks during learning, which indicate local optimum in the decision space and the system is nonlinear. After these peaks, the average system error continues its declining trend, which means that the learning does not get trapped at nonoptimal values. This figure also shows that the 3-node system has the fastest decreasing rate inaverage system error, but the 8-node system converges the fastest. Therefore, the fastest decreasing rate in average system error is not necessary the one that converges the fastest. However, readers should note that the learning performance reported here is highly problem-dependent. Different network configurations, input data, and training parameters may impact the performance.

Effects of example values in the training data base on learning and detection based on the 8-node net in the hidden layer system are shown in Table 7-2. The table shows that increasing examples may decrease the learning rate. In addition, more examples provide better detection accuracy because more data are available to characterize the possible decision space. It seems that detection accuracy depends on examples in the training data base rather than the number of nodes in the hidden layer.

It should be pointed out that these observations only apply to this particular problem. Caution should be exercised in attempting to generalize it to other applications and input data. Although an optimal number of nodes seems to exist for the hidden layer, for a particular problem domain and data set, a thorough analysis must be performed before reaching any conclusion.

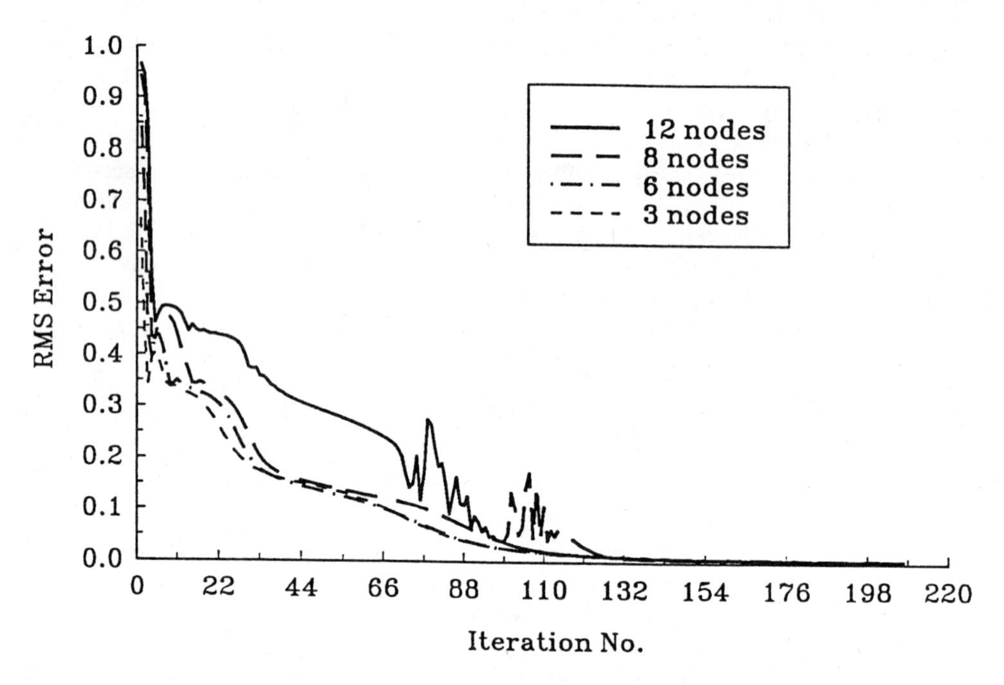

Figure 7-7: Network learning behavior with various numbers of nodes.

Table 7-2: Recognition results of various networks.

No. of examples	No. of iterations	No. of correct detection
10	131	40
9	159	35
6	211	35
3	341	26

Applications to Aperiodic Signals

In a second example, a machine that has an AC motor with a gear box is used to verify the feasibility of the proposed system on aperiodic process signals. Possible faults include motor spindle misalignment, broken or worn gear teeth, broken gear box, gear misalignment, and mechanical looseness. Accelerometers are mounted at the shaft and the gear box to collect vibration signals while the machine is operating. PCLAB [19] and a sampling circuit with an A/D converter are used. The sampling frequency is set at 400 Hz.

A vibration signal and its power spectrum for the AC motor in normal condition are shown in Figure 7-8. For comparison, the power spectrum of the motor status with a loose screw on the motor seal is shown in Figure 7-9. The difference between the the normal and abnormal conditions is clearly visible (while the significance is to be determined on further analysis). Since vibration signals of various machine faults have identifiable frequency response, machine status changes will be reflected in the power spectrum. Therefore, power spectrum is very informative for machine monitoring and diagnosis.

Instead of using the whole power spectrum as input to the neural net, some critical features must be extracted. In this case, the power spectrum was segmented into 15 equal intervals and the average value of each of the interval data points was used as an attribute. Assuming a power spectrum contains N data points, there are therefore n data ponits in each interval, where $n = INT(N/15)$. The analyzed power spectrum data stream is shown as follows:

$$(f_{1-1}, f_{1-2}, \ldots, f_{1-n}, \ldots, f_{i-1}, \ldots, f_{i-n}, \ldots, f_{15-1}, \ldots, f_{15-n})$$

Therefore, 15 attribute values can be obtained from the following equation:

$$t_i = \sum_{j=1}^{n} \frac{f_{i,j}}{n} \quad (13)$$

These 15 values, which are sufficient to represent the original power spectrum, are input to the back propagation net.

A test run of this system was then conducted with the following network parameters chosen based on our earlier discussion. In this network, there are fifteen nodes in the input layer and eight nodes in the hidden layer. Three nodes in the output layer, representing three machine conditions to be monitored, are used. There are eight examples in the training data set. The learning rate is 0.2, momentum 0.7, sacling factor 1.0, and threshold 0.1. Although the problem is much more complex than that of the periodic case, the computation based on the proposed monitoring system framework is still able to learn at a reasonably fast speed.

Learning behavior under various numbers of nodes in the hidden layer for this application is shown in Figure 7-10. Numbers of interations to convergence for the 5-node, 8-node, and 12-node systems are 1961, 321, and 328, respectively. This figure also shows that using more nodes in the hidden layer may increase learning iterations due to a more complex network structure and less nodes may also slow down learning because of reduced nonlinearity in the network. This observation is consistent with that in the periodic signal application.

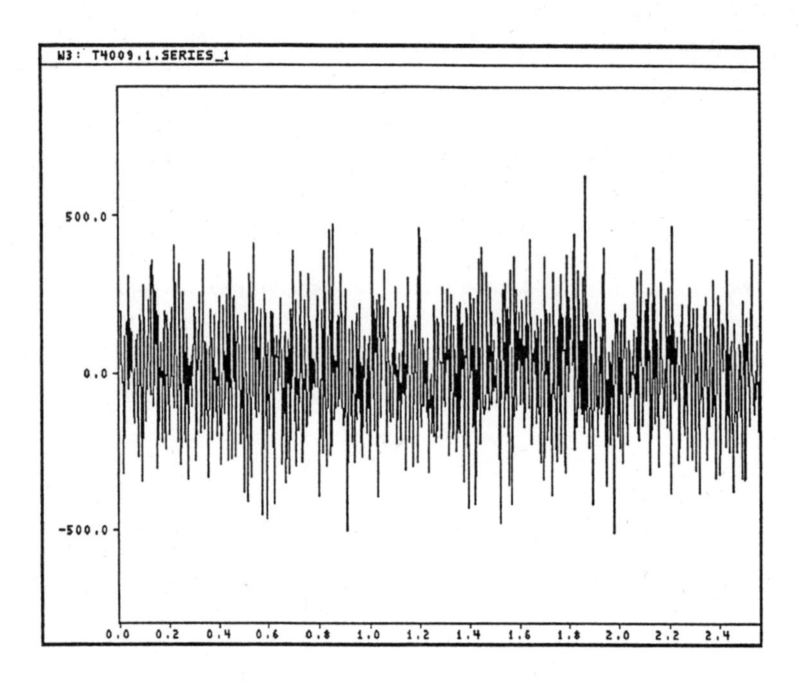

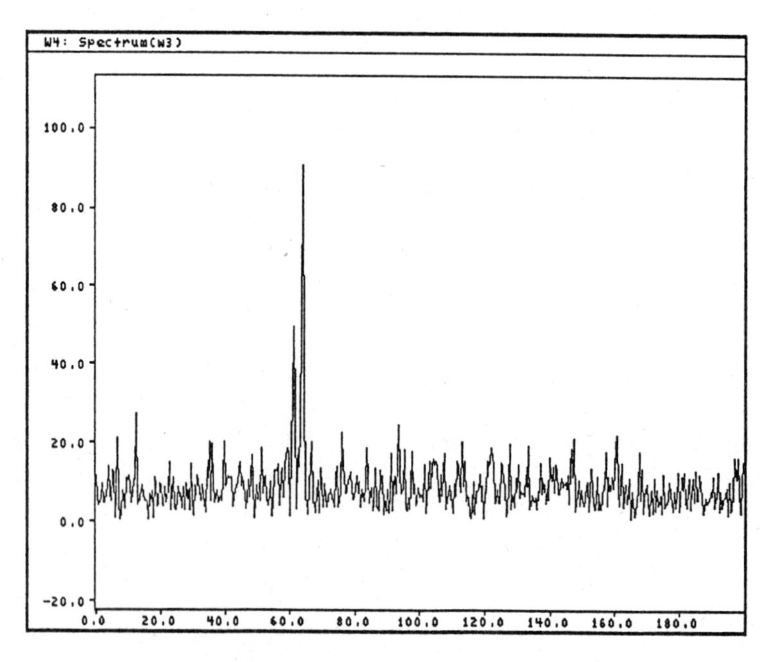

Figure 7-8: A vibration signal and its power spectrum.

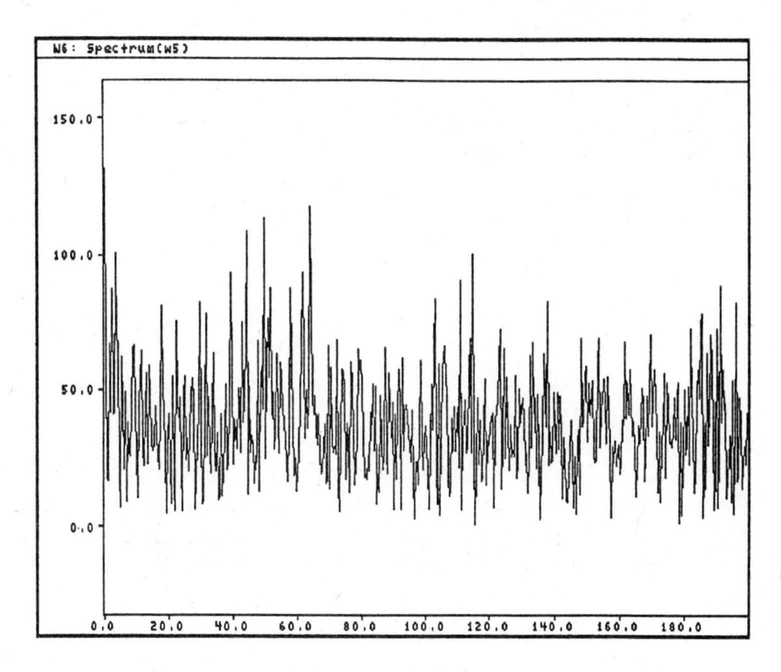

Figure 7-9: Power spectrum of a motor with a loose screw in the seal.

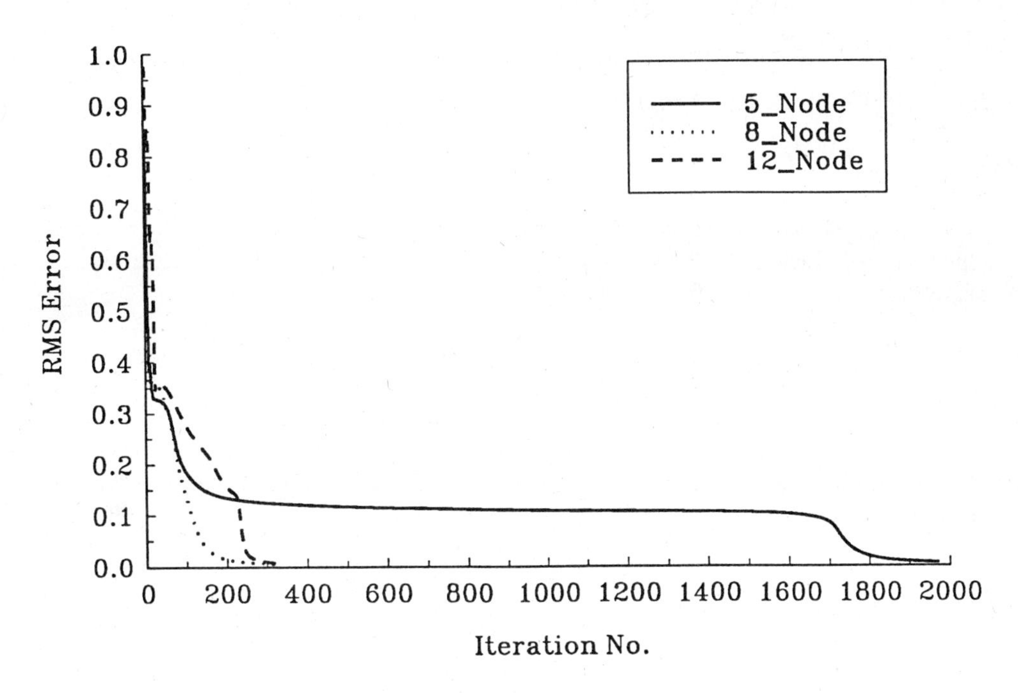

Figure 7-10: Learning behavior under various nodes.

7.5 CONCLUDING REMARKS

In this chapter, manufacturing process monitoring techniques and applications of neural network in this field were briefly reviewed, and the back propagation neural network algorithm and its computation logic were described in detail. By combining digital signal processing and neural networks, a generalized neural network based monitoring system for both periodic and aperiodic process signals was then proposed. At the end, two examples were presented to demonstrate the feasibility of the monitoring system and its recognition ability. The results show that the system is able to learn rapidly for both periodic and aperiodic signals. Number of nodes in the hidden layer may affect learning speed, but not necessarily detection accuracy. Instead, detection accuracy depends on characteristics of the training data set.

This work can be improved in several ways. First, the feature extraction process can be improved by using certain filters to extract specific frequency responses if the overall frequency response is very complex. The selection of the target frequencies may come from the domain knowledge about the process. For example, revolution speed, gear transmission ratio, and number of gear teeth should be used for monitoring a gear box. Second, expert systems can be used to interpret the monitoring results, diagnose faults, and suggest remedies [20]. Combining an expert system with neural network based pattern recognition [11] may lead to potential applications in manufacturing process monitoring and faults diagnosis. Finally, since it is normally difficult for the user to determine the appropriate number of nodes in the hidden layer, some specific rules or expert systems may be helpful in providing a general guideline.

ACKNOWLEDGEMENTS: This chapter is based on a paper by T.-H. Hou and L. Lin: "Manufacturing Process Monitoring Using Neural Networks," *Computers and Electrical Engineering*, Vol. 19, No. 2, pp. 129-141, copyright (1993), with kind permission from Elsevier Science Ltd., The Boulevard, Langford Lane, Kidlington 0X5 1GB, UK.

REFERENCES

[1] R.H. Lyon, The monitoring and diagnostics of machines and processes. Proceeding of the first International Machinery Monitoring and Diagnostics Conference and Exhibit, September, Las Vegas, Nevada, pp. xi-xiii, (1989).

[2] Y.B. Chen and S.M. Wu, A real-time automated diagnostic monitoring system for manufacturing processes. Proceedings of Manufacturing International, pp. 51-56, March (1990).

[3] Y.B. Chen and S.M. Wu, Machinery condition monitoring by prediction error analysis. Intelligent and Integrated Manufacturing Analysis and Synthesis, ed. C.R. Liu, A. Requicha and S. Chandrasekar. ASME, PED-Vol. 25, pp.129-139, (1987).

[4] P.L. Love and M. Simaan, Automatic recognition of primitive changes in manufacturing process signals. Pattern Recognition, Vol. 21, No. 4, pp. 333-342, (1988).

[5] J. Tranter, The fundamentals of, and the application of computer to, condition monitoring and predictive maintenance. Proceeding of the first International Machinery Monitoring and Diagnostics Conference and Exhibit, September, Las Vegas, Nevada, pp. 394-401, (1989).

[6] M. Serridge, Ten crucial concepts behind trustworthy fault detection in machine condition monitoring. Proceeding of the first International Machinery Monitoring and Diagnostics Conference and Exhibit, September, Las Vegas, Nevada, pp. 722-727, (1989).

[7] A. Bajpai, An expert system model for general-purpose diagnostics of manufacturing equipment. Manufacturing Review, Vol 1, No 3, pp. 180-187, (1988).

[8] G.S. Kumar and G.W. Ernest, An expert system architecture for predictive monitoring. Intelligent and Integrated Manufacturing Analysis and Synthesis, ed. C.R. Liu, A. Requicha and S. Chandrasekar. ASME, PED-Vol. 25, pp.121-128, (1987).

[9] P.G. Li and S.M. Wu, Monitoring drilling wear states by a fuzzy pattern recognition techniques. ASME Transactions, Journal of Engineering for Industry, Vol.110, PP. 297-300, (1988).

[10] W.T. Miller, III, Sensor-based control of robotic manipulators using a general learning algorithm. IEEE Journal of Robotics and Automation, Vol. RA-3, No. 2, pp. 157-165, (1987).

[11] D.A. Handelman, S.H. Lane, and J.J. Gelfand, Integrating neural networks and knowledge-based systems for intelligent robotic control. IEEE Control Systems Magazine, pp. 77-87, April (1990).

[12] S. Rangwala and D. Dornfeld, Integration of sensors via neural networks for detection of tool wear states. Intelligent and Integrated Manufacturing Analysis and Synthesis, ed. C.R. Liu, A. Requicha and S. Chandrasekar. ASME, PED-Vol. 25, pp.109-117, (1987).

[13] S.R. Naidu, E. Zafiriou, and T.J. McAvoy, Use of neural networks for sensor failure detection in a control system, IEEE Control Systems Magazine, pp. 49-55, April (1990).

[14] R.P. Gorman and T.J. Sejnowski, Analysis of hidden units in a layered network trained to classify sonar targets, Neural Networks, Vol. 1, pp. 75-89, (1988).

[15] R.P. Lippmann, An introduction to computing with neural nets. IEEE ASAP Magazine, Vol. 4, pp. 4-22, April (1987).

[16] Y.H. Pao, Adaptive Pattern Recognition and Neural Networks. Addison-Wesley Publishing Company, Inc. (1989).

[17] D.E. Rumelhart and J.L. McClelland, Parallel Distributed Processing. The MIT Press (1986).

[18] A.V. Oppenheim, A.S. Willsky and I.T. Young, Signals and Systems. Prentice-Hall Inc., (1983).

[19] PCLAB User Manual, Data Translation, Inc., 100 Locke Drive, Marlboro, MA 01752-1192, (1985).

[20] S.S. Wang, AI and expert systems for diagnostics. Proceeding of the first International Machinery Monitoring and Diagnostics Conference and Exhibit, September, Las Vegas, Nevada, pp. 516-522, (1989).

8

ARTIFICIAL NEURAL NETWORK APPROACH IN MODELING OF EDM AND WIRE-EDM PROCESSES

Gopal Indurkhya,Indotronix Int'l Corporation
K. P. Rajurkar, University of Nebraska-Lincoln
Ravi Thouti Reddy, University of Nebraska-Lincoln

Electro-Discharge Machining (EDM) and Wire Electro-Discharge Machining (WEDM) are used to machine high strength, temperature resistant (HSTR) alloys components in aeronautical and automotive industries. The selection of process parameters to achieve optimal performance needs adequate modeling of EDM and WEDM. The complexity and stochastic nature of the EDM process has defied numerous attempts of modeling it accurately. Therefore, this chapter reports an attempt of modeling of EDM and WEDM processes through an artificial neural network. For die-sinking EDM with an orbital electrode, a 9-9-2 size back-propagation neural network has been developed. Machining depth, tool radius, orbital radius, radial step, vertical step, offset depth, pulse ontime, pulse offtime, and discharge current are selected as input parameters. The material removal rate (MRR) and surface roughness (Ra) are output parameters for the model. For Wire-EDM a back-propagation neural network of 6-6-2 size with one hidden layer has been developed. The input parameters for the network are current, pulse duration, frequency, wire tension, wire speed, and flow rate. The material removal rate (MRR) and surface roughness (Ra) are output parameters for the network. The results for both processes have been compared with estimates obtained by multiple regression analysis. Experiments have also been performed to check the validity of the neural network models. The artificial neural network models for EDM and WEDM have been found to provide faster and more accurate results.

8.1 INTRODUCTION

Electro-Discharge Machining (EDM) and Wire-Electro Discharge Machining (WEDM) are used in machining high strength, temperature resistant (HSTR) alloys specifically in aeronautical and automotive industries. Successful integration of optimization techniques and adaptive control of EDM depends on the development of proper quantitative relationships between output parameters and controllable input variables. Numerous modeling approaches including the classical method involving the Fourier heat conduction equation[1-3], an innovative method involving stochastic modeling analysis by Data Dependent System[4], and a multiple regression model (based on experimental data)[5] have been reported in EDM literature. The complex and random nature of the EDM process, the large number of parameters, and a lack of understanding as well as influence of the ejection mechanism at the end of a pulse in quantitative terms have defied these valuable attempts in providing accurate and general models for EDM performance. Also, traditional self-tuning adaptive control techniques can only deal with linear systems or some special nonlinear systems. Orbital EDM and WEDM processes are more complex because the number of parameters increases, and the effect of process parameters of outputs are not linear in most of the cases.

For predicting the behavior of EDM and WEDM processes, new techniques of artificial intelligence can be used. The most recent technique of artificial intelligence is the artificial neural network. Neural network technology can provide solutions to a wide variety of science and engineering problems that involve extracting useful information from complex or uncertain data. The back-propagation neural networks have the capability to learn arbitrary nonlinearity and show great potential for adaptive control applications [6,7].

A neural network adjusts its internal structure according to the complexity of the process. Neural networks have been used for process modeling and control for turning[8], milling[9], welding[10] and other conventional processes. Some of the results of neural network approach in modeling orbital EDM process have been reported in [11]. This paper presents artificial neural network modeling of WEDM process. For the sake of completeness and continuity, artificial neural network modeling of orbital EDM and related additional results are also included in this paper. Two different hetero-associative type, back-propagation neural networks have been developed for orbital EDM and WEDM processes to study the effect of various input parameters on output parameters. In the second section orbital EDM and WEDM processes are explained, and the background of artificial neural networks with respect to back-propagation is explained in the third section. In the fourth and fifth sections experimental procedures and machine details are explained. In these sections models were developed for orbital EDM and for WEDM. The estimated output from artificial neural network models was compared with the actual output and absolute error and % error were calculated for each model. Further multiple regression models have been developed and outputs from these models are compared with the actual output and artificial neural network output. In the sixth section results are analyzed, and the last section presents the conclusion of this study.

8.2 ELECTRO-DISCHARGE MACHINING

The EDM process involves controlled erosion of electrically conducting materials by the initiation of rapid and repetitive electrical spark discharges between the tool electrode and the workpiece. This erosion process takes place while the tool and workpiece are immersed in a dielectric fluid medium and are separated by a small gap known as the spark-gap. A schematic diagram of an EDM system is given in Figure 8-1. When moving wire is used as a tool in EDM, the process is called wire electro-discharge machining (WEDM). When the tool in die-sinking EDM moves in planetary orbits with respect to the workpiece, the process is called orbital EDM.

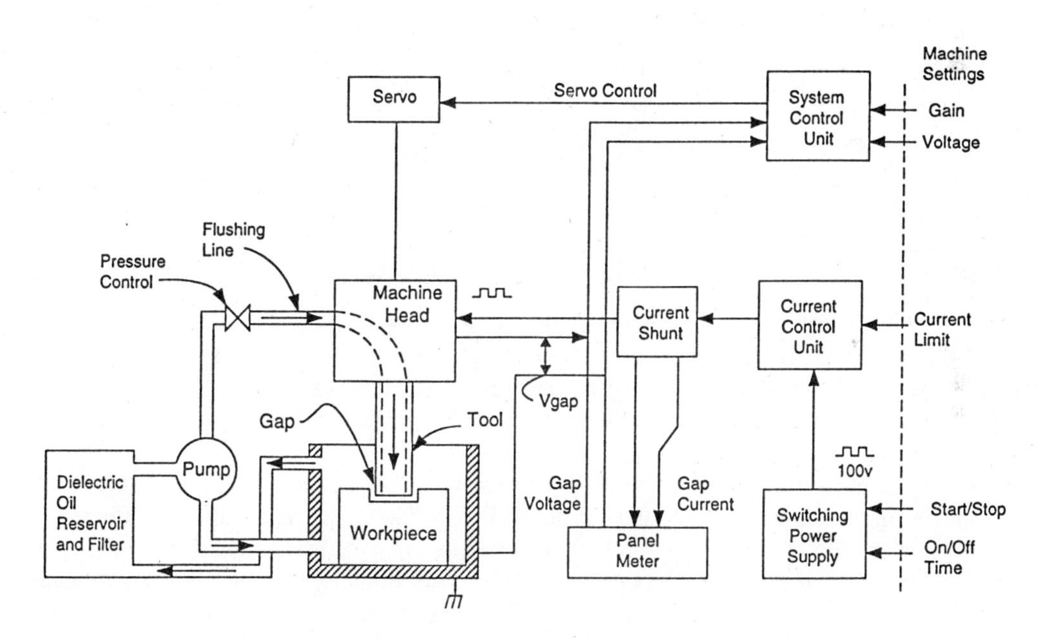

Figure 8.1 Schematic diagram of an EDM system

Orbital motion was introduced to EDM during the early 1970's. The goal of employing orbital motion was to lower production costs through increased productivity. Orbital motion improves the accuracy of the machined surface, compensates for differences in the spark gap by introducing relative motion between the tool and the workpiece, and improves the efficiency of EDM [12]. Figure 8-2 shows the vertical and radial step of material removal during orbital motion of a cylindrical tool in EDM. It has also been noted that orbital motion is particularly useful when it is difficult to flush the machining gap in an EDM sinking operation. Orbital motion induces a flushing effect and lowers the average roughness of the finished workpiece [13].

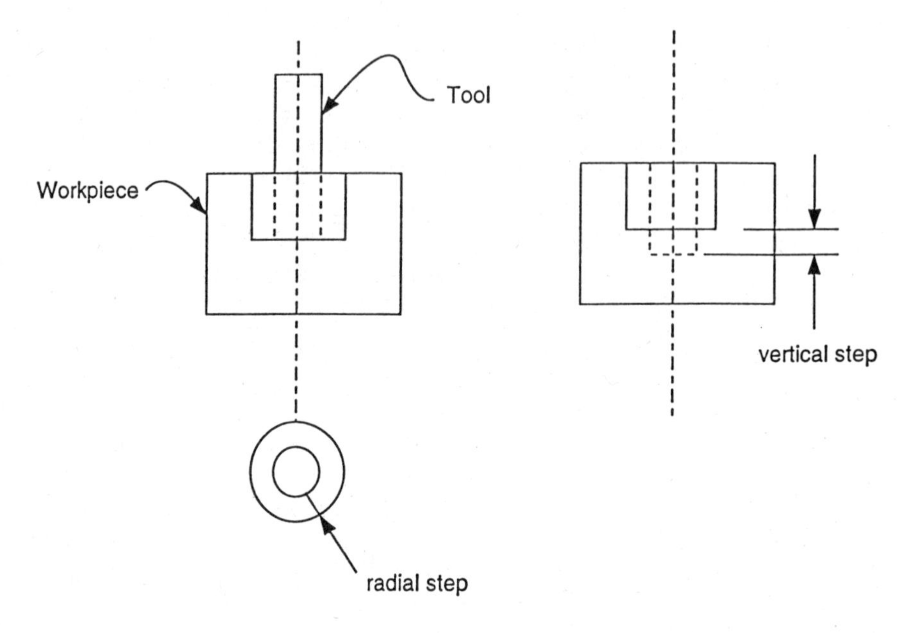

Figure 8.2 Material removal stages in Orbital EDM

Electric sparking takes place in the minute gap between the tool and workpiece. The temperature of the spot hit by the spark causes the work surface to melt, vaporize, and ultimately take the form of a crater which is quenched by the surrounding dielectric medium. This machining process is extremely complicated and stochastic in nature. Due to the randomness of the process it is very difficult to establish a quantitative relationship between controllable input parameters such as current, ontime, offtime, etc., and output parameters.

WEDM is a spark erosion process used to produce complex two and three dimensional shapes in electrically conductive workpieces. WEDM uses a thin (0.05-0.3 mm) diameter wire as the tool electrode. The high frequency pulses of electricity generate sparks between the wire and workpiece. The gap between the wire and the workpiece is filled with a liquid dielectric medium (deionized water). A series of discrete spark discharges between the wire and workpiece erodes material ahead of the traveling wire. Figure 8-3 presents a detail of the WEDM machining gap. The high degree of accuracy and fine surface finishes obtainable make WEDM attractive for applications involving the manufacture of press stamping dies, extrusion dies, and prototype parts [14].

The main goals of WEDM manufacturers and users are to achieve a better stability and higher productivity of the WEDM process, i.e., a higher machining rate with the desired accuracy, reduced wire breakage, and minimum surface damage. However, due to a large number of variables and the stochastic nature of the process, even a highly skilled operator working with a state-of-the-art WEDM machine is unable to achieve optimal performance and avoid wire rupture and surface damage as machining progresses.

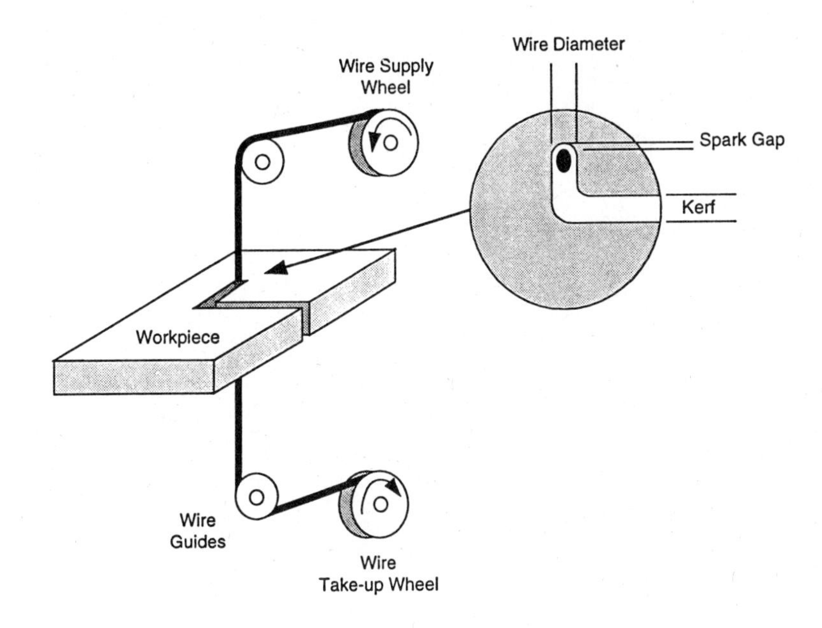

Figure 8.3 Details of wire-EDM cutting gap

The characteristics of electro-discharge machining by means of a wire electrode depend on the frequency and energy of the pulses, the wire electrode diameter and its tension, the length of cut, the wire speed, flow rate, and other parameters.

Some improvement in machining performance has been noticed by using a fuzzy controller for EDM [15,16] but an artificial neural network has not yet been used for modeling of EDM. Chen, et. al., analyzed the stability of the machining process of HS-WED and described an active fuzzy control system which can improve the process stability and increase the rate of HS-WEDM [17].

8.3 ARTIFICIAL NEURAL NETWORK

The main component of a neural network is the processing element, which is shown in figure 8-4. The input signals X_i are modified by the interconnection weights W_{ji} and summed to a single result. This result is then modified by an activation function, which can be a sigmoid, hyperbolic tangent, or sinh, etc. The activation function response is transmitted to the output path, which may be input to other nodes, or sent to external sources for interpretation.

The feed forward network is shown in figure 8-5. The input units are shown in the bottom layer i and output units are the top layer k. There can be many layers of hidden units in between, but every unit must send its output to a higher layer and must receive its input from lower layers. For a given input vector, the output vector is computed by a forward pass

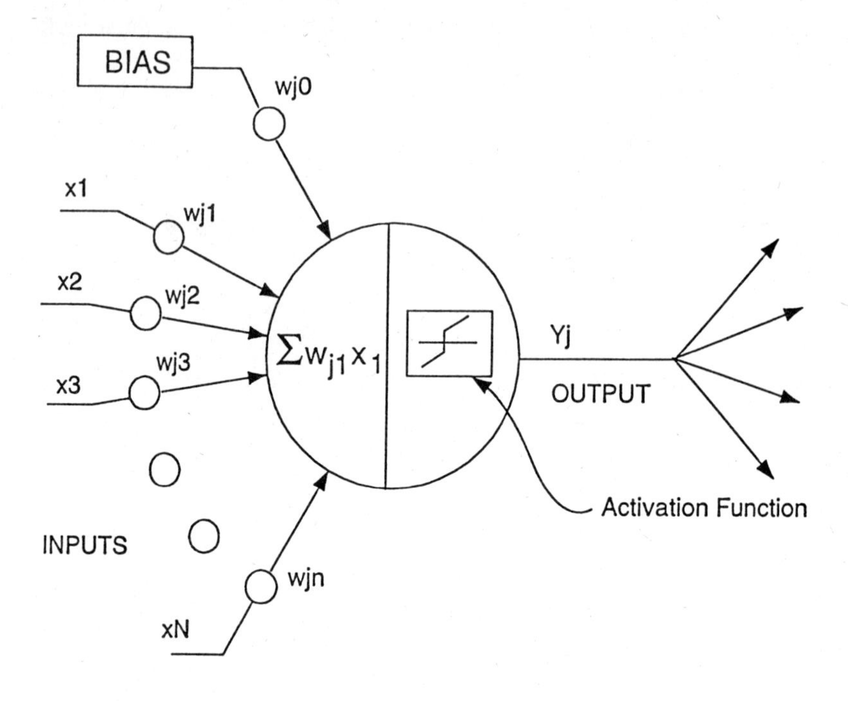

Figure 8.4 Neural network processing element

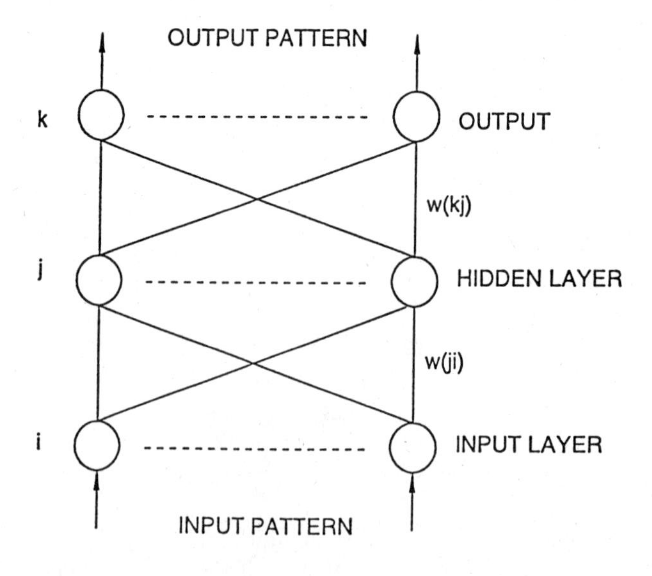

Figure 8.5 Multilayer semilinear feedforward network

which computes the activity levels of each layer, in turn, using the already computed activity levels in the earlier layers. The hidden units in hidden layer j with a linear activation function provide no advantage, so we have to use a non-linear activation function like sigmoid, tanh or sinh.

For the learning phase of the network, an input-output pattern pair is selected from the training samples randomly. The input is presented to the input layer and the input is fed forward up through the network. The difference between the desired output and the output from the output layer is used to compute an equivalent error, which is then fed back into the top layer. In this backward pass the backpropagating of errors, update the connection weights and also the thresholds in each layer. Once the network accomplishes this learning we present another set of the input-output pair from training samples, and ask the network to "learn" that association. This learning process is continued until a single set of weights and biases are obtained which can satisfy all the input and output pairs [18].

In general, the outputs will not be the same as the target or desired value. If t_k is the target or desired value at node k, the error E, is expressed in terms of the outputs o_k,

$$E = \frac{1}{2}\sum (t_k - o_k)^2 \quad (1)$$

The outputs o_k can be expressed as

$$o_k = f(net_k) \quad (2)$$

where net_k is the input to the kth node, and by definition is the weighted linear sum of all the outputs from the previous layer:

$$net_k = \sum w_{kj} o_j \quad (3)$$

From which we obtain

$$\delta_k = (t_k - o_k)\, f_k'(net_k) \quad (4)$$

where $f_k'(net_k)$ is the derivative of the nonlinear activation function or squashing function, f_k, for the kth unit, evaluated at the net input (net_k) to that unit.
For any output layer node k, we have

$$\Delta w_{kj} = \eta (t_k - o_k)\, f_k'(net_k) o_j = \eta \delta_k o_j \quad (5)$$

If the weights do not affect output nodes directly, we have

$$\Delta w_{ji} = \eta \delta_j o_i \quad (6)$$

$$\delta_j = f_j'(net_j)\sum_k \delta_k w_{kj} \tag{7}$$

Equations (4) and (7) give a recursive procedure for computing δ's for all units in the network. In other words, the δ's at an internal mode can be evaluated in terms of the deltas at an upper layer. Starting at the highest layer (the output layer), δ_k can be evaluated using equation (4), and then the errors are propagated backward to lower layers. Further, these δ's can be used to compute the weight changes in the network according to equation (6).

The learning procedures require only that the change in weight be proportional to -∂E/∂w. The constant of proportionality is η - the learning rate in our procedure. The larger this constant, the larger the changes in the weights. Rumelhart, et. al. [19], suggested that equations (5) and (6) might be modified to include the momentum term

$$\Delta w_{ji}(n+1) = \eta(\delta_j o_i) + \alpha \Delta w_{ji}(n) \tag{8}$$

where the subscript n indexes the presentation number and α is a constant which determines the effect of past weight changes on the current direction of movement in weight space. The second term in equation (8) is used to specify that the change in w_{ji} at the (n+1)th step should be somewhat similar to the change undertaken at the nth step. In this way some inertia is built in, and momentum in the rate of change is conserved to some degree.

8.4 ORBITAL EDM MODELING

Die-sinking EDM has been numerically controlled for automation, and its performance has been remarkably enhanced by orbital motion of tools [20]. At the same time, however, it has become very difficult to machine workpieces while fully utilizing the functions of the machine due to the increase in the number of parameters which have to be set by an operator. There are two approaches available for solving this problem: an application of artificial intelligence to the parameters which must be determined off-line, and an application of adaptive control to the parameters which must be controlled on-line.

To solve this problem, the artificial neural network approach in modeling of orbital EDM process has been used. For orbital EDM modeling, a back-propagation neural network was developed [11]. The machining depth, tool radius, orbital radius, radial step, vertical step, offset depth, pulse ontime, pulse offtime, and discharge current were selected as input parameters to the network. The output parameters were MRR and surface roughness. For comparison to different modeling techniques a multiple regression model was also developed.

Experiments: The experiments were performed with an Eltee pulsitron EDM machine, model TRM20/EP-30, using a low viscosity BP 180 dielectric. The orbital motion movement was provided by means of an Eltee Pulsitron Vector Machining Control (VMC). The design of the experiments was based on the matrix experiment of the standard L_{27} orthogonal-array Taguchi method [21]. The control factors and their levels are shown in table 8-1 [11]. The workpiece material was 4150 steel, and the cylindrical tool was made of copper. There was no flushing used during any of the experiments. The experimental procedure is described below:

a) Weigh the tool and workpiece using a Sortorius precision scale type E-1200S (maximum capacity = 1210 gm, resolution = 0.001 gm) before and after machining the workpiece;
b) Machine the workpiece up to 2 mm depth;
c) Clean the tool and workpiece with a solvent (wipe off the tool and workpiece with soft cloth and allow them to dry in air);
d) Measure the surface roughness at the bottom of the machined hole with a Mahr Perthen model C3A profilometer (stylus radius: 5 microns). The cut off length was kept at 0.03 in.

Table 8-1 Factors and their levels for orbital EDM [11]

CONTROL FACTORS	LEVEL I	LEVEL II	LEVEL III	UNIT
A. Machining depth	2	2.5	3	mm
B. Tool radius	3	5	7	mm
C. Orbital radius	1	2	3	mm
D. Radial step	0.01	0.02	0.254	mm
E. Vertical step	0.2	0.5	1	mm
F. Offset depth	1	1.5	1.8	mm
G. Pulse On-time	180	210	240	sec
H. Pulse Off-time	180	210	240	sec
I. Discharge current	12	18	24	amps

Neural Network Modeling : To train the neural network model the experimental results were used as the training set of data, which are shown in table 8-2. For 9 input and 2 output parameters different sizes of networks were tried with a greater or lesser number of hidden units in the hidden layer. The 9-9-2 size back-propagation neural network was found to be the most suitable network. The 9-9-2 size neural network is shown in figure 8-6 [11]. One hidden layer with 9 hidden units was used. The hyperbolic tangent function was used as a transfer function for the processing elements. The hyperbolic tangent function is shown in figure 8-7. The hyperbolic tangent function is simply a bipolar version of the sigmoid function.

For the hyperbolic tangent function, we have

$$o_j = \frac{e^{(net_j+\theta_j)} - e^{-(net_j+\theta_j)}}{e^{(net_j+\theta_j)} + e^{-(net_j+\theta_j)}} \tag{9}$$

Taking the partial derivative with respect to net_j, we obtain
The bias, θ_j, is treated as the weight of the link from a virtual node whose output value is

$$\frac{\partial o_j}{\partial net_j} = (1+o_j)(1-o_j) \tag{10}$$

unity. The bias or threshold offsets the origin of the logistic function, producing an effect that is similar to adjusting the threshold of the perceptron neuron, thereby permitting more rapid convergence of the training process. It should be noted that the derivative , $\partial o_j/\partial net_j$, reaches its maximum at $o_j = 0$ and, since $-1 \leq o_j \leq 1$, approaches its minima as o_j approaches -1 or +1. The deltas are given by the following two expressions:

$$\delta_{pk} = (t_{pk}-o_{pk})(1+o_{pk})(1-o_{pk}) \tag{11}$$

$$\delta_{pj} = (1+o_{pj})(1-o_{pj})\sum_k \delta_{pk} w_{kj} \tag{12}$$

The network was trained for 6000 iterations. The normalized cumulative delta rule was used as learning rule for training the network. Momentum was kept at 0.5. The learning coefficient was 0.9 for the hidden layer and 0.1 for the output layer. After completion of training, the network was tested for the training set of data for estimating MRR and Ra outputs. The additional experimental results are shown in table 8-3 which were used for verification of the neural network model [11]. The percentage error and absolute error shown in table 8-4 were calculated for both MRR and Ra. The maximum percentage error for estimating the MRR from the neural network model was -33.33%, and the absolute error was found to be 0.001 gm/min. The percentage error in the estimated Ra was 1.706%, and the absolute error was -5.8 micro inches.

Multiple Regression Approach: Statistical Analysis Software (SAS) was used to fit a multiple regression model for this approach. For orbital EDM output parameters MRR and Ra vary nonlinearly for most of the process parameters. A logarithmic transformation was incorporated as follows for MRR and Ra :

$$Y_1 = -\log_{10} (\text{ reciprocal of square of MRR}) \tag{13}$$

$$Y_2 = -\log_{10} (\text{ square of Ra}) \tag{14}$$

The best fitted multiple regression models for MRR and Ra are given as below:

$$E(Y_1) = \beta_0+\beta_1X_1+\beta_2X_2+...+\beta_9X_9 \tag{15}$$

$$E(Y_2) = \beta_0+\beta_1X_1+\beta_2X_2+...+\beta_9X_9 \tag{16}$$

where x1,...x9 are input parameters.

The estimated MRR and estimated Ra from the multiple regression models were compared with neural network estimates. The actual output parameters, neural network estimates, and multiple regression model estimates for MRR and Ra are given in table 8-5. The comparison indicates the superiority of the artificial neural network model over the multiple regression model.

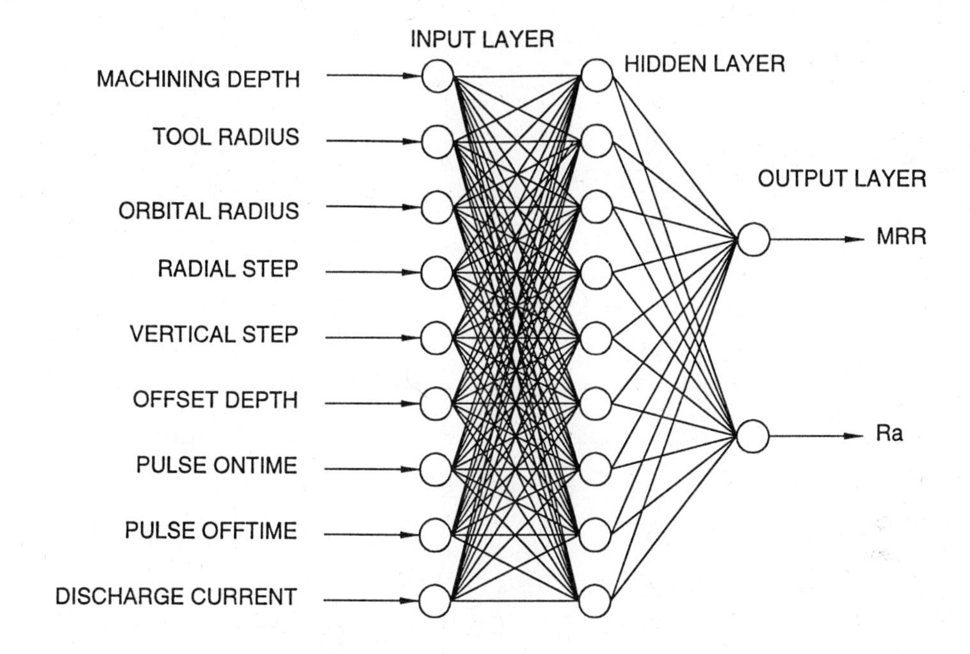

Figure 8.6 Neural network for Orbital EDM

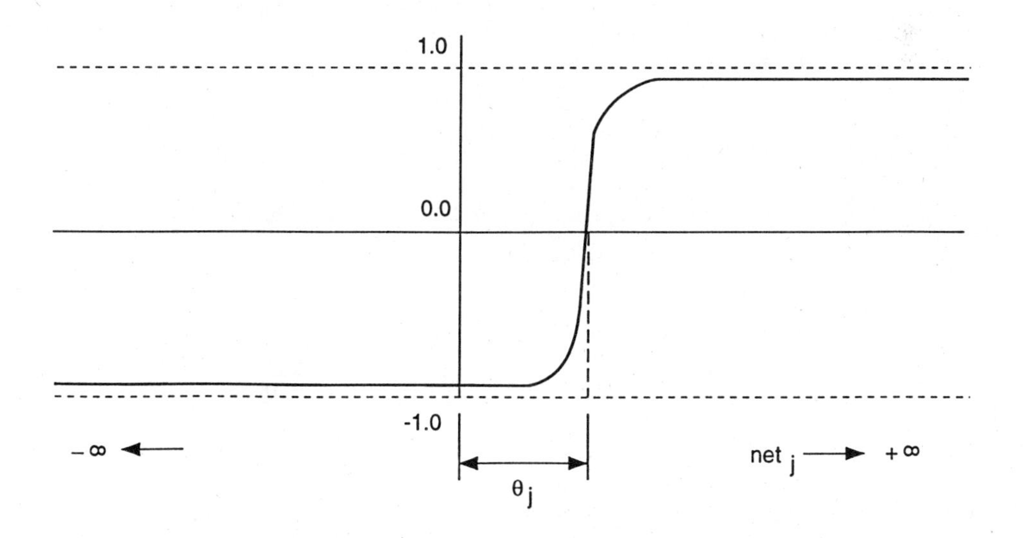

Figure 8.7 Hyperbolic tangent function

Table 8-2 Training data for ANN model for orbital EDM

S.#	Machining Depth	Tool radius mm	Orb. rad. mm	Radial step	Vert. step	Offset depth mm	On-time μsec	Off-time μsec	Current amp	MRR gm/min	Ra micro inch
1	2	3	1	0.010	0.2	1.0	180	180	12	0.005	237
2	2	3	1	0.010	0.5	1.5	210	210	18	0.016	378
3	2	3	1	0.010	1.0	1.8	240	240	24	0.018	390
4	2	5	2	0.020	0.2	1.0	180	210	18	0.008	319
5	2	5	2	0.020	0.5	1.5	210	240	24	0.023	458
6	2	5	2	0.020	1.0	1.8	240	180	12	0.019	347
7	2	7	3	0.254	0.2	1.0	180	240	24	0.043	376
8	2	7	3	0.254	0.5	1.5	210	180	12	0.047	211
9	2	7	3	0.254	1.0	1.8	240	210	18	0.053	340
10	3	3	2	0.254	0.2	1.5	240	180	18	0.027	292
11	3	3	2	0.254	0.5	1.8	180	210	24	0.031	305
12	3	3	2	0.254	1.0	1.0	210	240	12	0.028	198
13	3	5	3	0.010	0.2	1.5	240	210	24	0.004	313
14	3	5	3	0.010	0.5	1.8	180	240	12	0.007	297
15	3	5	3	0.010	1.0	1.0	210	180	18	0.007	277
16	3	7	1	0.020	0.2	1.5	240	240	12	0.029	224
17	3	7	1	0.020	0.5	1.8	180	180	18	0.052	326
18	3	7	1	0.020	1.0	1.0	210	210	24	0.068	397
19	3	3	3	0.020	0.2	1.8	210	180	24	0.003	382
20	3	3	3	0.020	0.5	1.0	240	210	12	0.006	189
21	3	3	3	0.020	1.0	1.5	180	240	18	0.008	150
22	3	5	1	0.254	0.2	1.8	210	210	12	0.023	211
23	3	5	1	0.254	0.5	1.0	240	240	18	0.065	326
24	3	5	1	0.254	1.0	1.5	180	180	24	0.084	398
25	3	7	2	0.010	0.2	1.8	210	240	18	0.010	373
26	3	7	2	0.010	0.5	1.0	240	180	24	0.014	426
27	3	7	2	0.010	1.0	1.5	180	210	18	0.022	224

Table 8-3 Verification Experiment for ANN model for orbital EDM

S.#	Mach depth mm	Tool radius mm	Orb. Rad. mm	Radial step	Vert. Step mm	Offset depth mm	On-time μsec	Off-time μsec	Current amp	MRR gm/min	Ra micro in
1	2.5	5	2	0.010	0.5	1.5	210	210	18	0.007	-
2	2.5	5	2	0.020	0.5	1.5	210	210	18	0.009	-
3	2.5	5	2	0.254	0.5	1.5	210	210	18	0.050	-
4	2.5	5	1	0.254	0.5	1.5	210	210	18	0.069	-
5	2.5	5	3	0.254	0.5	1.5	210	210	18	0.016	-
6	2.5	5	2	0.254	0.5	1.5	210	210	12	0.032	195
7	2.5	5	2	0.254	0.5	1.5	210	210	24	0.055	405
8	2.5	3	2	0.254	0.5	1.5	210	210	18	0.027	-
9	2.5	7	2	0.254	0.5	1.5	210	210	18	0.069	-

Table 8-4 Estimates and errors from neural net model for orbital EDM

Serial	ActMRR gms/min	EstMRR gms/min	% Error	Abs Error gm/min	Act R_a micro	Est R_a micro	% Error	Abs Error micro in
1	0.005	0.005	0.000	0.000	237.0	234.0	-1.266	3.0
2	0.016	0.015	-6.250	0.001	378.0	377.6	-0.106	0.4
3	0.018	0.018	0.000	0.000	390.0	393.2	0.821	-3.2
4	0.008	0.009	12.500	-0.001	319.0	319.1	0.031	-0.1
5	0.023	0.023	0.000	0.000	458.0	455.8	-0.480	2.2
6	0.019	0.019	0.000	0.000	247.0	249.2	0.891	-2.2
7	0.043	0.043	0.000	0.000	376.0	378.9	0.771	-2.9
8	0.047	0.047	0.000	0.000	211.0	213.4	1.137	-2.4
9	0.053	0.053	0.000	0.000	340.0	345.8	1.706	-5.8
10	0.027	0.028	3.704	-0.001	292.0	292.5	0.171	-0.5
11	0.031	0.031	0.000	0.000	305.0	306.3	0.426	-1.3
12	0.028	0.028	0.000	0.000	198.0	197.6	-0.202	0.4
13	0.004	0.005	25.000	-0.001	313.0	316.0	0.958	-3.0
14	0.007	0.008	14.286	-0.001	297.0	296.6	-0.135	0.4
15	0.007	0.007	0.000	0.000	277.0	278.1	0.397	-1.1
16	0.029	0.029	0.000	0.000	224.0	224.1	0.045	-0.1
17	0.052	0.052	0.000	0.000	326.0	326.0	0.000	0.0
18	0.068	0.068	0.000	0.000	397.0	398.3	0.327	-1.3
19	0.003	0.002	-33.333	0.001	382.0	381.9	-0.026	0.1
20	0.006	0.005	-16.667	0.001	189.0	188.4	-0.317	0.6
21	0.008	0.008	0.000	0.000	150.0	151.6	1.067	-1.6
22	0.023	0.023	0.000	0.000	211.0	209.7	-0.616	1.3
23	0.065	0.065	0.000	0.000	326.0	324.5	-0.460	1.5
24	0.084	0.083	-1.190	0.001	398.0	398.4	0.101	-0.4
25	0.010	0.011	10.000	-0.001	373.0	372.9	-0.027	0.1
26	0.014	0.014	0.000	0.000	426.0	426.6	0.141	-0.6
27	0.022	0.022	0.000	0.000	224.0	223.6	0.179	0.4

8.5 WEDM MODELING

For Wire-EDM modeling a back-propagation neural network model and a multiple regression model were developed. The input parameters for the network were current, pulse duration, frequency, wire tension, wire speed, and flow rate. The material removal rate (MRR) and surface roughness (Ra) were output parameters for the network.

Experiments: A factorial design model was used to predict the measure of performance for a variety of control settings. The matrix experiment used was a standard L_{18} orthogonal-array of the Taguchi method [21]. The 6 factors and their levels are shown in table 8-6.

The initial experiments were performed on a Robofil 100 WEDM machine (manufactured by Charmilles technology) by Shreedhar [22]. The machining circuit on this machine is of the relaxation type. The workpiece material was D2 tool steel. The height of the workpiece was 25 mm and the length of cut was 10 mm. Thus, the cross section of the cut was 10 mm X 25 mm. A 0.25 mm diameter zinc-coated stratified wire was used as a tool. The following factors were held constant during machining:

Table 8-5 Results from ANN and multi reg models for orbital EDM

Serial	Act MRR gms/min	Est MRR neural net	Est MRR mutli reg	Act R_a micro in	Est R_a neural net	Est R_a multi reg
1	0.005	0.005	0.008	237.0	234.0	244.0
2	0.016	0.015	0.013	378.0	377.6	320.9
3	0.018	0.018	0.024	390.0	393.2	408.6
4	0.008	0.009	0.010	319.0	319.1	313.9
5	0.023	0.023	0.016	458.0	455.8	412.7
6	0.019	0.019	0.018	247.0	249.2	249.8
7	0.043	0.043	0.040	376.0	378.9	387.3
8	0.047	0.047	0.037	211.0	213.4	242.0
9	0.053	0.053	0.071	340.0	345.8	308.2
10	0.027	0.028	0.018	292.0	292.5	286.8
11	0.031	0.031	0.028	305.0	306.3	344.6
12	0.028	0.028	0.034	198.0	197.6	186.3
13	0.004	0.005	0.006	313.0	316.0	386.9
14	0.007	0.008	0.007	297.0	296.6	210.5
15	0.007	0.007	0.010	277.0	278.1	263.7
16	0.029	0.029	0.023	224.0	224.1	281.1
17	0.052	0.052	0.030	326.0	326.0	354.4
18	0.068	0.068	0.053	397.0	398.3	423.0
19	0.003	0.002	0.003	382.0	381.9	336.8
20	0.006	0.005	0.004	189.0	188.4	185.8
21	0.008	0.008	0.007	150.0	151.6	221.4
22	0.023	0.023	0.039	211.0	209.7	234.7
23	0.065	0.065	0.058	326.0	324.5	285.8
24	0.084	0.083	0.090	398.0	398.4	357.3
25	0.010	0.011	0.014	373.0	372.9	316.6
26	0.014	0.014	0.017	426.0	426.6	404.5
27	0.022	0.022	0.023	224.0	223.6	218.3

1. Product shape (rectangle)
2. Location of workpiece on working table
3. Temperature of dielectric liquid (50°F)
4. Angle of cut (vertical)
5. Workpiece thickness = length of cut (10 mm)
6. Height of work piece (25 mm)
7. Machining voltage with no load on circuit (120 volt)
8. Reference average voltage under servo control (50%)
9. Servo control stability (0.957)
10. Wire type (stratified, Copper, diameter 0.25 mm)
11. Dielectric flow rate (lower guide = 2 bars).

The experimental procedure was as follows:

a) Weigh the workpiece using a Sortorius precision scale type E-1200S (maximum capacity = 1210 gm, precision = 0.001 gm) before and after machining the workpiece;

b) Each time an experiment was performed, a particular set of input parameters were chosen for the initial experiments or verification experiment;

c) The workpiece was cut through for 10 mm length;

d) The surface finish value (μ-inches) was obtained by measuring the Ra value (mean absolute deviation from the average surface level) using a model C3A Mahr Perthen profilometer.

Table 8-6 Factors and their levels for WEDM

CONTROL FACTORS	LEVEL I	LEVEL II	LEVEL III	UNIT
A. Discharge Current	16	24	32	amp
B. Pulse Duration	3.2	6.4	12.8	micro sec
C. Pulse Frequency	40	50	60	kHz
D. Wire Speed	7.6	8.6	9.2	m/min
E. Wire Tension	1000	1100	1200	gms
F. Dielectric Flow Rate	1.2	1.3	1.4	bars

Neural Network Modeling: For WEDM, the back-propagation neural network was also selected as the network paradigm. For 6 input parameters and 2 output parameters, there were 6 nodes in the input layer and 2 nodes in the output layer. There were 6 nodes selected in the hidden layer. The hyperbolic tangent function was used as the transfer function or activation function. Different sizes of networks were tried with a greater or lesser number of hidden units in the hidden layer. The sigmoid activation function was also tried as a transfer function, but the 6-6-2 size artificial neural network with a hyperbolic tangent function was found to be the best possible network, which is shown in figure 8-8.

The experimental results were used as the training set of data. The training set of data is shown in table 8-7. The additional experiments were performed to check the adequacy of the model. The results of these experiment are given in table 8-8. The normalized cumulative delta rule was used as the learning rule. The momentum was kept at 0.4 and the learning coefficients were kept at 0.9 and 0.1 for the hidden layer and output layer. The network was trained for 12000 iterations. After the accomplishment of the training task the network was tested for the same set of data of input parameters which was used for training. The percentage error and absolute error were calculated for the estimated MRR and estimated Ra as shown in table 8-9. The maximum percentage error in MRR was 0.275 and the absolute error was -.0003 gm/min. For estimated Ra model provides error free results.

Multiple Regression Approach: In order to establish a relationship between the input and the output parameters, a second order nonlinear multiple regression model was selected for the WEDM process for comparison with neural network modeling. MRR and Ra vary nonlinearly for most of the process parameters. The best fitted nonlinear multiple regression model for the WEDM process is shown below:

$$E(Y)=\beta_0+\beta_1X_1+\beta_2X_2+...+\beta_6X_6+\beta_7X_1^2+\beta_8X_2^2+...+\beta_{12}X_6^2 \quad (17)$$

where x1,...,x6 are input parameters.

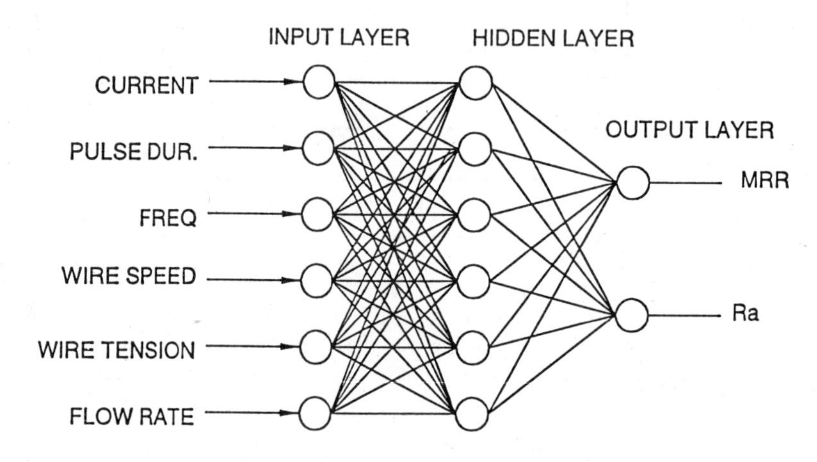

Figure 8.8 Neural network for WEDM

The output parameters MRR and Ra are affected by the complex mechanism of EDM but experimental data for WEDM also have their own level of inaccuracies. The estimated MRR and estimated Ra from both the artificial neural network model and the multiple regression model were compared with actual outputs. The comparisons are shown in table 8-10. It is clear from this table that neural network modeling is more accurate than the multiple regression approach.

8.6 RESULTS AND DISCUSSION

For the orbital EDM and WEDM processes the effect of various input parameters on MRR and Ra were analyzed by using the artificial neural network and multiple regression models. The effects of significant parameters on outputs were analyzed. The additional experiments were performed to check the adequacy of models. It was observed that the artificial neural network approach works very well for modeling such complex processes. It was noticed that obtaining an adequate model by multiple regression analysis is more difficult and time consuming. Also, the multiple regression models had more noise.

Table 8-7 Training data for ANN model for WEDM

S.no	Current Amp.	Pulse Dur usec	Freq, kHz	W speed m/min	Wire Tension .gm	Flow Rate Bars	Act MRR gm/min	Act Ra u in
1	24	3.2	40	7.6	1000	1.2	0.190	137.4
2	24	6.4	50	8.6	1100	1.3	0.1817	149.8
3	24	12.8	60	9.2	1200	1.4	0.1595	147.6
4	32	3.2	40	8.6	1100	1.4	0.1355	143.4
5	32	6.4	50	9.2	1200	1.2	0.2120	156.6
6	32	12.8	60	7.6	1000	1.3	0.1993	158.6
7	16	3.2	40	7.6	1200	1.4	0.1026	140.2
8	16	6.4	50	8.6	1000	1.2	0.1632	150.6
9	16	12.8	60	9.2	1100	1.3	0.1757	153.0
10	24	3.2	40	9.2	1100	1.2	0.1381	151.8
11	24	6.4	50	7.6	1200	1.3	0.1298	152.0
12	24	12.8	60	8.6	1000	1.4	0.1343	148.8
13	32	3.2	40	9.2	1000	1.3	0.1614	139.6
14	32	6.4	50	7.6	1100	1.4	0.2471	154.6
15	32	12.8	60	8.6	1200	1.2	0.1459	157.0
16	16	3.2	40	8.6	1200	1.3	0.1177	143.4
17	16	6.4	50	9.2	1000	1.4	0.1204	147.6
18	16	12.8	60	7.6	1100	1.2	0.1903	145.2

Table 8-8 Verfiication Experiment for ANN model for WEDM

S.no	Current Amp	Pulse Dur. usec	Freq. kHz	Wire speed m/min	Wire Tension .gm	Flow rate Bars	Act. MR gm/min	Act. Ra micro in.
1	16	6.4	50	8.6	1100	1.3	0.1673	-
2	32	6.4	50	8.6	1100	1.3	0.2192	-
3	24	3.2	50	8.6	1100	1.3	0.1332	140
4	24	12.8	50	8.6	1100	1.3	0.1609	148
5	24	6.4	40	8.6	1100	1.3	0.1237	-
6	24	6.4	60	8.6	1100	1.3	0.2030	-
7	24	6.4	50	8.6	1000	1.3	0.1654	-
8	24	6.4	50	8.6	1200	1.3	0.1708	-

Orbital EDM: Figure 8-9 shows the effect of current on surface roughness [11]. The surface roughness increases with increasing current. This can be explained by the effect of current on the crater depth. As current increases crater depth increases, hence, surface roughness increases. When the current was increased beyond 24 amp, the Ra value increased by 30 μinches. The Ra value increased rapidly between the 18 to 24 amp current range. The multiple regression model shows an almost constant increase in Ra as current is increased.

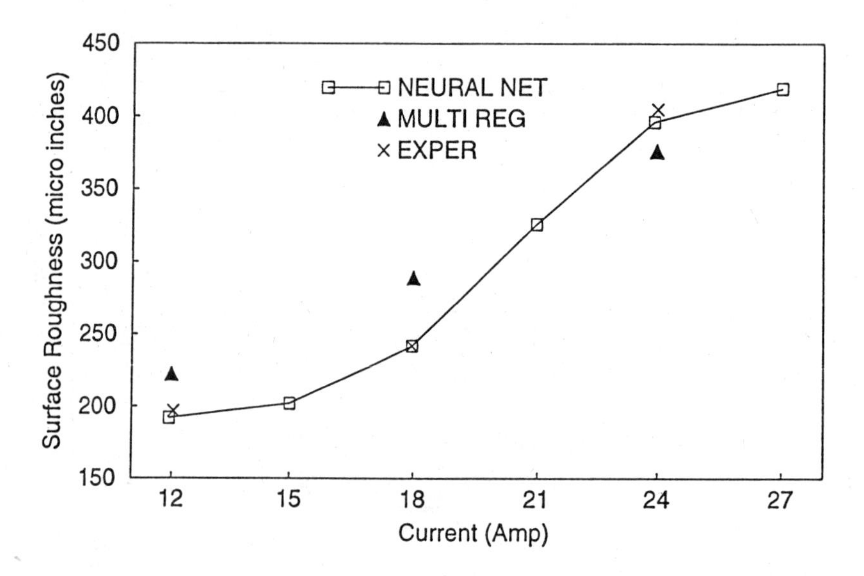

Figure 8.9 Effect of discharge current on surface roughness

Figure 8-10 shows the effect of current on MRR [11]. MRR increases with increasing current. MRR does not increase much above 22 to 24 amps of current, and remains almost constant, because increasing the current above a particular level makes the process unstable. The effect of an orbital radius on MRR is shown in figure 8-11 [11]. As the orbital radius is increased, while keeping other parameters at constant level, the MRR decreases. It was observed during experimentation that when the orbital radius is increased the orbital speed is reduced to avoid arcing. This tends to reduce MRR very quickly. The effect of "radial step" on MRR can be explained from the figure 8-12 [11]. MRR increases with increasing radial step.

Because 0.01 and 0.02 mm radial steps are very small when compared to a 0.254 mm radial step, it is obvious that if orbital speed is not increased very fast for small radial steps, MRR will not be as high as it is for a larger radial step.

Table 8-9 Estimates and errors from neural net model for WEDM

Sl. #	Act. MRR gm/min	Est MRR gm/min	% Error	Abs. Error gm/min	Act. Ra micro inch	Est. Ra micro inch	% Error	Abs. Err micro inch
1	0.1090	0.1093	0.275	-0.0003	137.4	137.4	0.0	0.0
2	01817	0.1819	0.110	-0.0002	149.8	149.8	0.0	0.0
3	0.1595	0.1593	-0.125	0.0002	147.6	147.6	0.0	0.0
4	0.1355	0.1357	0.148	-0.0002	143.4	143.4	0.0	0.0
5	0.2120	0.2121	0.047	0.0001	156.6	156.6	0.0	0.0
6	0.1993	0.1993	0.000	0.0000	158.6	158.6	0.0	0.0
7	0.1026	0.1026	0.000	0.0000	140.2	140.2	0.0	0.0
8	0.1632	0.1632	0.000	0.0000	150.6	150.6	0.0	0.0
9	0.1757	0.1757	0.000	0.0000	153.0	153.0	0.0	0.0
10	0.1381	0.1380	-0.072	0.0001	151.8	151.8	0.0	0.0
11	0.1298	0.1296	-0.154	0.0002	152.0	152.0	0.0	0.0
12	0.1343	0.1344	0.074	-0.0001	148.8	148.8	0.0	0.0
13	0.1614	0.1613	-0.062	0.0001	139.6	139.6	0.0	0.0
14	0.2471	0.2472	0.040	-0.0001	154.6	154.6	0.0	0.0
15	0.1459	0.1460	0.069	-0.0001	157.0	157.0	0.0	0.0
16	0.1177	0.1178	0.085	-0.0001	143.4	143.4	0.0	0.0
17	0.1204	0.1204	0.00	0.0000	147.6	147.6	0.0	0.0
18	0.1903	0.1904	0.053	-0.0001	145.2	145.2	0.0	0.0

Table 8-10 results from ANN and multi reg. models for WEDM

Sl. #	Act. MRR gm/min	Est. MRR gm/min	Est. MRR multi. reg	Act. Ra micro inch	Est. Ra neural net	Est. Ra multi. reg
1	0.1090	0.1093	0.0917	137.4	137.4	140.1
2	01817	0.1819	0.1824	149.8	149.8	150.7
3	0.1595	0.1593	0.1515	147.6	147.6	152.9
4	0.1355	0.1357	0.1368	143.4	143.4	144.5
5	0.2120	0.2121	0.2038	156.6	156.6	155.1
6	0.1993	0.1993	0.2093	158.6	158.6	155.3
7	0.1026	0.1026	0.1090	140.2	140.2	136.8
8	0.1632	0.1632	0.1611	150.6	150.6	151.7
9	0.1757	0.1757	0.1642	153.0	153.0	151.5
10	0.1381	0.1380	0.1547	151.8	151.8	146.7
11	0.1298	0.1296	0.1378	152.0	152.0	151.4
12	0.1343	0.1344	0.1330	148.8	148.8	145.6
13	0.1614	0.1613	0.1602	139.6	139.6	143.2
14	0.2471	0.2472	0.2365	154.6	154.6	155.6
15	0.1459	0.1460	0.1532	157.0	157.0	156.1
16	0.1177	0.1178	0.1105	143.4	143.4	144.4
17	0.1204	0.1204	0.1313	147.6	147.6	146.8
18	0.1903	0.1904	0.1925	145.2	145.2	148.8

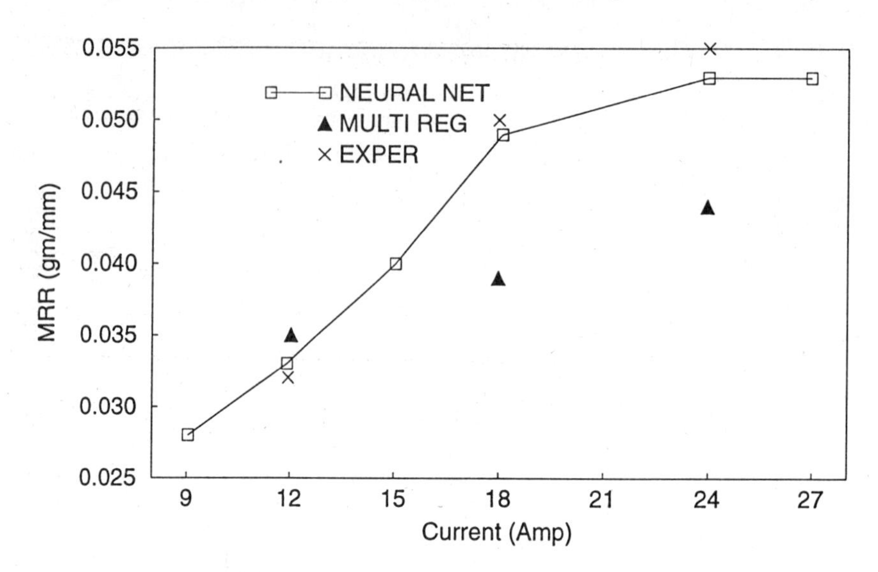

Figure 8.10 Effect of discharge current on MRR

The neural network estimates are closer to actual experimental results, which can be concluded from all the figures. Similar results from table (5) were noticed when output parameters estimated from the neural net and from multiple regression models were compared to actual experimental results. For example, in table (5) for serial number 24, actual MRR was 0.084 gm/min whereas the neural net model estimated .083 gm/min, and the multiple regression model estimated .090 gm/min. Similarly, the actual surface roughness (Ra value) for the same serial number was observed at 398.0 μin. The estimated Ra values by the neural network and multiple regression models were 398.4 μin and 357.3 μin, respectively. For all other serial numbers it was noticed that the neural net model is more accurate compared to the multiple regression model. It was also evidenced from figures 8-9 and 8-10 that the neural network estimated the output parameters more accurately over a wider range of process parameters.
WEDM: The effect of current on MRR is shown in figure 8-13. It was observed that with increasing discharge current the MRR increases because just as increasing the current the crater-size will increase. MRR increases rapidly when current is increased from 16 to 32 amps. When MRR was estimated for an extended range of current, it was noticed that MRR does not increase rapidly. Figure 8-14 shows the effect of pulse duration on MRR. First, the MRR increases with increasing pulse duration up to 6.4 μsec, but it decreases with any further increase in pulse duration.

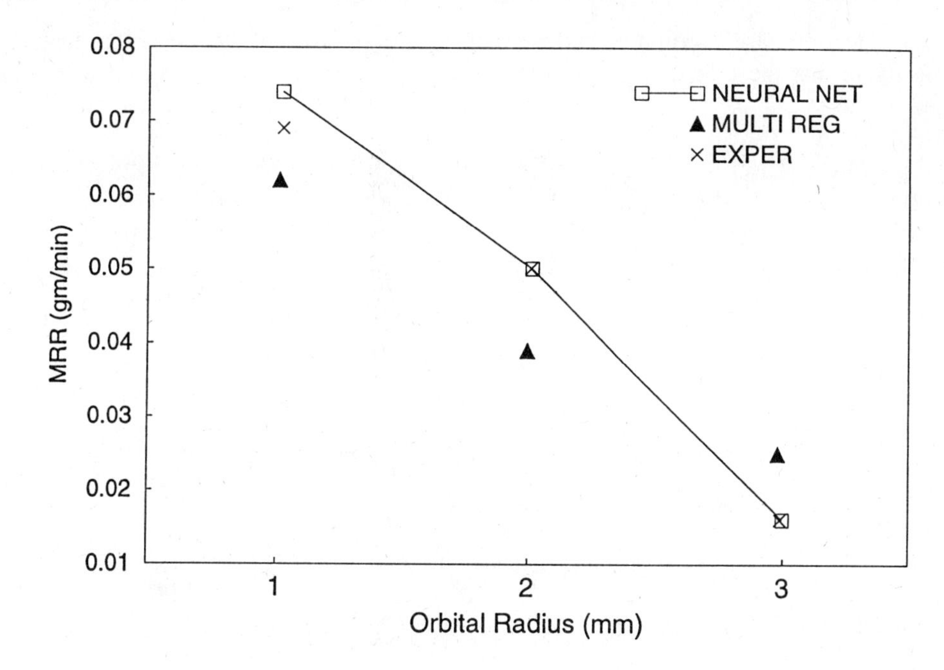

Figure 8.11 Effect of orbital radius on MRR

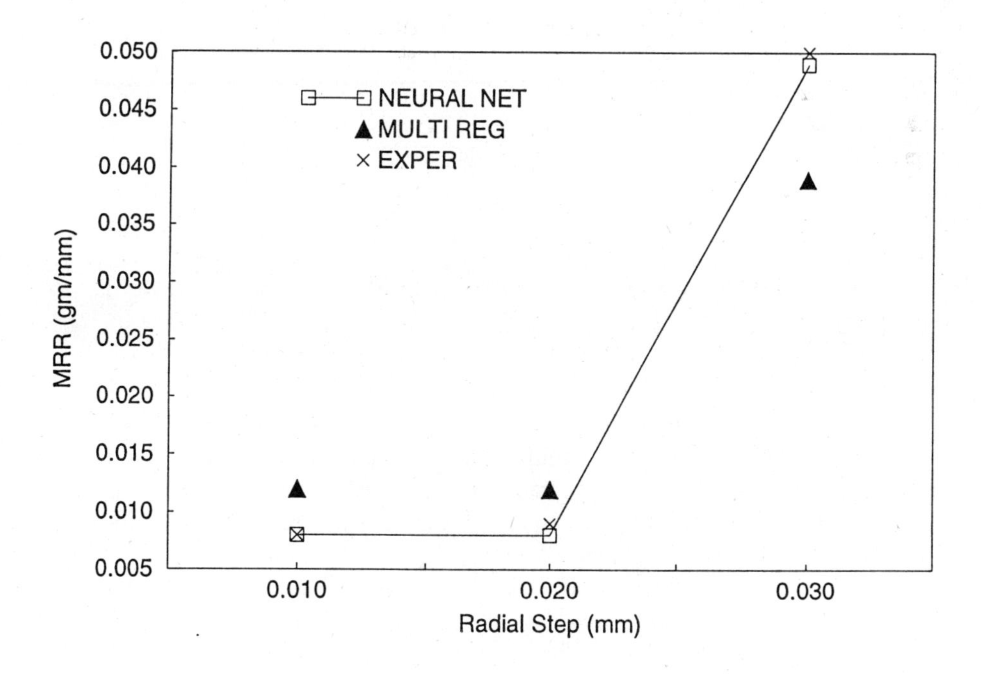

Figure 8.12 Effect of radial step on MRR

The effect of pulse duration on surface roughness is shown in figure 8-15. It is analyzed from the figure that the effect of pulse duration on Ra is similar to its effect on MRR. Ra increases by increasing the pulse duration up to 6.4 μsec, but decreases by further increasing the pulse duration.

Figure 8-16 shows the effect of pulse frequency on MRR. The neural network shows the significant effect of pulse frequency on MRR. This is also confirmed by the experiments as shown in the figure. The MRR increases faster for a 40 to 55 Khz pulse frequency but for further increases in pulse frequency the increase is relatively little.

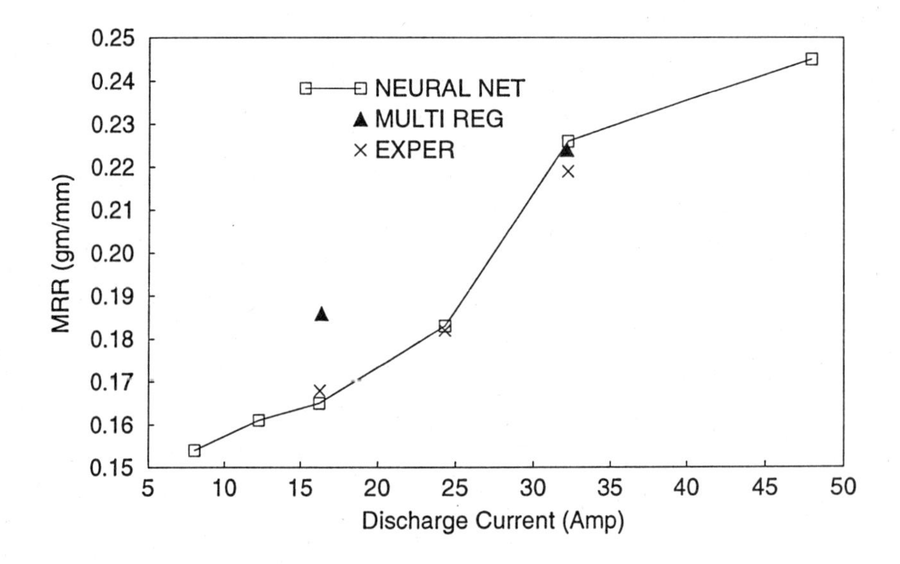

Figure 8.13 Effect of current on MRR

It was determined from all the figures that the artificial neural network models predicted the output parameters more accurately as confirmed by experimental results. Similar results from table (10) were noticed when output parameters estimated from the neural net and from multiple regression models were compared to actual experimental results. For serial number 16 the actual MRR was 0.1177 g/min, whereas the neural network estimated 0.1178 gm/min, and the multiple regression model estimated 0.1105 gm/min. Similarly, the actual surface roughness for the same serial number was 143.4 μin. The estimated Ra values, by the neural net and multiple regression models were 143.4 μin and 144.4 μin, respectively. For all other serial numbers it was noticed that the neural network model is more accurate than the multiple regression models.

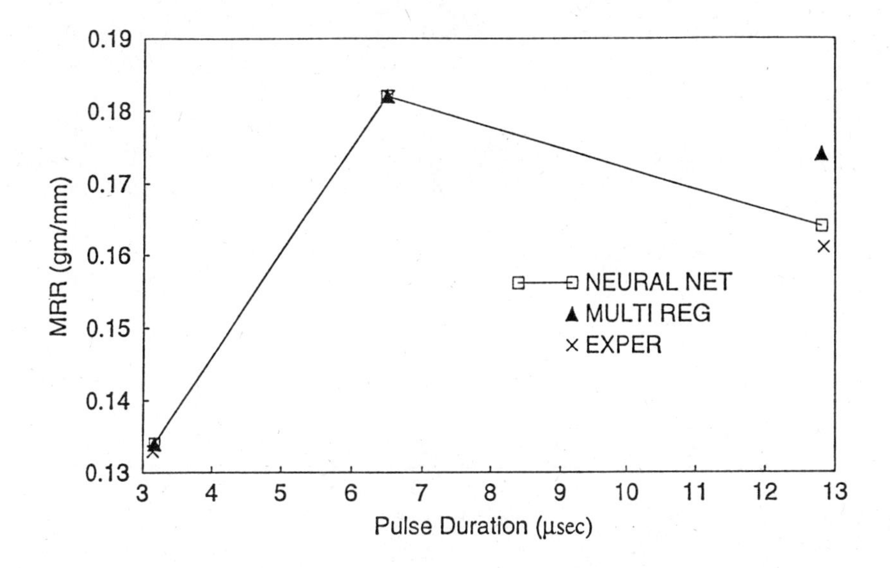

Figure 8.14 Effect of pulse duration on MRR

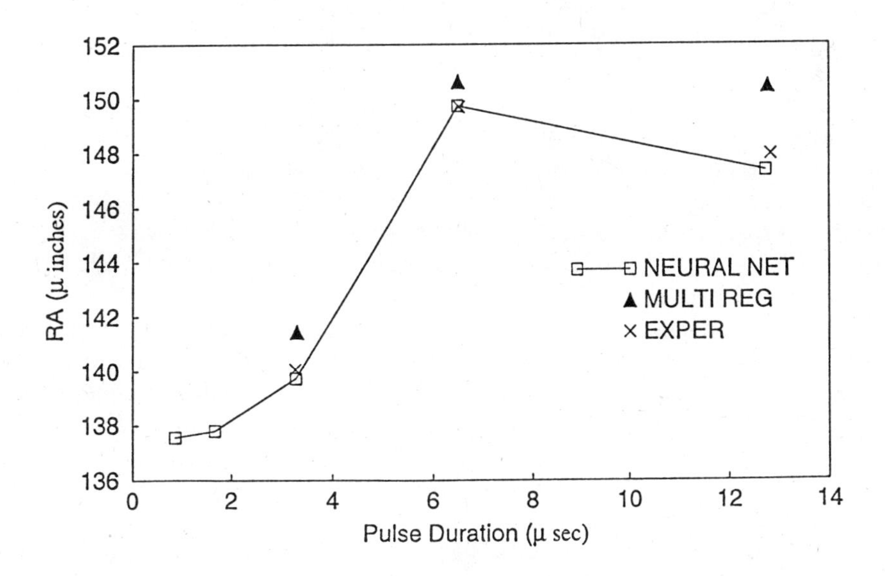

Figure 8.15 Effect of pulse duration on surface roughness

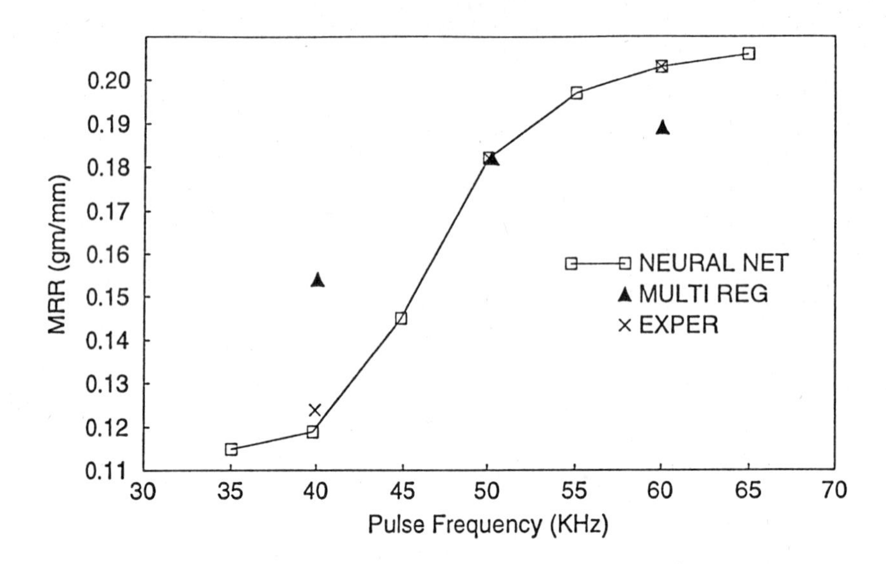

Figure 8.16 Effect of pulse freqency on MRR

8.7 CONCLUSIONS

In this study 9-9-2 size and 6-6-2 size backpropagation neural network models were developed for orbital EDM and WEDM, respectively. The following conclusions can be drawn from this research:

(1) The quantitative relationships between input and output parameters for these complex processes were successfully established by modeling with backpropagation neural networks.

(2) The neural network models have been found to be smooth and have less errors than multiple regression models.

(3) Additional experiments confirmed the validity of neural network models.

(4) For orbital EDM the effect of current, orbital radius and radial step on MRR was found to be significant. The effect of current on Ra was also significant.

(5) For WEDM the effect of pulse duration on both outputs (MRR and Ra) were found to be significant. The pulse frequency and current were also significantly affect MRR.

(6) When multiple regression models were used for estimating the output parameters for orbital EDM and WEDM processes, they were found to have much more noise than the neural network models.

(7) Finally, we can conclude that the neural network can model very complex processes like EDM and WEDM, and is less sensitive to noise included in the data. These models predict accurate results and can be very useful on shop floors for operators in predicting the random process performance.

ACKNOWLEDGEMENT

Support from the National Science Foundation (Grant # DDM-9215298) and the State of Nebraska (Nebraska Research Initiative Fund) is gratefully acknowledged.

REFERENCES

[1] Konig, W., Wertheim, R., Zvirin, Y., and Toren, M., "Material Removal and Energy Distribution in Electrical Discharge Machining, " Annals of the CIRP Vol. 24/1, pp. 95-100, 1975.

[2] DiBitonto, D. D., Philip, T. E., Patel, M. R., and Barrufet, M. A., "Theoretical Models of the Electrical Discharge Machining Process. I. A Simple Cathode Erosion Model, " Journal of Applied Physics, 66 (9), November, pp. 4095-4103, 1989.

[3] Snoeys, R. and Van Dijck, F., "Physico-Mathematical Analysis of the EDM Process, "Proceedings of the North American Metal Working Research Conference- Mc Master Univ., Canada, pp. 181-199, 1973.

[4] Rajurkar, K. P. and Pandit, S. M., "Prediction of Metal Removal Rate and Surface Roughness in Electrical Discharge Machining," Transactions of NAMRI/SME, pp. 444-450, 1982.

[5] Indurkhya, G., Jain, V. K., and Rajurkar, K. P., "Experimental Investigations into EDD: A Response Surface Approach, " AMT III Conf., IQ90-241, Sept., 1990.

[6] Chen, F., "Back-Propagation Neural Networks for Nonlinear Self-Tuning Adaptive Control, " IEEE Control Systems, pp. 44-48, 1990.

[7] Nguyen, D. H. and Widrow, B., "Neural networks for Self-Learning Control Systems", IEEE Control Systems, pp. 18-23, 1990.

[8] Chryssolouries, G. and Guillot, M., "A Comparison of Statistical and AI Approaches to the Selection of Process Parameters in Intelligent Machining, " Journal of Engineering for Industry, Transaction of the ASME, Vol. 112, pp. 122-131, May, 1990.

[9] Okafor, A. C., Marcus, M., and Tipirneni, R., "Multiple Sensor Integration Via Neural Networks for Estimating Surface Roughness and Bore Tolerance in Circular End Milling,"Transactions of NAMRI/SME, pp. 128-136, 1990.

[10] Andersen, K., Cook, G. A., Gabor, K., and Ramaswamy, K., "Artificial Neural Networks Applied to Arc Welding Process Modeling and Control, " IEEE Transactions on Industry Applications, Vol. 26, No. 5, pp. 824-830, Sep./Oct., 1990.

[11] Indurkhya, G., and Rajurkar, K. P., "Artificial Neural Network Approach in Modeling of EDM Process, " Proceedings of the Artificial Neural Networks in Engineering (ANNIE '92) Conf., pp. 845-850, Nov., 1992.

[12] Staelens, F. and Kruth, J. P., "A Computer Integrated Machining Strategy for Planetary EDM, " Annals of the CIRP, Vol. 38, No. 1, pp. 187-190, 1989.

[13] Rajurkar, K. P., Williams, R. E., and Royo, G. F., "Process Control of EDM Gap Conditions Using Radio Frequency and Orbital Motion, " Japan-U.S.A. Symposium on Flexible Automation, A Pacific Rim Conference, Kyoto, Japan, pp. 763-769, 1990.

[14] Benedict, G. F., Non Traditional Manufacturing Processes, Marcel Dekker, Inc, New-York, 1987.

[15] Morita, A., et al., "Fuzzy Controller for EDM, " Proceedings of the International Symposium for Electro Machining (ISEM-9), Nagoya, Japan, pp. 240-243, 1989.

[16] Xiong, Y., Qu, S. and Mao, S., "Fuzzy Pattern Recognition and Fuzzy Control in EDM,"Proceedings of the International Symposium for Electro Machining (ISEM-9), Nagoya, Japan, pp. 357-360, 1989.

[17] Chen, K. and Xi, S., "The Stability Analysis and the active Fuzzy Control System of HS-WEDM, " Proceedings of the International Symposium for Electro Machining (ISEM-9),Nagoya, Japan, pp. 90-93, 1989.

[18] Pao, Y., Adaptive Pattern Recognition and Neural Networks, Addison-Wesley Publishing Company, Inc, Massachusetts, 1989.

[19] Rumelhart, D. E., Hinton, G. E., and Williams, R. J., Learning internal Representations by Error Propagation. In Rumelhart, D. E. and McClelland, J. L., Parallel Distributed Processing: Explorations in the Microstructure of Cognition, Vol. 1: Foundations, MIT Press, Cambridge, MA, pp. 318-362, 1986.

[20] Kishi, M., Suzuki, Y., and Araya, S., "Optimization of Multi-Stage Planetary Machining Parameters in NC Die Sinking EDM, " Proceedings of the International Symposium for Electro Machining (ISEM-9), Nagoya, Japan, pp. 34-37, 1989.

[21] Douglas C. Montgomery, Design and Analysis of Experiments, John Wiley & Sons.

[22] Scott, D., Boyina, S., and Rajurkar, K. P., "Analysis and optimization of parameter combinations in wire electrical discharge machining, " International Journal of Production Research, Vol. 29, No. 11, pp. 2189-2207, 1991.

9

A KNOWLEDGE-BASED EXPERT SYSTEM FOR SELECTION OF INDUSTRIAL ROBOTS

Shobha Shashikumar, University of Missouri
Rolla

Ali K. Kamrani, University of Michigan
Dearborn

9.1 INTRODUCTION

As research fields in AI accelerate and a greater number of experts are demanded by industry, Expert Systems play an important role in meeting the technological sophistication required in today's competitive world. Industries are demanding the assistance of human experts to solving complicated problems. However, there is a shortage of experts due to this demand. Expert Systems are rapidly becoming one of the major approaches to solve engineering and manufacturing problems. They have been implemented for several practical applications in many decision making problems. Expert Systems are helping major companies to diagnose processes in real time, schedule operations, maintain machinery and to design service and production facilities.

The area of robotics is a fertile one for the application of Expert Systems. Robots are an integral part of today's manufacturing environment. New tasks are being defined for robots in order to meet the challenges of Flexible Manufacturing Systems. Robots are entering every facet of manufacturing. Along with this growth there is an increasing variety of robots to choose from. One of the major problems facing the potential robot user will be his/her choice of an optimum robot for a particular task. Various parameters should be considered and the user should choose an industrial robot whose characteristics satisfy the requirements of the intended task.

This work will present a viable solution to the problem of selecting an optimum robot by building a Knowledge-Based Expert System using a LEVEL5 shell. LEVEL5 is an Expert System software created by Information Builders Inc., which runs on the IBM Personal Computer, XT and AT. The system will ask the user several questions regarding the usage and requirements of the desired robot. It uses the knowledge base and the rules to determine an optimum robot. If this analysis leads to more than one robot, then a test for economic feasibility of the suggested robots is performed and ranked. Based on this, the robot which is the most economical will be chosen.

9.2 AUTOMATION

Automation is a technology concerned with the application of mechanical, electronic, and computer-based systems to operate and control production. Industrial robots are major contributors to and factors in the automation technology. An industrial robot is a general purpose, programmable machine, possessing certain anthropomorphic characteristics. However, RIA, The Robot Institute of America, has defined a robot as ***"A re-programmable multi-functional manipulator designed to move material, parts, tools, or special devices through variable programmed motions for performance of a variety of tasks"***. According to this definition, robots can be classified as a programmable automation device, where the production equipment is designed with the capability to change the sequence of operations to accommodate different product configurations. Computers have had dramatic impact on the development of production automation technologies. Today, almost all modern production systems are computer-based or use a computer as the control and supervisory system [1].

There is no industry standard for the classification of robots. Although in general robots have been classified based on the basic configuration of the manipulator, which includes the body, arm and the wrist, they have also been categorized by their reach, the controller capability and its intelligence, load-carrying capacity, applications, drive systems, and the operation's ease and speed of programming. The direct impact of technology on robotics has introduced other means of classification for robotic systems. These may include on-board computers, on-board memory storage devices, and devices for outside interaction and interlocks which add to the robot's intelligence.

As manufacturing turns more toward automation in order to reduce costs and improve productivity, there will be more aggressive implementations of computers and Artificial Intelligence (AI) capabilities. Expert Systems (ES) have begun to prove their potential for solving important problems in engineering and manufacturing environments. Many applications of ES can be found in manufacturing areas. ES can provide assistance during the scheduling of activities in a job shop, detailed planning of machine operations and facility layouts, the monitoring of thousands of process variables, maintenance, and the fault diagnosis of equipment. The area of robotics is a fertile one for the application of ES. Some potential areas for using ES in robotics may include kinematic system design, robotic system selection, and robotics cell layout.

The real reason to introduce robots in a factory is to assign them a certain job and utilize them in a way that increases productivity and quality. The selection of a particular robot from among the wide variety of robots available should be well planned. The selection process may have to accommodate multiple conflicting objectives. An ES can provide the potential for solving these types of situations. This work describes the application of a knowledge-based expert system for the selection of industrial robots for manufacturing systems operations and automation.

9.3 INDUSTRIAL ROBOTS

9.3.1 Robot's manipulator and anatomy

The basic robot's manipulator consists of the body, arm, and wrist. An extension to the wrist of the manipulator is used for the orientation of the end-effector which consists of either grippers or tools. The volume of the area which these components will cover is known as the "work envelope," and it consists of the area in which the robot is capable of performing the required task. The shape of this area varies with the robot's configuration. This difference in shape is important in the selection of the appropriate robot for a given task. A description of an industrial robot's anatomy focuses on the physical structure of the robot, which includes the size of the robot, type of joints, and links. A joint in an industrial robot provides relative motion between two parts of the body. Each joint will create a degree of freedom for the robot and provide the robot with the capacity to move its end-effector to the desired position [2]. Robots are often classified according to the total number of degrees of freedom which they possess. The mechanical joints which are commercially available for robots fall into four categories.

Linear and Orthogonal Joints: The linear joint involves a sliding motion or translational motion of the connecting links. The axes of the input and output are parallel in linear joints. The motion of the axes is achieved by a telescoping mechanism or relative motion along a linear track or rail.

Rotational Joint: In a rotational joint, the axis of rotation is perpendicular to the axis of the two connecting links.

Twisting Joint: This type of joint requires a twisting motion between the input and the output links. The axis of rotation is parallel to the axes of both links.

Revolving Joint: In a revolving joint, the input link is parallel with the axis of rotation of the output link, and the output link is perpendicular to the axis of rotation.

The robot's wrist provides the end-effector the required orientation of the part and consists of three degrees of freedom. These are: *Roll* (Rotation of the wrist about the arm axis), *Pitch* (Up and down rotation of the wrist), and *Yaw* (Right and left rotation of the wrist).

The shape of the work space (work envelope) depends on the type of robot's configuration. Industrial robots are classified into five categories based on their physical configurations [3].

- Cartesian or Rectangular Configuration
- Articulated or Jointed Arm Configuration
- SCARA Configuration
- Cylindrical Configuration
- Spherical Configuration.

Cartesian or Rectangular Configuration: All the motions performed by this type of robot are along three linear and orthogonal axes. These robots are also known as pick-and-place robots. Due to their configuration, they have been used for assembly operations which require high accuracy for the positioning of components. The work volume of these robots describes a rectangular box within which work is performed.

Articulated or Jointed Arm Configuration: A series of revolving and rotary motions is used in order to move the manipulator in this type of robot. This robot's configuration is similar to the human arm and its motion. The work volume of robots with this configuration is irregular, which makes it possible for these robots to perform a variety of tasks. Due to the flexibility provided by this type of configuration, the manipulator can also reach over and under objects to perform tasks. Robots of this type are used for process operations such as coating and welding. In some cases, they have also been used for assembly and material handling.

SCARA (Selective Compliance Assembly Robot Arm) Configuration: This configuration is similar to the jointed arm with the exception that the shoulder and the elbow rotational axes are in the vertical position. This robot is used for insertion-type assembly operations in which the insertion is made from above.

Cylindrical Coordinates: These robots have two linear and orthogonal motions with a rotary base. This provides the robot with rapid motion for operations. The work envelope of this robot is cylindrical, and the maximum and minimum reach of the robot determines the maximum and minimum size of the work volume. This type of robot has been used for material handling operations mostly in the area of machine loading and unloading. They have also been used for applications such as assembly and pick and place.

Spherical Configuration: The work volume generated by this robot describes a sphere. These robots consist of two rotational axes and one linear axis. Robots of this type have the lowest weight and the shortest joint travel compared to the other forms of robot configuration. They have been used for material handling operations due to the large area which they can cover.

9.3.2 Drive systems

Joints of industrial robots are moved by actuators powered by a particular form of drive systems. The drive systems are of three types: a) pneumatic, b) electric, and c) hydraulic.

Pneumatic Drives: These are the simplest and cheapest of the drive systems used in robotics. Due to their lack of accuracy and repeatability, and lack of control within intermediate positions, pneumatic drives are used in the simplest robotics applications, such as pick-and-place operations. The advantages of using pneumatic drive systems are their low cost and low maintenance requirements.

Electric Drives: DC-Servo (Direct-Current) and stepper motors are the two types of electric drive systems used in robots. These drive systems can provide precise positioning, with high torque. These motors are reliable and clean. Due to the low cost associated with electrical devices, these drives will be the principal system used for the robotics of the future.

Stepper motors are manipulated by a series of electric pulses which control the number of revolutions made by the motor's shaft. This type of motor is not as sophisticated as the dc-servo motor and has limited performance capabilities. The advantages of electric drive systems are their high accuracy, repeatability, low cost, smaller size, and ease of maintenance.

Hydraulic Drives: These types of drive systems are designed for heavy duty tasks since they can provide high power which results in high strength and speed. Hydraulic drive systems were used in early robots. They are still in use, but they are criticized for their high noise level, large size, and high maintenance cost. After time, these drive systems tend to leak and the chemical compound in the fluid may be hazardous to the operator and the working environment.

9.3.3 Programming Techniques

In order for the robot to perform a desired task, information such as operating coordinates and conditions must be provided. This information must be taught by a programmer and recorded by the controller. Most robots are equipped with a teaching module which assists the programmer during this task. Two methods of programming are used.

On-line Programming Approach: The on-line approach, also known as teaching by showing, is a process in which the robot is manually manipulated by the operator through the desired task while recording the coordinates and conditions. Teaching using this approach will require the programmer to take the robot off the production line, which results in loss of production time. The on-line programming technique is further divided into manual and lead-through programming.

In the manual technique, the manipulator is maneuvered through the desired points using a teach box, also known as teach pendent. A series of buttons and joysticks is used to control the robot's motion and to direct the end-effector through the desired path.

Lead-through programming requires the operator to physically grab the robot's arm and maneuver the robot through the desired path. In this form of teaching, the path of the required task is being recorded by the controller. This method of programming has been used for applications such as spray painting and welding, where the operator has the knowledge and the experience of the proper path and operating conditions. The speed of operation is controlled during the teaching phase in order to provide the required safety for the operator.

Off-line Programming Approach: In this approach, the program is written and developed without the presence of the robot. Programs are developed while the robot is still in operation, therefore the down-time of the robot is reduced. This method of programming is beginning to receive the attention of system designers and programmers, due to the increased complexity of the robot's controllers and tasks.

9.3.4 The Controller Unit

The main function of the controller unit is to direct the manipulator and the end-effector through a defined position, while maintaining the required orientation and speed. This unit is capable of controlling the robot's motion either in point-to-point or point-to-point with continuous path [4].

The trajectory of the robot's path in the point-to-point method is defined and planned by the controller. The programmer cannot define the trajectory, although a set of intermediate points can be used to define an approximate path. Speed of operation relative to accuracy and repeatability is an advantage of using this form of control.

In point-to-point with continuous path control, the trajectory of the robot's motion can be defined by the operator. This technique is also known as walk-through continuous path, in which the robot emulates the motion of the operator's hand. This technique is required for tasks such as painting and arc-welding. The controller's path is achieved by recording while the operator is leading the end-effector through the desired path.

The advent of technology has enhanced the capability of the robots by providing new functions with which the operator can both teach the coordinates and define the trajectory of the motion. This method is known as the controlled-path, where a series of mathematical tools are provided to the robot's controller. In this situation, the operator defines a series of points, and the required path of motion is generated by the computer.

9.3.5 Precision of Movement

As a measure of performance for industrial robots, the precision of movement is used [5]. The precision of movement is classified and defined as the following functions.

Accuracy: Accuracy is defined as the robot's ability to position its wrist end at a desired target point within the work volume. It relates to the robot's capability to be programmed in order to achieve a given targeted point.

Repeatability: Repeatability of a robot is defined as the robot's ability to position its wrist or an end-effector attached to the wrist at a point in space that had previously been taught to the robot. Repeatability is usually considered more important than accuracy.

Load Capacity: Load carrying capacity of a robot is dependent on its physical size, configuration, construction, and drive system. This capacity ranges from 1 pound to several hundreds of pounds. Robots usually work with tools or grippers attached to their wrist, hence the net load carrying capacity of the robot is reduced by the weight of the gripper. The manufacturer's specification of this feature is the gross weight capacity.

Speed: The speed determines how quickly the robot can accomplish a given work cycle. The speed capabilities of current industrial robots range up to a maximum of 5 ft/sec.

Other factors which are considered important in robotic systems are their hardware capabilities which include:

Memory Capacity: Some robots have slow access time memory devices such as cassette tapes and disk drives. The advent of new technology has provided robotics systems with on board processors and high speed memories such as ROM, RAM, and EPROM. Use of high level languages and operating systems could provide very efficient utilization of available memory capacity.

Interface (Input/Output) Capacity: Interfaces are the robot's connections to the external world for integration and communication. Signals must be received from auxiliary equipment, computers, and sensors. Often, the robot is required to signal the completion of a task or motion, so other actions can take place. The major provision in I/O communication is the provision of correct interfaces to ensure the compatibility of signals and control information to and from the device attached to the system. A standard interface widely used in robotics is the

RS-232 line.

9.4 ROBOT APPLICATIONS AND THEIR CHARACTERISTICS

Industrial robots can be considered as substitutes for human labor under the following conditions [6]:

- ***Hazardous work environment for human operator***: Robots can substitute for the operator when the work environment is unsafe and hazardous for people. Hazardous work situations may include spray painting, welding, and forging.
- ***Repetitive work cycle***: Another situation which normally promotes the use of industrial robots is a repetitive work cycle. If the steps involved in the work cycle are simple, the robot is often more suitable for the task than is a human operator.
- ***Difficult handling***: Handling operation tools and heavy loads are other situations in which an industrial robot can be considered as a good substitute for the human operator.
- ***Multi-shift operation***: When a firm operates on multi-shifts, finding qualified operators is often a time-consuming task. An industrial robot can easily be used for these situations.

Robotic applications are classified into the three categories. These include material handling, processing operations, and other applications such as assembly and inspection.

9.4.1 Material Handling Operations

The material handling task can be divided into two categories: material transfer, and machine loading/unloading. Material transfer applications are defined as operations in which the primary objective is to move parts from one location to another location.

Pick-and-Place: In Pick-and-Place operations the robot picks up the part at one location and moves it to another. These operations are usually considered as the most straightforward robot applications.

Palletizing and related operations: In palletizing operations, the robot picks up individual cartons or other containers and places them on a pallet. The pallets are then handled mechanically within the plant using some form of material handling device.

Depalletizing: In this operation, the robot picks up cartons and places them onto a conveyor. The operation of inserting parts into cartons from a conveyor is very similar to palletizing, and the operation of removing parts from cartons is similar to depalletizing.

Stacking and unstacking: In this operation the objects, usually flat pieces such as metal sheets, are stacked on top of each other.

Machine loading: In machine loading applications, the robot loads the raw material into the machine, but the part will be ejected from the machine by some other means. An example of this situation is the press working operation.

Machine unloading: In this group, the machine is loaded without robot assistance, and the robot is used only for unloading the machine. Examples for this group may include die casting and plastic molding operations.

9.4.2 Processing Applications

In processing applications, robots perform some form of operation on the part. For this group of applications, robots are equipped with tools rather than grippers. The processing operations are divided into the following categories:

Spot welding: In this operation two sheet metal parts are fused together at localized points by passing a large electric current through the parts where the weld is to be made.

Continuous arc welding: Arc welding is a continuous welding process. In this form of welding, continuous weld joints are provided rather than the individual welds at specified contact points as in spot welding. The long welded joints made in this operation often form an airtight seal.

Spray coating: Spray coating is the most common form of robot application. The spray coating process makes use of a spray gun to apply the coating to the object. The fluid flow through the nozzle of the gun is applied over the surface of the object. The use of robots with such characteristics and features as continuous-path control, hydraulic drive system, manual leadthrough programming method, and multiple program storage is recommended for the spray-coating operations.

Besides spot welding, arc welding, and spray coating, other processing applications in which industrial robots can be utilized may include:

- **Drilling, routing, and other machining operations**
- **Grinding, polishing, and deburring operations**
- **Waterjet cutting**
- **Laser drilling, cutting, and welding**
- **Assembly & inspection operations.**

9.5 ECONOMIC JUSTIFICATION FOR ROBOTICS

To perform an economic analysis for an industrial robot, information must be gathered by engineers and managers. This information must include the type of project, the costs associated with installing the robot, and the project life cycle. The installation of robotic systems in a company can be of two cases: a) new application (this is the case in which no facility exists, and there is a need for a new facility. The use of industrial robots may be considered as one alternative approach to satisfy this need); and b) substituting the existing production method with one or several robots and automated systems (this is the case in which operations are performed manually and the use of robots is somehow considered to be desirable. The usual approach in this case is to demonstrate that the current method is costly and economically inefficient).

To perform the economic justification and analysis, the analyst should gather two types of cost data. These costs include investment costs and operating costs. A list of investment and operating costs which are encountered in robot installation projects is summarized in Tables 9-1 and 9-2 [3].

TABLE 9-1. LIST OF INVESTMENT COSTS IN A ROBOT PROJECT

Cost	Characteristics
Robot Purchase	*The basic price of the robot needed to perform the desired application (the end effector must be excluded from this price).*
Engineering	*The costs of planning and design by the user company's engineering staff to install the robot.*
Installation	*This includes the labor and materials needed to prepare the installation site.*
Special Tooling	*This includes the cost of the end effector, parts positioners, and other fixtures and tools required to operate the work cell.*
Miscellaneous	*This covers the additional investment costs not included by any of the above categories.*

TABLE 9-2. A LIST OF THE OPERATING COSTS IN A ROBOT PROJECT

Cost	Characteristics
Direct Labor	*The direct labor cost associated with the operation of the robot cell. Fringe benefits are usually included in the calculation of direct labor rate, but other overhead costs are excluded.*
Indirect Labor	*The indirect labor costs that can be directly allocated to the operation of the robot cell. These costs may include supervision, setup, programming, and other personnel costs not included in category 1.*
Maintenance	*This includes the indirect labor (maintenance crew), parts, and service calls. It is recommended that annual maintenance cost in the absence of better data should be estimated on the basis of 10 to 15% of the purchase price.*
Utilities	*The cost of utilities to operate the robot cell. This may include electricity and air pressure.*
Training	*The cost of employees' training may be included in investment costs table. It is often said that training should be a continuing activity, and hence, it shall be included in operation costs.*

Several methods are currently used in industry to justify and evaluate the economic feasibility of robot installation and application. These methods may include Net Present Value

Analysis (NPV), Internal Rate of Return (IRR), Benefit Cost Ratio (B/C), and the Payback Period (PP). From these methods, the payback period is still considered an effective approach and used by many industries to justify the feasibility of their projects. In this method, the length of time required for the net accumulated cash flow to equal the initial cost is determined. A project is considered acceptable if the payback period is equal to or less than some predetermined time set by the management. The following illustrates the formulation which can be used for the determination of the payback period for robotic projects:

$$PP = (TCRA - ITAXC) \div (DLCS + MCS - MC\&OC)(1-ITAXC) + (Dep * TR) \qquad (1)$$

where:

PP = Payback Period
TCRA = Total Cost of the Robot and Associated Accessories
ITAXC = Investment Tax Credit due to advance technology application
DLCS = Direct Labor Cost Saving annually
MCS = Material Cost Saving due to high quality products development
MC&OC = Maintenance and Operating Costs of robot annually
Dep = Annual Depreciation and *TR* = Tax Rate

9.6 ARTIFICIAL INTELLIGENCE

Artificial Intelligence (AI) is no longer simply an emerging technology that is *in* one day and *out* the other. It is, in fact, a reality which not only has the ability to provide a major impact on industry, but is doing so on a daily basis. As AI assumes characteristics of intelligence which involve reasoning, understanding, learning, and deduction, it also assumes the ability to replace the human decision maker. This aids in realizing levels of automation and CIM as well as providing solutions to specific tasks and problems.

There are many definitions of AI, but one which has been commonly accepted is that "*AI deals with programming computers to carry out tasks that would require intelligence if carried out by humans.*" Another definition of AI could be that it deals with tasks which are considered to require knowledge, perception, reasoning, learning, understanding, and other cognitive abilities. The goal of AI is a qualitative expansion of computer capabilities, and regardless of its precise meaning, it has been agreed that AI's applications and its usefulness to industry have a definite and growing impact.

Programs which characterize AI are considered mostly symbolic processes that involve complexity, uncertainty, and ambiguity. On the other hand, conventional programming consists of algorithmic solutions that are primarily numeric. AI deals with more qualitative issues that are typical of human problem solving and decision making. In AI architecture a heuristic search approach is taken in order to arrive at the correct solution, and the solution steps are implicit; this is due to the large number of solution paths. However, in conventional programming, the solution steps are explicit.

One important element of AI is its ability for heuristic search which encompasses the idea that once a decision has been made, the situation has changed, thus giving rise to new opportunities for further decision analysis and making. This process is modeled using decision trees as representing the decision making process with an initial condition and a subsequent branch for every decision thereafter. By further analyzing this chart, it is apparent that as one continues down the various branches of the tree, the number of decision possibilities increases greatly. For problems that require many solution steps, the number of branches can be enormous.

The problem solving task is a set of actions taken in order to achieve a goal. The elements of a problem solver are the initial situation, a goal or a desired situation, a control strategy or generalized actions that can be used for changing situations, and a control strategy applied to the procedure for achieving the desired goal.

Another important element of AI is knowledge representation. Researchers have determined that intelligent behavior depends more on the knowledge one has to reason with, rather than the method of reasoning. Therefore, methods to model knowledge efficiently are important. One method which has been practiced in developing and representing the knowledge is use of the production rules which provide a simplified modular technique to represent knowledge. Production rules have been considered as the basis for expert systems. The entire production system consists of the production rules, a global database which represents the system status, and a rule interpreter or control structure for choosing the rules to execute. The production rules consist of the domain facts as well as the heuristic for search. Knowledge is critical to high-performance intelligent systems. Two methods of knowledge representation are declarative or object-oriented (fact and representation) and procedural (actions).

9.7 KNOWLEDGE-BASED EXPERT SYSTEMS

In the manufacturing environment, the need for expert systems arises due to the limitations offered by conventional programming and decision making. Even though computers are a power source of information, they cannot provide human-like expertise. Expert systems offer an environment where the right and proper capabilities of humans and the power of computers can be incorporated to overcome many limitations offered by conventional decision making tools. The major advantage of choosing expert systems lies in the fact that the manipulation of knowledge is possible using the expert systems, whereas the manipulation of data is possible with conventional programming.

Expert systems derive their name from the fact that the system contains the expertise required to solve specific, domain related problems. The role of the program is to act as an intelligent consultant in the field of interest, capturing the knowledge of one or more experts. The non-experts can then interface with the expert to answer questions, solve problems, and make the required decisions. The major strength lies in the fact that the presence of the expert is not needed. Like all artificial intelligence software, expert systems are knowledge-base systems containing facts, data, and relationships that are applied to solve a problem. What distinguishes expert systems from other AI software is heuristic knowledge and real-world data. Expert systems are organized in four unique components: knowledge-base, inference engine, database and the user interface.

The knowledge-base is the core of the system. The knowledge-base is not to be confused with the database. This module is created by the knowledge engineer, who translates real knowledge into rules and strategies. Depending on the problem scenario, these rules and strategies will change accordingly. For expert systems, knowledge can be represented through problem solving rules, facts, predicate calculus, lists, frames, scripts, semantic nets, or intuitions that a human expert might use in solving a given problem domain. Through considerable experience, it has been determined that production rules are the most effective method used for representing knowledge for the expert systems. The IF-THEN rule format facilitates development of the rules in the knowledge-base and in turn creates an impressive knowledge-base quickly. Moreover, by applying production rules, modification and additions can be easily done. Once all the rules have been developed, they are stored in the computer memory, and the knowledge-base becomes the search space in which the inference engine works.

The inference engine is a software that implements a search and pattern-matching operation. The inference engine is also known as the rule interpreter due to its operation which behaves similarly to a software interpreter. However, major differences exist. A language interpreter looks at each line of code in a program and then implements the operations specified, whereas the rule interpreter examines the rule in a specified sequence looking for matches to the initial and current conditions given in the database. As rules matching these conditions are discovered, the rules are executed, in turn initiating the actions specified by the rules.

As the execution of the rules continues, the rules will reference one another to form an inference chain. Each time a new rule is examined, it is checked against the current status of the problem solution stored in the database, new information is added, and the next rule is selected until a solution is reached. The inference engine also functions as a hypothesis testing station. When a hypothesis is given to the expert system, the inference engine first checks to see if the hypothesis is stored in the database. If it is, the hypothesis is considered to be proven fact and no further operation or action is required. However, if the hypothesis is not found, which is usually the case, it must be proven by inferencing. Two methods used by the rule interpreter to search for answers are forward chaining and backward chaining. Forward chaining starts with axioms and definitions and derives as many conclusions as possible. It is a data or fact driven approach. The rules in forward reasoning are of "IF condition THEN conclusion" format. Backward reasoning, however, starts with a goal and tries to accumulate enough facts to substantiate the goal. Therefore, it is a goal driven approach. The rules in backward reasoning are of the form "conclusion IF condition."

Another important part of an expert system is the database. The database is referred to as a global database due to its broad range of information about the current status of the problem being solved. Known facts are stored in the database initially. New facts are added to the database as the problem solving process proceeds. Other data such as the initial conditions are also stored here. The inference engine begins its search by matching the rules in the knowledge-base against the information in the database.

The final element of expert systems is the user interface. User interface is a software that allows the user to communicate with the system. With the use of questions and menus, the user interface allows the user to enter the initial information in the database. In general, the user interface serves as a communication channel between the user and the system.

The problem of selecting an industrial robot using expert systems has been analyzed using a number of approaches. Some examples are:

> ***ROBOTEX - Choosing the Right Robot for the Job*** : This technique requires that the user have an expert knowledge of the manufacturing processes to be automated using robots. This is an interrogative system which requires no knowledge of robotics. This system asks the user a series of questions about the manufacturing environment regarding subjects such as temperature, access, power supply, safety, and then the specific process itself. The system will evaluate the feasibility of robot usage, and then make the selection [7].

TOWILL - A Production Engineering Approach to Robot Selection : This methodology presents the production engineering approach to robot selection with the emphasis on the role of production engineer as systems designer. The types of available robots and the robot time and motion study concept are provided along with the pitfalls which face the production engineers in their specifications. This system further recommends an approach for throughput time estimation [8].

ROSE - Robot Selection Expert : This method requires the user to construct the environment in which the robot will be operating and conform to it by using 3-D modeling techniques. The user will choose from a menu and place the various objects with which the robot will have to interface. The system will query the user as to the characteristics of the desired robot and the expert systems will choose an optimum robot from the available choices in the database [9].

ROBOSPEC : This expert system is developed in OPS5 and designed for the selection of industrial robots based on the production system architecture. The system asks the user to select the proposed implementation of the robot and then searches through the encoded knowledge and provides the user with a broad outline of the specifications which a particular robot requires to perform a specific task. The system carries out the analysis and then provides the desired robot, based on the specifications set by the user [10].

SPECIFICATION & ROBOT SELECTION : This model presents a methodology which addresses the robot selection problems by proposing a two-phase decision model that promotes the analysis and matching of robot technologies to tasks. An integer programming model and a rule-based expert system are developed for this task. The first phase (analysis) identifies the best set of robotics technologies that are required to perform the considered task. In the second phase (matching), a particular robot and accessories are chosen from a set of given options. The goal of this phase is to determine the best available robot to fulfill the required technologies [11].

In general, most of the developed systems are prototypes and have a limit of 25 robots. In contrast, the developed expert system has a database which supports information and specifications on 250 commercially available robots. The system development is discussed in the next section.

9.8 A KNOWLEDGE-BASED EXPERT SYSTEM FOR SELECTION OF INDUSTRIAL ROBOTS

A knowledge-based expert system is designed and developed to assist the expert user in the selection of an industrial robot for a particular task. This system takes into account the parameters specified by the user and performs the economic analysis, if required. Payback period formulation has been used to perform this analysis. The expert system is linked to the economic analysis software implemented for this task. The knowledge-based system is implemented using LEVEL5, a software developed by Information Builder Incorporated, which runs on an IBM/PS2 model 70. An overview of the system architecture is illustrated in Figure 9-1.

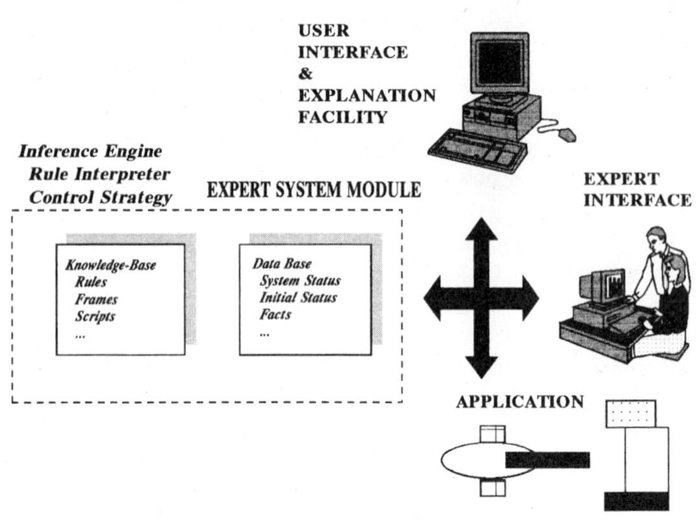

Figure 9-1. KBESR Overview

The system architecture is as follows.

9.8.1 Knowledge acquisition

The first step in the process of creating a knowledge base is to evaluate the issues involved in choosing an industrial robot. It is necessary to consult an expert and to review the literature in order to accomplish this task. Much of the information for this work was obtained from "International Robotics Industry Directory" and "Robotics Engineering Handbook"[12,13]. As a result, the prototype expert system knowledge base consists of robots from companies all over the world. An important objective in data collection is ensuring that the information collected is accurate. When, for a given robot, a given piece of data could not be obtained, the robot was excluded from the relevant analysis.

9.8.2 Decision tree

The next step in creating a knowledge base is prioritizing the necessary criteria involved in choosing a robot. General information should be evaluated first, such as the type of application and load carrying capacity, followed by more specific information, such as the memory capacity and type of programming requirements. The order of criteria has been summarized into a decision tree. The parameter "Application" is the general criterion and thus it is placed at the top of the decision tree. Branching down in the tree will increase the number of specifications, which in turn leads to a series of results which corresponds to the applications with the required specifications. A parameters listing used in a decision tree for palletizing and stacking tasks is illustrated in Figure 9-2.

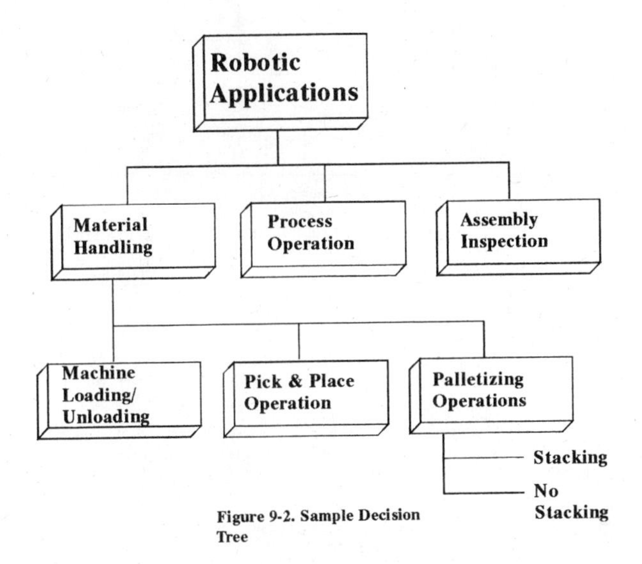

Figure 9-2. Sample Decision Tree

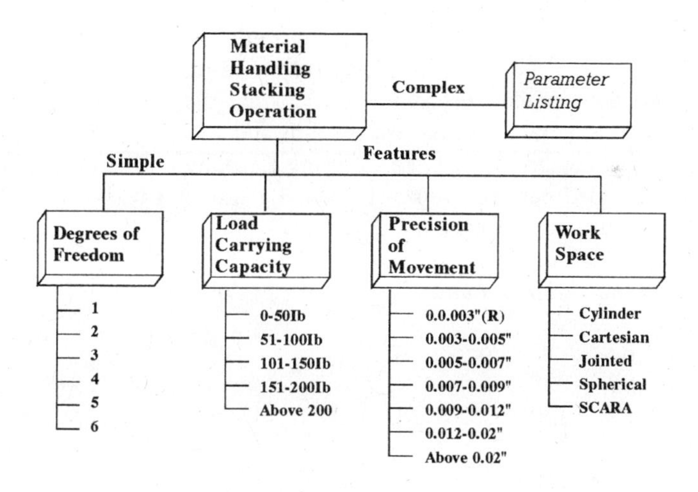

Figure 9-2. Sample Decision Tree (Continued)

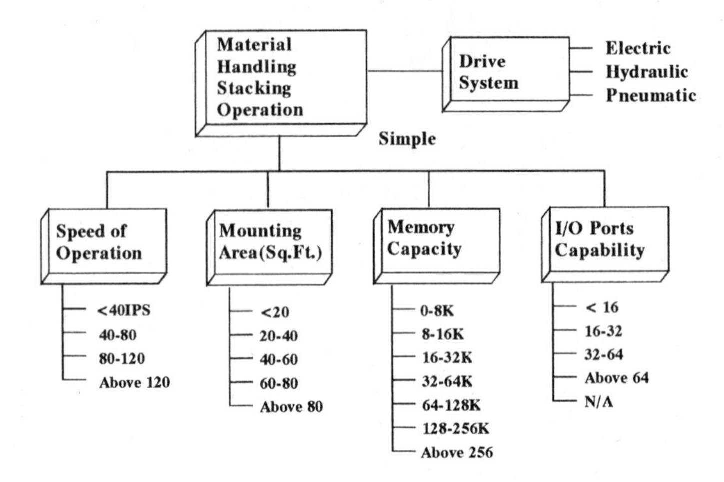

Figure 9-2. Sample Decision Tree (Continued)

9.8.3 Knowledge base development

The third step in creating a knowledge base is translating the hierarchy of information from the decision tree into a language that the expert system's inference engine understands. This task is accomplished through the development of rules and parameters. Accompanying each parameter are possible choices that correspond to a characteristic of an industrial robot. Thus, the expert system will ask the user a series of questions and will be provided with a selection of choices. An example is the parameter "Load Carrying Capacity." The system will ask the user to specify the required load carrying capacity, which requires a user response. The list of choices provided to the user is as follows:

Please Select the Desired Load Carrying Capacity

→ Below 50 Pounds
50 - 100
100 - 150
150 - 200
Above 100

Rules are lists of if-then-else statements that define the criteria that lead to the choosing of an industrial robot. Following directly from the ordering of the parameters contained in the rule, the user is prompted with the question that corresponds to the first parameter (in this case, application is the first parameter). Then, based on the user's response, the system scans all the rules and eliminates those rules that do not contain this response. The expert system then asks the question related to the next parameter from the remaining rules. This process terminates once all of the if-then-else conditions for the specific problem have been satisfied, and displays the recommended industrial robot to the user. In some cases, the analysis may not lead to the recommendation of a robot. In such a case, the user is informed of the blockage and is requested to provide a different value for the factor where the blockage occurred and thus the analysis continues. In other instances, the analysis recommends more than one robot that will satisfy the user needs. In this case, an economic feasibility analysis is performed to select the most feasible robot.

The parameters discussed in the overview of robot technology in this article are used by the expert system for the selection of industrial robots. For this work, performance data for 250 robots was obtained. The chart shown in Figure 9-3 illustrates the characteristic accuracy and repeatability exhibited by the population of industrial robots surveyed. Some 24% of the robots had repeatability ranges of 0 - 0.003 inch, and another 24% had repeatability ranges of 0.02 inch and above. A good percentage (18%) of the robots had repeatability ranges of 0.003 - 0.005 and 0.007 - 0.009 inch each.

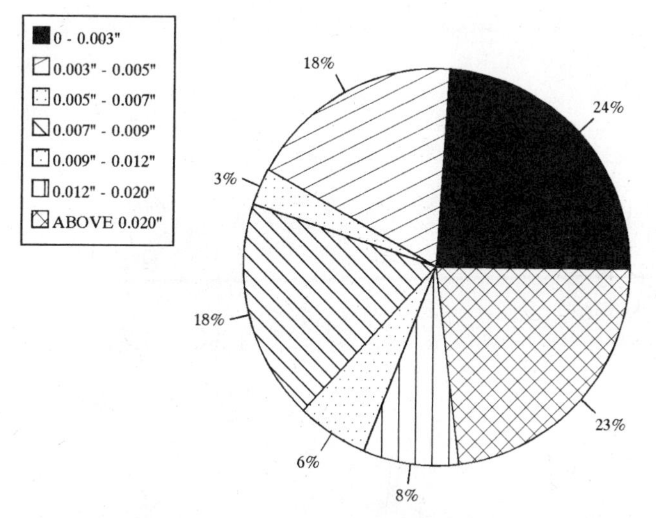

Figure 9-3. Accuracy & Repeatability Ranges for Surveyed Robots

The load carrying capacity of the robots examined varies between a few pounds to several hundred pounds. Figure 9-4 displays the maximum load capacities of the robots. Clearly the 0 - 50 pounds range of load carrying capacity was found to be the most popular, with approximately 64% of the robots falling in that range and approximately 12% of the robots falling in the 200 pounds and above range. From this analysis, shown in Figure 9-5, it was also seen that small robots requiring mounting areas in the range of 0 - 20 square feet, were found to predominate (64%), followed by 10% of the robots requiring an area in the range of 20 - 40 square feet. Figure 9-6 illustrates the analysis for the operation speed.

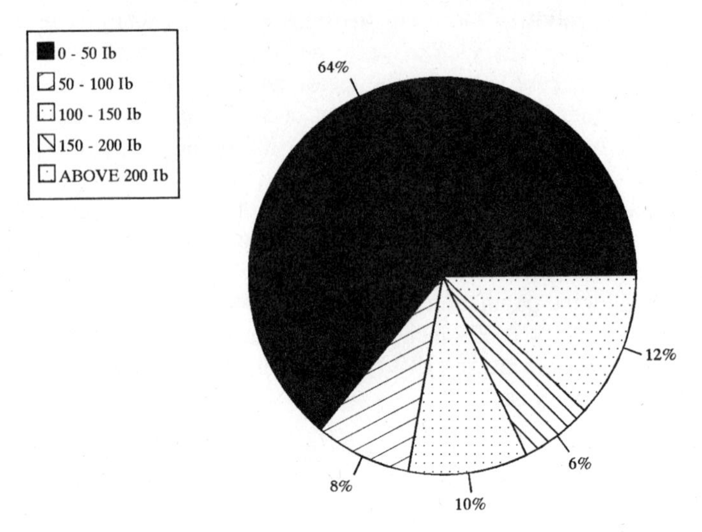

Figure 9-4. Load Carrying Capacity Ranges for Surveyed Robots

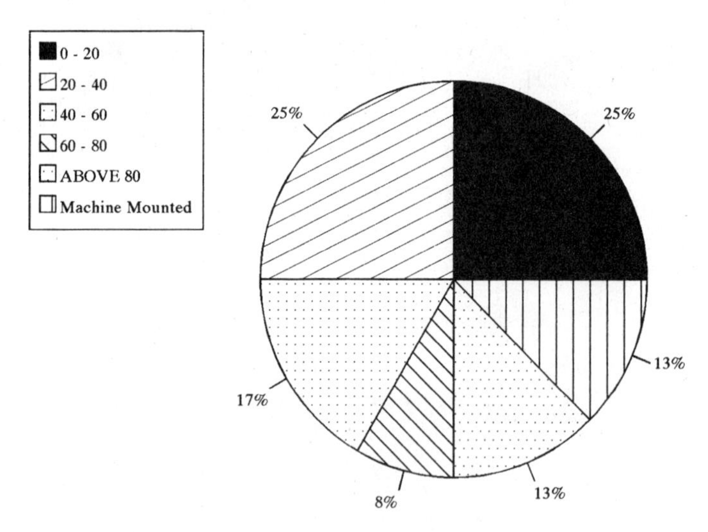

Figure 9-5. Mounting Area Ranges for Surveyed Robots (Square Inch)

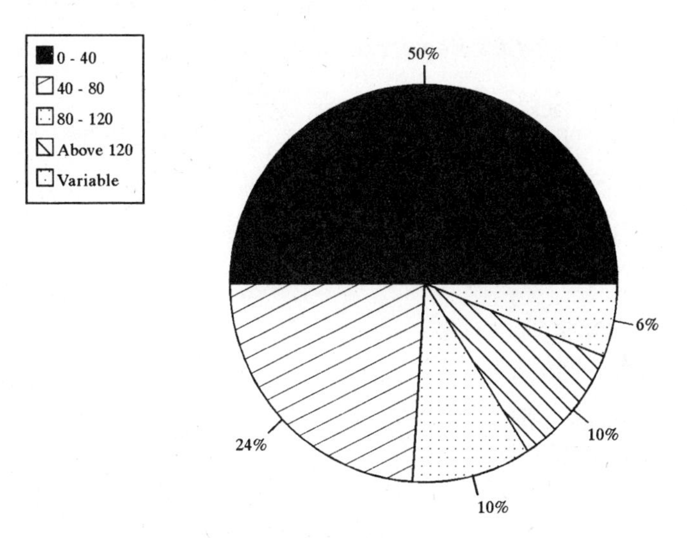

Figure 9-6. Speed Ranges for Surveyed Robots (Inch/Seconds)

9.8.4 Sample scenario

Once the user has decided on the type of application, the knowledge-based system can be used as an essential tool to assist in selecting the appropriate robot. For example, consider a situation in which the user intends the robot for the palletizing task. The following illustrates the sequence of questions and corresponding responses as they appear on the screen. The first screen is the required application.

Select the Type of Application

→ Material Handling
Processing Operation
Assembly
Inspection

The user selects the first option, material handling, and the system responds by asking the next question which addresses the specific job within the material handling application.

Select the Specific Job Under the Material Handling Application

Pick-&-Place
→ Palletizing
Machine Load/Unload

The user is then prompted with the specific characteristics associated with the palletizing task and the type of product which is to be palletized.

Select the Palletizing Task Requirements

No Stacking Required
→ Stacking Required

and

Select the Product Geometry Specification

→ Simple Geometry
Complex Geometry

At this stage, the following rule is fired, which is the indication of the minimum number of requirements for the degrees of freedom necessary.

{ Rule 1 }

IF	**Question appl \ Choice MH**
AND	**Question job \ Choice PAL**
AND	**Question stck \ Choice STACK**
AND	**Question part \ Choice Simple**
THEN	**min num of DOF is 4**

This result which is generated by this rule eliminates a number of rules from further analysis. The remaining parameters can assist the user in the selection of a specific robot. The user is prompted with the following set of questions which will collect the required specifications on the load-carrying capability and repeatability performance.

Select the Desired Load Carrying Capability

Below 50 Pounds
50 - 100
100 - 150
150 - 200
→ Above 200

and

Select the Desired Repeatability

Below 0.003
0.003 - 0.005
0.005 - 0.007
0.007 - 0.009
0.009 - 0.012
0.012 - 0.20

→ Above 0.020

The following rule is used to capture these specifications:

{ Rule 21 }

IF	**min num of DOF is 4**
OR	**min num of DOF is 5**
OR	**min num of DOF is 6**
AND	**Question load \ Choice VERY HIGH**
AND	**Question repy \ Choice VLOW**
THEN	**we have ROBOTS4**

ROBOTS4 is used as a temporary parameter which illustrates the result of the fired rule, based on the set of specifications determined by the designer.

The next stage is the selection of the work space configuration, speed, and the required mounting area. The user is prompted with a list of options which illustrates these specifications, and the result from the selected options is the fired rule which also uses a temporary parameter.

Select the Work Space Configuration

Cartesian
Articulated
Spherical
→ Cylindrical
SCARA

Select the Desired Speed Range

→ Below 40
40 - 80
80 - 120
Above 120

and

Select the Desired Mounting Area

Below 20
20 - 40
→ 40 - 60
60 - 80
Above 80

The following rule is used to capture these specifications.

{ Rule 49 }

```
IF      we have ROBOTS4
AND     Question space \ CYLIN
AND     Question speed \ SLOW
AND     Question area  \ SMALL
THEN    recommended ROBOTS 4a
```

The system continues with the list of questions regarding memory capacity, input/output capabilities, and programming technique. The following rule illustrates this situation, which results in the identification of robots capable to perform the required application.

{ Rule 58 }

```
IF      recommended ROBOTS 4a
AND     Question memy  \ Choice a (0 - 8k)
AND     Question inout \ Choice 1 (0 - 16)
AND     Question prog  \ Choice tp
THEN    Robot is Found
AND     DISPLAY ROBOTS 4a
```

At this point, the expert system displays the robots model(s) and provides descriptions of their features. The specifications of the selected robots for this sample are illustrated in Appendix 9-A. The selected robots are models FB and FC by Prab Robotics Incorporated. These robots possess all the characteristics required by the user, but the price for the FB model is $80,000, while the price for the FC model is $125,000. This price contrast clearly indicates the advantage of using this system for robotic system implementation. Further economic analysis is possible using the software interface to the knowledge base.

The rest of the applications follow the same order of questions, with the exception that the criteria used for the selection of the robot are different. For example, in the above sample scenario, the drive system was not a deciding factor, but in a spray painting application, the drive system would be pivotal.

9.9 CONCLUSION

A viable solution to the problem of selecting an optimum robot by building a Knowledge-Based Expert System is presented. The system is designed to ask the user questions regarding the usage and requirements of a desired robot and then by using the knowledge base and knowledge rules provided, a solution for the optimum robot(s) is given. If this analysis leads to more than one robot, then a test for economic feasibility of the suggested robots is performed and the result is ranked. Based on the ranking, the robot which is the most economical will be selected.

REFERENCES

1. Kamrani, A.K., "**Computer Application in Manufacturing Environment : A**

Planning, Justification, & Implementation Handbook," University of Missouri-Rolla, 1993.

2. Amirouche, F.M., "**Computer Aided Design & Manufacturing**," Prentice Hall Publishing, New Jersey, 1993.
3. Groover, M.P., "**Automation, Production Systems & Computer Integrated Manufacturing**," Prentice Hall Publishing, New Jersey, 1987.
4. Fu, K.S.,R.C. Gonzales, and C.S.G. Lee, "**Robotics : Control, Sensing, Vision, & Intelligence**," McGraw Hill Publishing, New York, 1987.
5. Groover, M.P., et al, "**Industrial Robotics : Technology, Programming, & Applications**, McGraw Hill Publishing, New York, 1986.
6. Parsaei H.R., A.K. Kamrani, "**Design & Management of Automation Handbook**," University of Louisville, 1988.
7. Rogers, "**Choosing the Right Robot for the Job**," The Industrial Robot, 13 (1), 1986, pp. 33-34.
8. Towill, D.R., "**A Production Engineering Approach to Robot Selection**," OMEGA, 12 (3), 1984, pp. 261-272.
9. Mauceri, J.G., "**The Robot Selection Expert - ROSE**," Journal of Automated Reasoning, 1 (4), 1985, pp. 357-390.
10. McGlennon J.M., G. Cassaidy, and J. Brown, "**ROBOSPEC - A Prototype Expert System for Robot Selection**," A. Kusiak, Editor, Artificial Intelligence : Implications for Computer Integrated Manufacturing, IFS Publications, Kempston, U.K. and Spring Verlag, New York, pp. 505-515.
11. Fisher E.L. and O.Z. Maimon, "**Specifications & Robot Selection**," A. Kusiak, Editor, Artificial Intelligence : Implications for Computer Integrated Manufacturing, IFS Publications, Kempston, U.K. and Spring Verlag, New York, pp. 162-187.

APPENDIX 9-A: Sample Robots Specifications

MODEL : FB

Company	Prab
Control System	microcomputer
Coordinate System	cylindrical or rectilinear
Power	n/s
Weight	2,300
Number and type of axis	3-7
Resolution	0.012
Accuracy	n/s
Repeatability	0.050"
Load carrying capacity	600 lbs
Velocity range	0 - 36 ips
Velocity programmable	yes
Floor space required	48" * 48"
End effectors	mechanical, vacuum, magnetic

Sensors	As required
Synchronized operation	yes
Mass storage available	yes, cassette, RS 232
Standard input devices	computer, cassette, I/O
Standard memory size	128 programs / 7000 points
Memory devices	varies with control option
Number of steps or points	7000
Actuators available	servo-hydraulic or electric
Control inputs supported	hand held teach unit, cassette recorder
Languages supported	n/a
Language	n/a
Applications supported	spot welding, die casting, investment casting, forging machine tool load/unload, parts transfer, plastic molding, machining, palletizing, stacking/unstacking

Cost - $80,000

<u>MODEL - FC</u>

Company	Prab
Control System	Same as FB
Coordinate System	Same as FB
Power	220/440
Weight	2000
Number and type of axis	5
Resolution	0.012
Accuracy	n/s
Repeatability	0.080"
Load carrying capacity	2000 lbs
Velocity range	Same as FB
Velocity programmable	yes
Floor space required	60" * 60"
End effectors	Same as FB
Sensors	Same as FB
Synchronized operation	yes
Mass storage available	Same as FB
Standard input devices	Same as FB
Standard memory size	Same as FB
Memory devices	Same as FB
Number of steps or points	7000
Actuators available	Same as FB
Control inputs supported	Same as FB
Languages supported	n/a
Language	n/a
Applications supported	forging, investment casting, parts transfer, palletizing, welding, stacking/unstacking, spot welding

Cost - $125,000

10

A CASE-BASED KNOWLEDGE SYSTEM EMPLOYED TO TROUBLE SHOOT FAILURES IN A MANUFACTURING ENVIRONMENT

Edward C. Chung, Ohio University

C. A. Vassiliadis, Ohio University

10.1 INTRODUCTION

Troubleshooting newly installed computer equipment can be an extremely time-consuming and frustrating process. Contentions between different hardware settings and conflicts in memory usage by various software products often complicate the installation process thus causing unnecessary down-time. This is especially true for personal computer products that are intended to be field-upgraded or field-installed by end-users. It is the responsibility of the manufacturer to provide technical support to assist end-users in determining and resolving problems during installation. A competent technical staff is needed to handle the job but this approach requires a considerable amount of training and resource allocation. This work describes a case-based knowledge system called ShootDem-Ks which effectively troubleshoots product failures without the need of specially trained personnel. ShootDem-Ks is based on a problem-solving paradigm called Case-Based Reasoning[1] (CBR). CBR solves new problems by retrieving from its case library solutions that have solved similar problems in the past and then adapts these solutions to evaluate the new problems. The development of ShootDem-Ks is presented here with a comparison to its rule-based counterpart. The performance of ShootDem-Ks is evaluated and discussions on future potentials for this system are included.

End-users often find themselves stranded for hours after attempting to install a new piece of computer hardware or software. Even though the user's manual tries to be as comprehensive as possible, it is very difficult to anticipate all the different types of hardware and software that the end-users might have already installed in their systems and which can create adverse effects on the newly installed product. It becomes the responsibility of the manufacturer to provide technical support to assist end-users in determining the correct combinations of switch-settings on different peripheral boards. It is equally important for the technical staff to identify faulty products instead of blaming the customer's existing computer system being incompatible.

A capable technical staff that can provide prompt and accurate answers to end-users often wins customer satisfaction. This not only helps building the company's image but frequently increases sales through users referrals. Nevertheless a competent technical staff is hard to acquire and maintain. Training requires a considerable amount of time and as the product line grows, the demand on the technical staff being familiar with every detail of each product becomes overwhelming. Striving to alleviate the burden on the technical support staff, an expert system called ShootDem (which stands for troubleShoot moDem) was developed. This system was designed first as a rule-based expert system to assist a manufacturer to troubleshoot five different IBM personal computer (PC) add-on peripherals: four different types of data/fax modems and one video graphics adapter. ShootDem analyzed the information reported by troubled customers and made appropriate suggestions on solving the problem. With approximately two hundred and eighty rules stored in the knowledge base, the rule-based ShootDem was capable of locating the causes of product failures and providing accurate solutions within a few seconds.

As technology changes product designs must change to satisfy the consumer marketplace. Recently the manufacturer has added four new modems and dropped two old modems to make up a new line of seven data/fax modems. The old video adapter has also been replaced with three new high speed, true-color models. In order to keep up with the rapid changes in the product line, knowledge engineers must constantly revise and update the rules in ShootDem. Soon, the company discovered that ruled-based expert systems have their weaknesses. Four major drawbacks associated to a rule-based approach have been recognized. First, the knowledge engineer must painstakingly acquire the troubleshooting knowledge from the product design engineers. Not all knowledge is easily translated into rules. In fact the product design engineers often tell the knowledge engineers that they solve problems by past experience or even by instinct and they rarely follow any set of rules. This makes knowledge acquisition and rules formulation exceedingly difficult. Second, building and maintaining a rule base requires special programming skills. This means a special person must be hired and trained to do just this job; this increases the cost of the process. Third, maintenance is time-consuming even if all the necessary rules are available, therefore, rule-based expert systems cannot easily keep pace with changes. As the number of rules increases, the maintenance process becomes even more tedious. Last but not least, the execution speed of the rule-based expert system significantly reduces as the number of rules exceeds six hundred.

The shortcomings of rule-based expert systems have motivated the manufacturer to seek a more appropriate and intelligent methodology that can improve the performance and lower the maintenance time of ShootDem. The potentials of a relatively contemporary AI (Artificial Intelligence) approach called Cased-Based Reasoning (CBR) is being investigated first. Unlike traditional rule-based methodologies, CBR organizes knowledge in terms of examples of previous problems and their solutions. These previous examples are called cases. Case-based knowledge systems "remember" past cases and adapt old solutions to new problems. Old solutions can provide almost-correct answers and can warn of potential mistakes or failures. In other words, case-based knowledge systems can generalize when new problems do not match exactly to any past cases. Furthermore, new cases can be added into the system with ease to increase the case-based knowledge. The ability to increase knowledge by adding new cases to the system as it runs, makes a CBR system especially suitable for a dynamic business environment such as the computer and electronics industry. CBR systems not only reduce maintenance time significantly, but they also lessen the burden of training special personnel to maintain such systems. With successes in the early experiments with CBR methodologies, a new cased-based knowledge system called ShootDem-Ks (pronounced as ShootDem Case) is developed to replace the old rule-based version. This paper describes the development of ShootDem-Ks and compares this case-based system with its rule-based counterpart. Performance evaluation of ShootDem-Ks and future potential of the system will be further discussed.

10.2 A RULE-BASED EXPERT SYSTEM

The premise of expert systems, which are rule-based systems in practice, is that knowledge of a human expert can be embodied in a set of rules [2]. This premise was historically derived from Newell and Simon's pioneering work on the General Problem Solver (GPS). GPS was one of the earliest AI programs and came to realization in the DENDRAL project, one of the first rule-based expert systems [3]. In this belief, AI becomes an attempt to translate all the domain knowledge that can be found into rules that will be used to program an expert system. This belief is of great commercial value since the claim is that human expertise and knowledge are collections of rules that can be extracted, coded into a machine, and executed systematically. Thus the claim suggests that by programming more and more rules into machines, they can be made more and more intelligent.

Newell and Simon's idea of a general problem solver is also appealing because it implies that human thought depends upon a set of reasoning principles that are independent of any given domain of knowledge. Under this premise, humans reason the same way regardless what the problem domain is. Furthermore, a computer can be programmed to reason by giving it general principles and methods of applying those principles to any domain. This premise was also the motivation behind the development of the first ShootDem expert system, a ruled-based system to troubleshoot product failures encountered by end-users during installation.

ShootDem was initially designed to troubleshoot five PC peripherals including a video graphics adapter and four different models of data/fax modems. Since the PC's system bus design utilizes a rigid addressing and interrupt request scheme, add-on peripherals can easily occupy the same address location or interrupt request line that can cause hardware contentions. This leads to hardware malfunctions and often arouses the user's suspicion of poorly manufactured products. To further complicate matters device drivers that are needed to control special hardware, such as a hi-resolution video display driver, must be loaded into memory before the hardware becomes operational. Under the architecture of PC-DOS and MS-DOS, certain device drivers may need to be loaded into memory in a specific order. Neglecting the proper procedures for loading these device drivers can lead to memory conflicts, system halt, loss of data, or even damage of hardware.

With only five products to handle, ShootDem acquired approximately two hundred and eighty rules to implement. The general operation block diagram of ShootDem is given in Figure 10-1. The whole system was implemented with the Automated Reasoning Tool for Information Management (ART-IM) from Inference Corporation[1]. ART-IM is a collection of tools which includes mainly an interactive development environment, an inference engine based on the RETE algorithm[4], a debugger, and an enhanced CLIPS-like language[5] compiler. By employing the data-driven RETE algorithm, ART-IM rents itself to a subclass of rule-based systems known as production systems[6]. The principle advantage of production systems, as identified by Rychener of Carnegie-Mellon in 1976, is that they provide a repository for conditional knowledge where each rule's conditions remained independent of any other rules and required no effort on the part of a programmer to organize a body of conditional knowledge into a procedural flow chart. In other words a production system, such as ART-IM, allows conditional knowledge to be acquired and refined incrementally without reorganizing the entire collection of production rules every time new conditions are realized. An example of some production rules used by ShootDem to troubleshoot a fax/data modem is shown in Figure 10-2.

Being built on a data-driven production system, the burden of organizing conditional knowledge into procedural flow charts was removed completely from the programmers who developed ShootDem. Furthermore, the system directly supported the incremental refinement and evolution of the knowledge base by modifying old or asserting new rules. Nevertheless, knowledge acquisition remained the major hurdle in delivering a truly robust system. Knowledge acquisition is a long and tedious process, as shown in Figure 10-3, that continues until it elicits enough problem-solving knowledge to enable the knowledge-based system to achieve expert performance. This process is long and tedious not only because it involves carefully planned interviews between human experts and knowledge engineers, translating and encoding answers from the interview into production rules, but because the main issue relates to human reasoning. As noted by Riesbeck and Schank[1], human reasoning is more than deriving a set of first-principle rules and applying them to solve problems. In fact, most experts agree that their

---

[1]Automated Reasoning Tool for Information Management and ART-IM are trademarks of Inference Corporation.

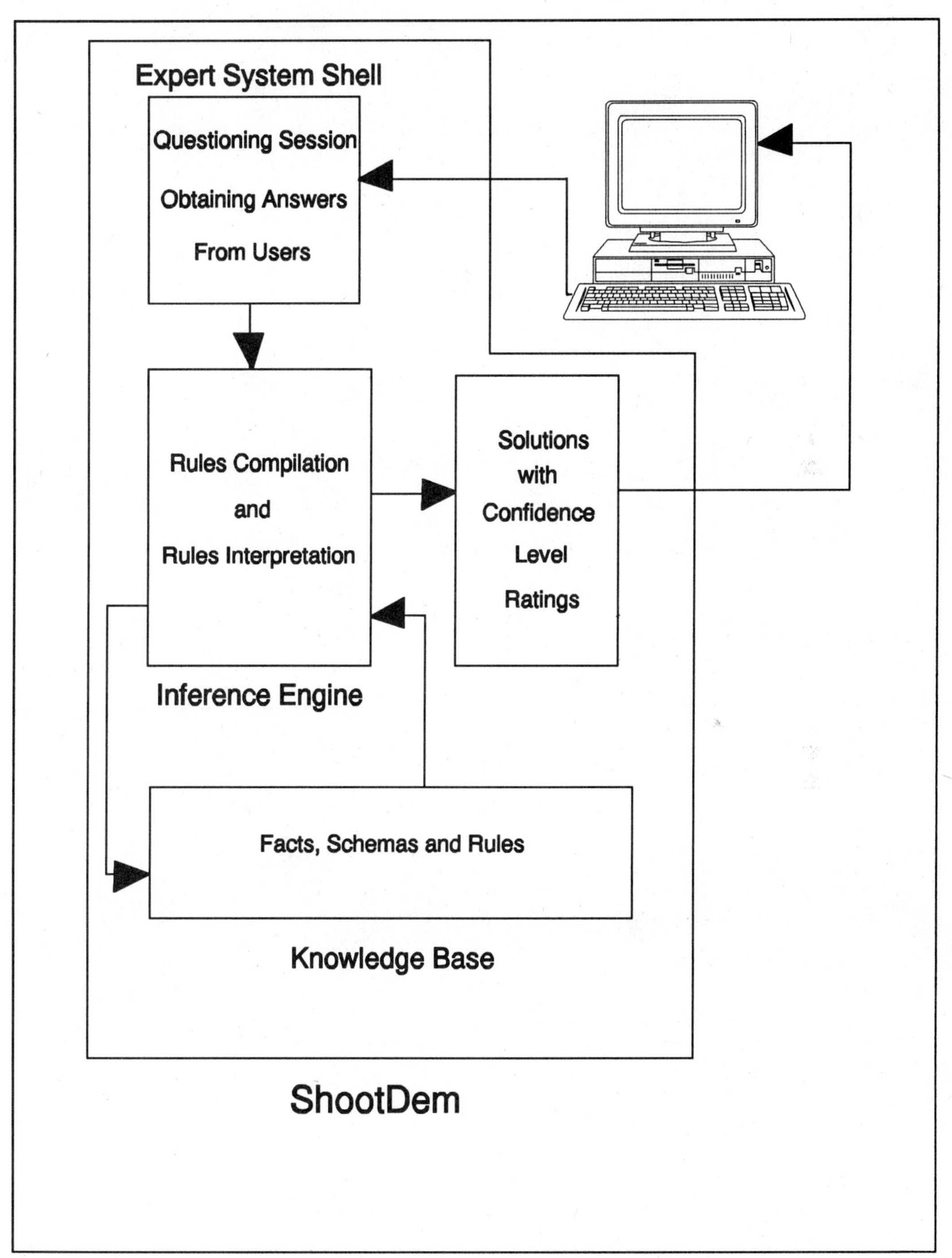

Figure 10-1 The general operation block diagram of rule-based expert system ShootDem.

```
(DEFRULE ASK4-TYPE
    (DECLARE
        (RULESET :DEFAULT))
    (LOGICAL
        (SCHEMA ASK-FOR
            (TYPE ?TYPE_USED)))
    =>
    (PRINTOUT T T "What product are you having problems with ?" T)
    (PRINTOUT T "       1 = 9624/AT Fax/Data Internal Modem     " T)
    (PRINTOUT T "       2 = 9696/MNP Fax/Data Internal Modem    " T)
    (PRINTOUT T "       3 = 9696/MNP Fax/Data External Modem    " T)
    (PRINTOUT T "       4 = 9696/MNP Fax/Data Pocket Modem      " T)
    (PRINTOUT T "       5 = 144/V.32 Fax/Data Internal Modem    " T)
    (PRINTOUT T "       6 = 144/V.32 Fax/Data External Modem    " T)
    (PRINTOUT T "       7 = 9696 Voice/Fax/Data Internal Modem " T)
    (PRINTOUT T "       8 = EISA Super VGA/True-Color Adapter   " T)
    (PRINTOUT T "       9 = ISA Super VGA/True-Color Adapter    " T)
    (PRINTOUT T "      10 = Hi-Color SVGA Windows Accelerator   " T)
    (PRINTOUT T "       0 = Return to Main Menu                 " T T)
    (PRINTOUT T "Select by the number in front of the product? ")
    (BIND ?PRODUCT_TYPE
        (STRING-TO-SYMBOL
            (STRING-UPPER-CASE
                (READ-LINE))))
    (WHILE
        (NOT
            (OR
                (= ?PRODUCT_TYPE 0)
                (= ?PRODUCT_TYPE 1)
                (= ?PRODUCT_TYPE 2)
                (= ?PRODUCT_TYPE 3)
                (= ?PRODUCT_TYPE 4)
                (= ?PRODUCT_TYPE 5)
                (= ?PRODUCT_TYPE 7)
                (= ?PRODUCT_TYPE 8)
                (= ?PRODUCT_TYPE 9)
                (= ?PRODUCT_TYPE 10))) DO
        (PRINTOUT T T "Invalid response! Choose from 0 to 10. Your choice =? ")
        (BIND ?PRODUCT_TYPE
            (READ-FROM-STRING
                (READ-LINE))))
    (MODIFY
        (SCHEMA ASK-FOR
            (PRODUCT_TYPE ?PRODUCT_TYPE))))


(DEFRULE IRQ-CONFLICT
    (DECLARE
        (RULESET :DEFAULT))
    (LOGICAL
        (SCHEMA ASK-FOR))
    (SCHEMA PRODUCT-TYPE
        (MODEL_CODE  ?MODEL))
    (SCHEMA INTERRUPT_SETTINGS
        (INTERRUPT-VALUE  ?IRQ-VALUE))
    (SCHEMA CONFLICTING-BOARD
        (CONFLICT ?CONFLICT-IRQ))
    ?*REPLACE_FACT* <- (SEQ ?INDEX  ?LOOKUP_IRQ_TABLE &:(= ?IRQ-VALUE ?CONFLICT-IRQ))
    =>
    (BIND ?*OLD_INDEX* ?INDEX)
    (RETRACT ?*REPLACE_FACT*)
    (ASSERT
        (INFORM CONFLICT-IRQ-EXIST)))


(DEFRULE I_O_ADDRESS-CONFLICT
    (DECLARE
        (RULESET :DEFAULT))
    (LOGICAL
        (SCHEMA ASK-FOR))
    (SCHEMA PRODUCT-TYPE
        (MODEL_CODE  ?MODEL))
    (SCHEMA INTERRUPT_SETTINGS
        (INTERRUPT-VALUE  ?I_O_ADDRESS-VALUE))
    (SCHEMA CONFLICTING-BOARD
        (CONFLICT ?CONFLICT-I_O_ADDRESS))
    ?*REPLACE_FACT* <- (SEQ ?INDEX  ?LOOKUP_I_O_TABLE &:(= ?I_O_ADDRESS-VALUE ?CONFLICT-I_O_ADDRESS))
    =>
    (BIND ?*OLD_INDEX* ?INDEX)
    (RETRACT ?*REPLACE_FACT*)
    (ASSERT
        (INFORM CONFLICT-I_O-EXIST)))
```

Figure 10-2 Some production rules from the expert system ShootDem.

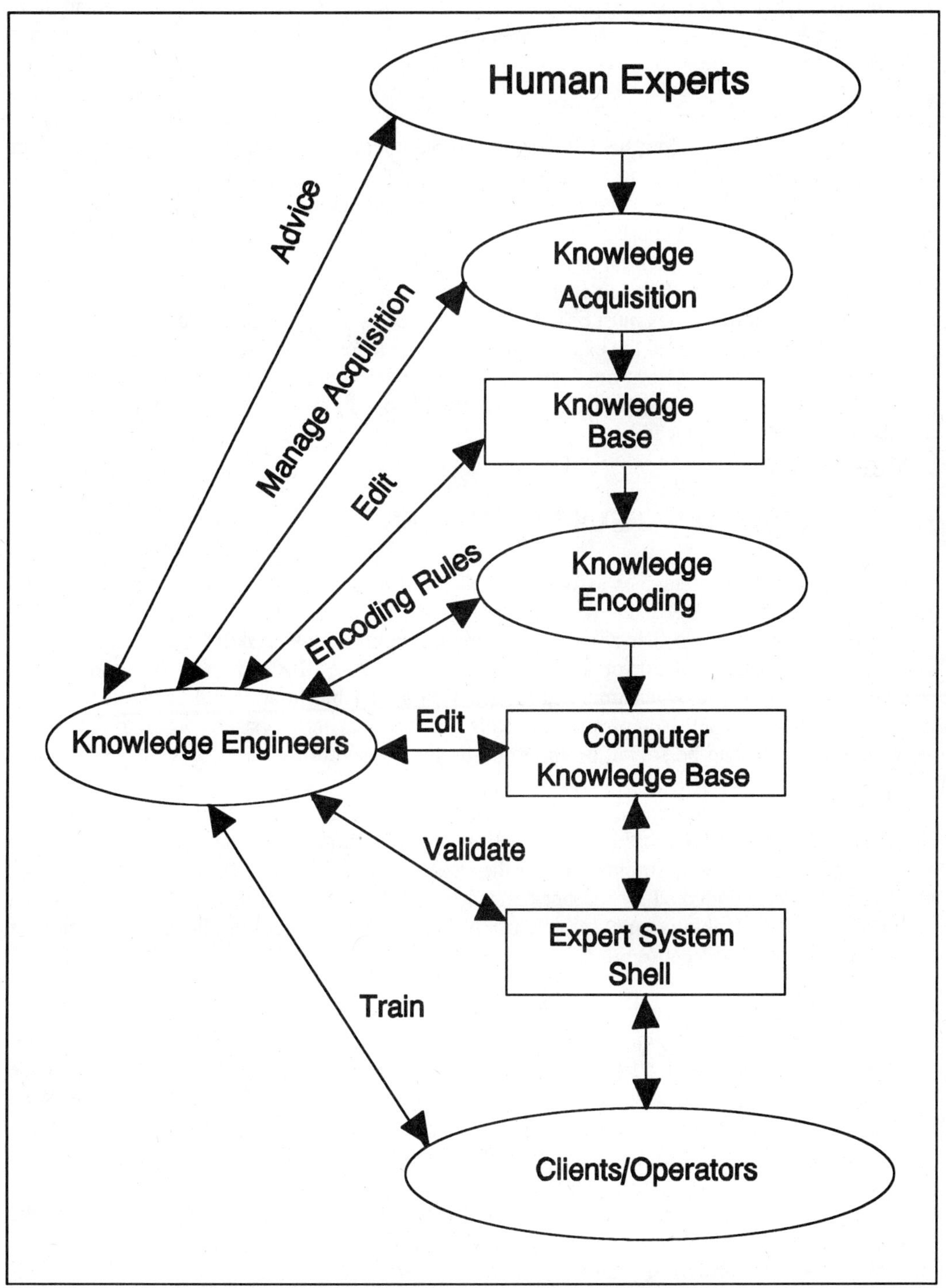

Figure 10-3 Knowledge engineers' roles in the process of knowledge acquisition.

experience is what makes them experts and experience is not something that can be formulated into rules. Certainly experts attempt to provide knowledge engineers with rules that they claim they use, but whether the experts actually use such rules when they reason is an entirely different question.

Knowledge acquisition, being the major problem with developing ShootDem, was not the only obstacle the project had to face. Specially trained knowledge engineers were needed to encode the acquired knowledge into rules. This requirement naturally meant higher development cost. The experience with ShootDem also indicated that maintenance of the rule-based ShootDem required extensive time for programming, debugging and testing. This problem was most evident when products were going through revision changes. To accommodate these changes, old rules must be refined and new rules must be added to keep the system up to date. This time-consuming maintenance process impaired the expert system's ability to keep pace with changes in a dynamic business environment. Last but not least, the knowledge engineers involved with this project pointed out that execution speed of the expert system was significantly reduced as the number of rules exceeded six hundred.

10.3 CASE-BASED VERSUS RULE-BASED

Realizing the deficiencies associated with rule-based expert systems, a manufacturer is compelled to seek a more efficient paradigm to cope with a dynamic business environment. Our goal is to develop a system that requires less time to maintain, lower skill level to operate, and most important of all, no effort for acquiring and encoding rules. Instead, the new system should be able to utilize previously accumulated information on troubleshooting old products to handle similar new designs. CBR is appropriate in satisfying these criteria because problems recur and when they do, they can be solved by the same or similar solutions.

CBR has been proposed by AI researchers, such as Riesbeck and Schank[1] and Hammond[7], as a more psychologically plausible model of human reasoning than the rule-based model. Figure 10-4 shows a summary of the theoretical comparison between CBR and rule-based reasoning that was conducted by Riesbeck and Schank[1]. Both researchers postulate that given a choice between thinking from scratch using first principles and recalling past events or experiences, people will choose the latter every time.

CBR organizes knowledge in terms of examples of problems encountered in the past and their solutions. These previous examples are called cases. Case-based knowledge systems "remember" past cases and adapt old solutions to new problems. Old solutions can provide almost-correct answers and can warn of potential mistakes or failures. In other words, case-based knowledge systems can generalize when new problems do not match exactly to any past cases. Furthermore, new cases can be added into the system with ease and increase the case-based knowledge. The ability to increase knowledge by adding new cases during the operation of the system makes a CBR system especially suitable for dynamic business environments such as the computer and electronics industry. CBR systems not only reduce maintenance time significantly, they also lessen the burden of training maintenance personnel.

Criterion	Rule-based Reasoning	Case-based Reasoning
Knowledge unit	Rule	Case
Granularity	Fine	Coarse
Knowledge acquisition Units	Rules in hierarchies	Cases in hierarchies
Explanation mechanism	Backtrack of fired rules	Precedent cases
Characteristic output	Answers with plus confidence measure	Answers with similarity scores plus precedent cases
Knowledge transfer across problems	■ High, if using backtracking ■ Low, if deterministic	Low
Execution speed as a function of knowledge base size.	■ Exponential, if using back-tracking ■ Linear, if deterministic	Logarithmic, if using a balanced index tree
Domain requirements	■ Domain vocabulary ■ Good set of inference rules ■ Rules apply sequentially ■ Domain mostly obeys rules	■ Domain vocabulary ■ Case library of past cases and their solutions ■ Manu exceptions to rules
Advantages	■ Flexible use of knowledge ■ Potentially optimal answers provided a good set of rules is employed	■ Rapid knowledge acquisition ■ Explanation by examples ■ Rapid response
Disadvantages	■ Computationally expensive ■ Long development time ■ Requires knowledge encoding into rules	■ Suboptimal solutions ■ Redundant knowledge base

Figure 10-4 Comparison of rule-based and case-based reasoning.

The process of CBR is depicted graphically in Figure 10-5 which is based on Riesbeck's and Schank's model[1]. As shown in the figure, the major components of a CBR system include a case library and a problem solver. The case library contains cases of past problems and their solutions. A case may describe the nature, circumstances, and causes of a problem. It may include complete, incomplete or no solutions for solving the problem. More details about the architecture of a case are given in the following section. The problem solver itself has two components: a case retriever and a case reasoner. Given a new problem, the case retriever searches the case library and identifies the most appropriate cases with the most matched features. These retrieved cases are then presented to the case reasoner. The effectiveness of the case retriever depends on the method of identifying the most appropriate neighbors. The simplest method to identify the similar cases is to use nearest-neighbor search[8]. However, this is computationally expensive especially when the size of the case library grows. To avoid an exhaustive case comparison while maintaining accuracy, cases in the case library are organized with indices. The retriever needs only to compare the indices with the new problem. Only those cases with the closest matching indices with the new problem are retrieved. Indices can be organized hierarchically, as discussed by Kolonder[9], to further improve the performance of the case retriever.

All retrieved cases are presented to the case reasoner for further examination. The case reasoner tries to solve the new problem by modifying the old solutions to conform to the new situation. Any modified solution is called a proposed solution. The proposed solution can be tested on whether it fits the new situation by the *Test* module inside the case reasoner. If the proposed solution succeeds, it is assigned with appropriate indices and then stored as a working solution for the presented problem. Otherwise, the *Explain* module tries to explain why the proposed solution failed and the repair module attempts to alter the solution to fit the new solution. After repairs are done, a new solution is returned to the *Test* module to conduct further testings until the solution is fit for the new situation.

Depending on the domain and the nature of the cases, CBR systems do not always need a case reasoner. A CBR system can retrieve almost-correct answers and present them to the human user. It is the human user who decides if any modifications and repairs are needed [10]. This function of "retrieving-and-suggesting" solutions is most suitable for customer support engineers to use it as a decision aid. During the development of the case-based ShootDem-Ks, it is discovered that a case reasoner is in fact not necessary for the problem domain that the system tries to manage. Furthermore, two major issues on making the system efficient have been identified: (1) a complete and compact case library, and (2) an effective indexing scheme. "Completeness" refers to the fact that after a reasonably short amount of time required to build the cases, the CBR system should be able to solve most of the problems in its domain without the need for additional cases to be added into the case library. "Compactness" refers to a small and manageable case library. An "effective indexing scheme" refers to the ability to match and retrieve similar cases and the successfulness of applying these cases to new problems.

The pilot system of ShootDem-Ks is developed with CBR Express[2] by Inference

[2] CBR Express is a trademark of Inference Corporation.

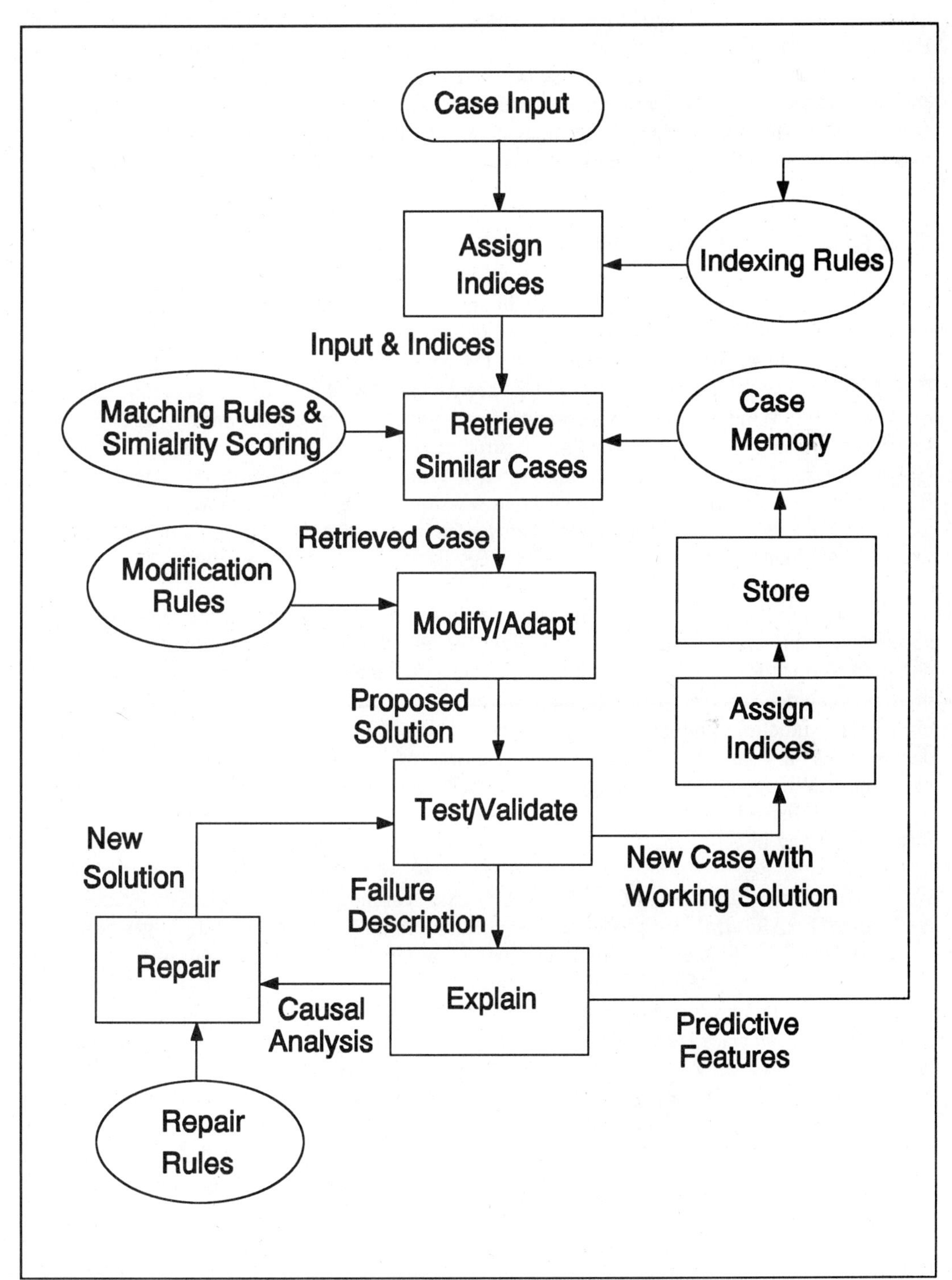

Figure 10-5 The process of case-based reasoning based on Riesbeck's and Schank's model.

Corporation. It is a knowledge base development tool that embodies the CBR methodology. CBR Express is a front-end expert system that is customized to handle diagnostics and expert knowledge query type of tasks. It is especially suitable for designing case-based systems in a customer service help desk environment, where customers confronted with technical problems will be calling for assistance. The general operation block diagram of the case-based system ShootDem-Ks is shown in Figure 10-6.

10.4 CBR EXPRESS BASIC ARCHITECTURE

CBR Express is an object-oriented implementation of, and procedural interface to, an associative retrieval mechanism for ART-IM objects[11]. The CBR interface is provided as a utility that can be combined with other ART-IM features to build case-based systems. Indeed, CBR Express is an ART-IM application in the area of case-based decision support that runs under Microsoft (MS) Windows 3.1[3]. It is composed of four sophisticated software packages operating together to produce a dynamic case based search environment. CBR Express is consisted of a Graphical User Interface written in ToolBook, a Windows application development software, from Asymetrix Corporation; multiple databases written in the Raima Data Manager, a database manager; and a case-matching search mechanism that is part of ART-IM. Some of the interactive screens, call "panels," that are provided by CBR Express are shown in Figure 10-7.

It is important to expose the user to a detailed description of the components of this tool, so that the capabilities of the system are fully understood. The three major components are designed to operate under MS Windows 3.1. ART-IM features a forward-chaining rule-based inference engine; an object-oriented data representation (schemas); a procedural language for user-defined functions and messages; a data-conversion utility for communication with external data sources; a complete procedural interface for integration with user-written C-language programs; a Windows 3.1 development environment; and kernel support for high-speed case base matching. All listed features are integrated into a single package. Thus ART-IM applications may make use of any or all of these features simultaneously.

The primary mission of CBR Express is case searching and/or matching. This function is supported by ART-IM's case base utilities. In addition, all of the artificial intelligence features of ART-IM are available for use in expanding or customizing a CBR Express developed application. The entire CBR Express System is based upon the concept of a client-server architecture. In this type of architecture, ART-IM operates as a server to the CBR ToolBook Graphical User Interface (GUI). The server is inactive until the user requests a search; a service provided by the server. At this point the server becomes active and proceeds to nominate a set of potential matches. This set of matching cases is returned to the ToolBook interface for display purposes. The Toolbook interface provides facilities to produce textual and graphical representations of the matching cases, solutions and their corresponding actions (even in animated form). Such a ToolBook developed application plays the role of the client which receives services from ART-IM.

---

[3] Windows 3.1, a Graphics User Interface (GUI)/operating system, is a trademark of Microsoft Corporation.

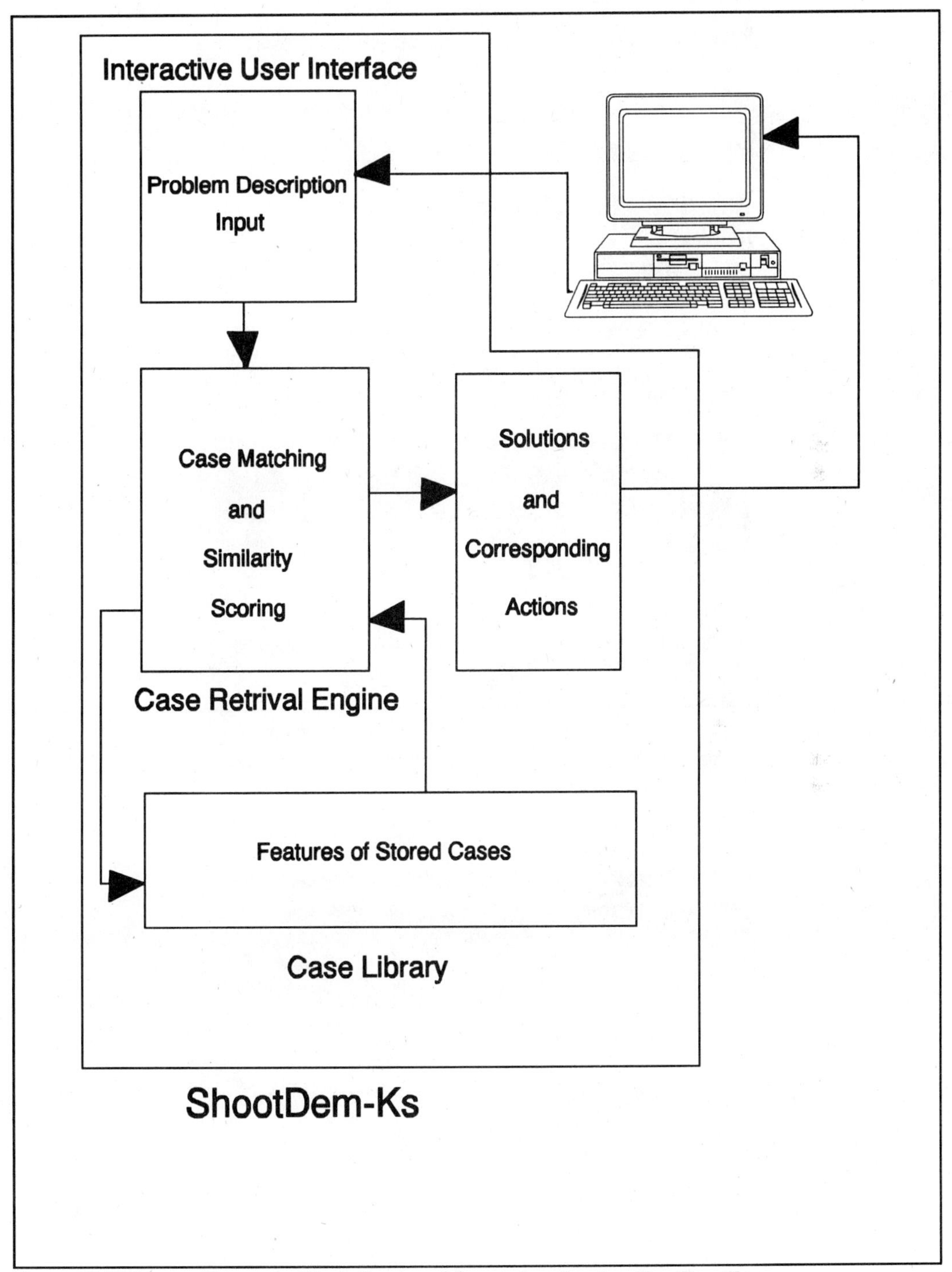

Figure 10-6 The general operation block diagram of the case-based system ShootDem-Ks.

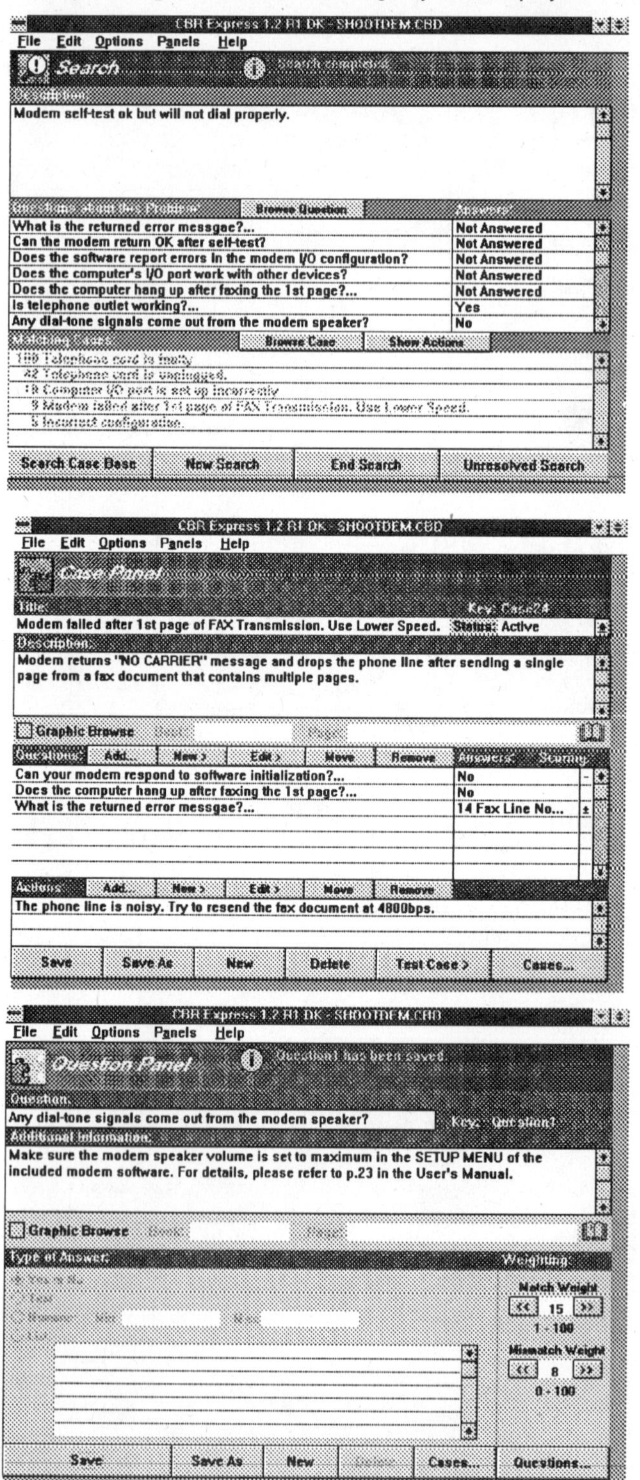

Figure 10-7 The *Search Panel*, *Case Panel* and *Question Panel* provided by CBR Express.

The default behavior of CBR Express is to run ART-IM as a "transparent" dynamic-link library (DLL), which is isolated from the user. However, some application programmers require direct access to ART-IM. This can be accomplished by executing the ART-IM/Windows development environment before running CBR Express. When CBR Express begins to execute, it will detect the presence of ART-IM and will not invoke the DLL version. This allows the programmer to directly access the internal functions of ART-IM.

10.4.1 ToolBook Interface

The CBR Express GUI is written in ToolBook from Asymetrix Corporation. Toolbook acts as an interface that manages the activities of both ART-IM and RDM. ToolBook can create graphical front-ends to many expert systems development kits such as ART-IM running under MS Windows 3.1. The CBR Express GUI consists of a title panel, five panels that form the public aspect of the interface, and a series of additional panels that serve to organize and store the operational scripts. The screen objects and scripts are organized to provide clean interfaces between modules that are user-configurable.

10.4.2 RDM Databases

RDM is a database development package that supports fields of unknown length in a record. This feature is essential for CBR Express, since it deals with open-ended descriptions of cases and answers that may run to any reasonable length.

CBR Express is designed to make it possible to use the ToolBook interface and ART-IM in conjunction with pre-existing data sources of interest to the individual client. To this end the ToolBook scripts that communicate between the ToolBook interface and RDM pass data through a small set of specific functions. By altering these few functions, the data may be redirected to other sources and sinks. These data sources and sinks will most likely be other relational databases and applications that execute CBR Express, by means of ToolBook pages or even text files as data repositories.

In a typical CBR Express session, the ToolBook interface maintains two opened RDM databases while ART-IM sometimes maintains a third one. ToolBook manages the databases for customer call tracking and for case-base records. The development version of ART-IM uses RDM to store its own internal knowledge base objects, such as the text of rules and schemas. RDM acts as a data server to both the ToolBook interface and to ART-IM. Communications between the various components of the CBR Express System is accomplished through the Windows dynamic-link libraries (DLL).

10.4.3 C Libraries as DLL's

In addition to ART-IM, ToolBook, and RDM, CBR Express also utilizes miscellaneous functions written in C and linked to ToolBook as DLLs or actually compiled into an executable form. In general, these C functions were written to supplement the inadequacies of ToolBook, where ToolBook scripts did not provide sufficient execution speed. In addition, recourse to C and the Windows API (Application Program Interface) to supply dialog boxes throughout the interface is also available.

10.5 THE FUNDAMENTAL ROLES OF CASE BASES

Unlike rule-based expert systems, case-based systems do not require a rigid structure in organizing accumulated knowledge. Building a case base is similar to a free-form documentation task rather than a programming job. Unfortunately for the developer, the above statement does not imply that the need for planning and organizing information can be ignored when designing case-based systems. On the contrary, careful planning allows the design of a compact set of questions and an efficient questioning strategy. As a result, more accurate solutions are retrieved after a minimal set of questions is answered.

Before constructing any CBR system, the relationship between the problem domain and the case base must be understood. The designed CBR system, ShootDem-Ks, concentrates on retrieving similar past cases followed by a conclusion on whether the retrieved cases are applicable to solve the new problem. The entire process involves two fundamental roles served by the case base: knowledge browsing and decision making.

To support knowledge browsing, a case base must be able to perform high recall searches. The ability to support high recall searching can be measured by the ratio between the number of relevant cases retrieved and the total number of relevant cases in the case base. In mathematical form, the recall factor can be expressed as:

$$\text{RECALL FACTOR} = \frac{\text{number of relevant cases retrieved}}{\text{number of all relevant cases in case base}} \qquad \text{(EQ.1)}$$

To support decision making, a case base must be able to perform high precision searches. The ability to support high precision searching can be measured by the ratio between the number of relevant cases retrieved and the number of all retrieved cases. In mathematical form, the precision factor can be expressed as:

$$\text{PRECISION FACTOR} = \frac{\text{number of relevant cases retrieved}}{\text{number of all retrieved cases}} \qquad \text{(EQ.2)}$$

It is important to visualize these two styles of searching as two distinct operations because they require different kinds of search behavior from CBR Express. The two fundamental case-base styles provided are:

1) **Inclusive** case bases, where the output of the system is a set of cases that are generally (or even weakly) similar to the search criteria. The operator of such a system is expected to browse the matching cases for information that contributes to a better view and understanding of the current situation. This should be the style chosen for designing systems that require high-recall searching as in the case of browsing knowledge.

2) **Exclusive** case bases, where the output of the system is expected to be a single case that is very similar, if not exactly equal, to the search criteria. This should be the style chosen for designing systems that require high-precision searching as in the case of decision making.

Both styles can be combined into a single case base. However, the weight of importance for each style must be balanced to achieve the desired behavior for the intended application. ShootDem-Ks, being a help desk application, is designed for operators who have only limited expertise on troubleshooting PC products, and must rely on the system to make a suggestion. In this environment high-precision searching is favored over high-recall searching.

10.6 CASES OF SHOOTDEM-KS

Speed is essential to ShootDem-Ks, and the rapid elimination of conflicting cases is a priority. The philosophy here is to employ a two-stage retrieval process: retrieval based on general or surface features regarding a failure, followed by a precision retrieval to pinpoint the most accurate solution.

Upon receiving a customer's call for technical assistance, the operator enters the description of the query or complaint into ShootDem-Ks Search Panel. In CBR Express, this original query for each case is called a "*case description.*" An initial search of the case base is automatically triggered in an effort to match the case description. If the search is inconclusive or imprecise, more questions are posed by the system. The search is refined iteratively as more questions are being answered. Ideally, questions should be used progressively to partition the population of retrieved cases into gradually smaller groups. Every question answered should lead to the separation of one group of cases from another. Any question that does not help to distinguish at least a few of the retrieved cases from the general population is redundant and useless. The best designed questions are the ones that can evenly partition up the case base into distinct groups of related cases from each other.

The strategy here is to represent the questions in a "space" where the questions themselves serve as coordinate axes and the cases may be separated along these axes into various subgroups. This concept is illustrated in Figure 10-8. All questions can be classified into two general groups: (1) Context questions, that separate groups of related cases from each other; and (2) Confirmation questions, that distinguish individual cases from a subgroup of similar cases. The case description initiated the first search in the case base which is then followed by context questions asked by the system. Context questions are needed to sharpen the focus of the search. To narrow down the search to a single solution, confirmation questions are later used. Confirmation questions serve to distinguish one neighboring case from another.

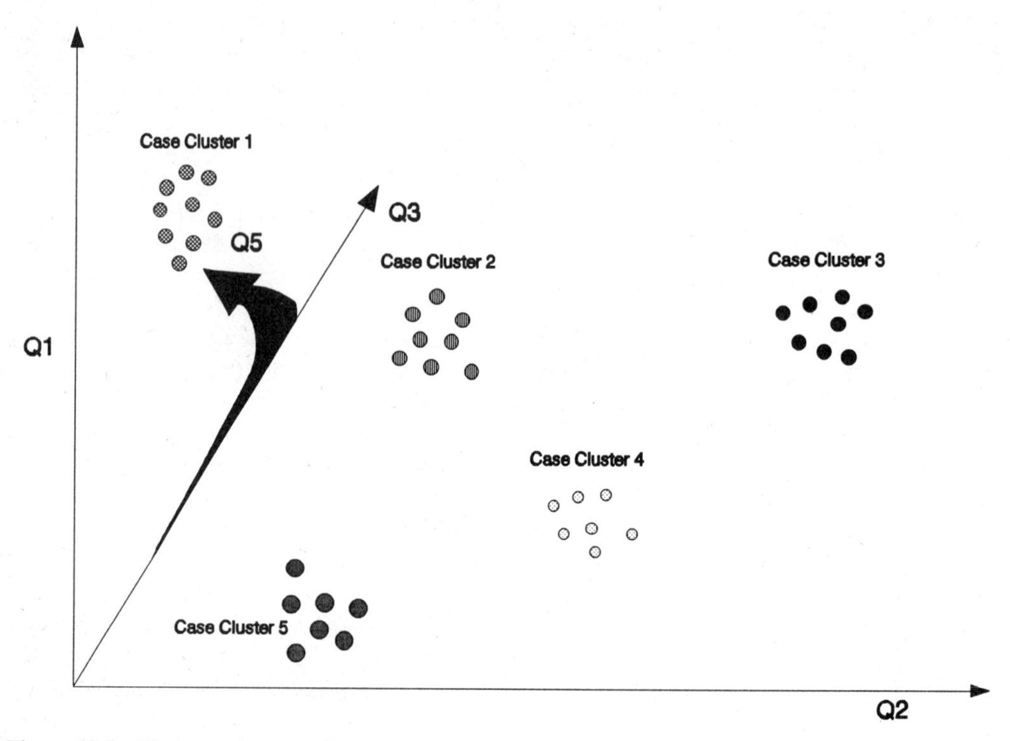

Figure 10-8a The "Question Space." Context questions (Q1, Q2 and Q3) steer the search to a cluster (Case Cluster 1) of similar cases.

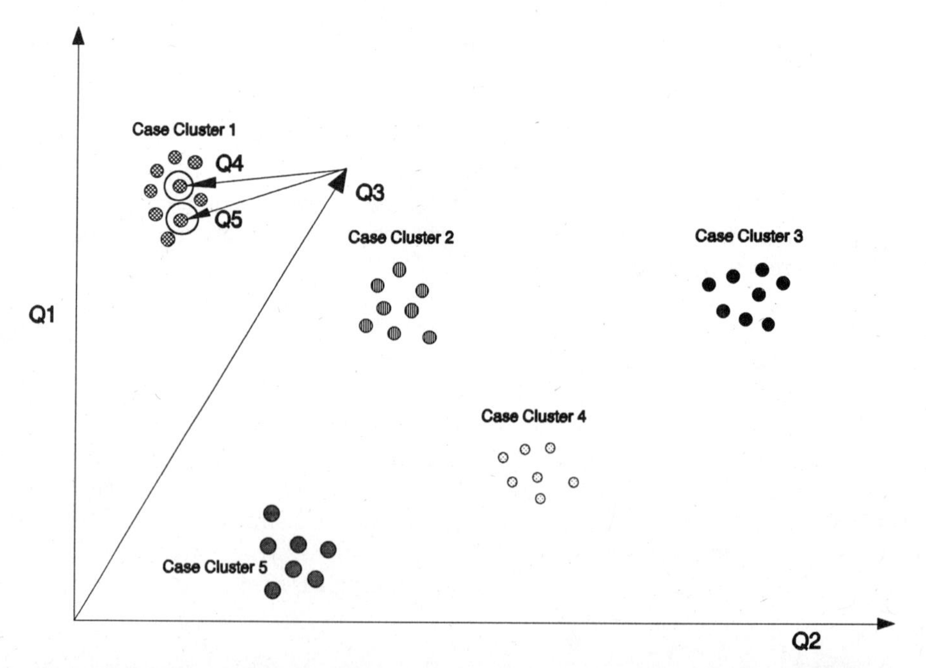

Figure 10-8b The "Question Space." Confirmation questions (Q4 and Q5) establish the relevance of a particular case. They serve to differentiate similar cases.

For example, the following case description contains several surface features of a possible hardware contention problem:

Modem failed to function with mouse device driver loaded.
Modem worked fine without mouse installed.

Since hardware contention may be caused by improper address selection or improper interrupt request (IRQ) line selection(or both), the system needs to ask the following context questions:

Does your mouse use the same IRQ level as your modem board ?
(refer to the Hardware Installation Guide for details on setting IRQ jumpers)

Does your mouse use the same COM port address as your modem board ?
(refer to the Hardware Installation Guide for details on setting port addresses)

Answers to this kind of questions can effectively divide the cases corresponding to the IRQ selection problem from the cases corresponding to the address selection problem. Suppose that the system identified that the problem is contributed to IRQ contentions. A confirmation question in this case may be:

Does you mouse use IRQ4?

The answer to this question is needed for the system to search for the optimum solution. Suppose there are three possible actions in the case base that correspond to the solution of this problem:

Action1 : *Reconfigure the modem board hardware to use IRQ3.*
Action2 : *Reconfigure the modem board hardware to use IRQ4.*
Action3 : *Disable the use of interrupt in the modem board hardware.*

If the mouse is currently using IRQ4, the proper action is to reconfigure the modem to use IRQ3. Action3 is less preferable since disabling the use of interrupts can degrade the performance of the modem.

Composing a case base using CBR Express is very similar to writing a hypertext document using natural-language. All case descriptions, context questions, confirmation questions and suggested actions are composed using plain English text. The system can also be configured to accommodate other languages such as Spanish or French. CBR Express is equipped with a sophisticated text-matching algorithm to compare the description of the new case in question with the descriptions already stored in the case base. This algorithm disregards most common words, such as collective nouns, articles and pronouns. CBR Express treats these words as "noise" text and keeps them in a list of words to be ignored. The algorithm concentrates on the more informative and substantive words in the descriptions of cases. It is worth noting that syntax and style used for composing these descriptions can affect the performance of the case base. A summary of the components that make up a case and the guidelines to compose each one effectively is given in Figure 10-9.

Components of a Case	Function	General Guidelines
Case Title	A one-line phrase that identifies the case	Include: ■ A domain code to help group similar cases. ■ The cause or the main symptom of the problem. ■ A brief description of the proposed solution. ■ The author's initial for identification purpose (optional).
Case Description	A critical paragraph of text that forms the basis for the initial search of the case library.	■ Generally summarizes the content of the most important questions and their answers. ■ The case title is automatically treated as part of the description. ■ Do not repeat information that has already been included in the case title. ■ Use the kind of language and terminology the end users used most. ■ Concentrate on standard key words and phrases.
Context Questions and Confirmation Questions	Context questions help the system to retrieve cases with similar features. Confirmation questions should distinguish individual cases from a subgroup of similar cases.	■ Each question should only ask for a single piece of information. ■ Define each question once and reuse it for other cases when applicable ■ Ask context questions before asking confirmation questions. ■ Combine "Yes/No" questions into List questions ■ Place questions in a logical and consistent order ■ Identify the context questions for the user ■ Use absolute scoring only where appropriate ■ Use numeric questions only when appropriate ■ Make sure the questions verify the case description ■ Review cases with fewer than four questions. These questions may be incomplete
Actions	An action is a solution to a particular case	■ Actions should be imperative and detailed ■ Supply additional information when possible for each action ■ Reuse existing actions where appropriate ■ Emphasize the differences between similar actions ■ Review cases that require more than four actions

Figure 10-9 A summary of the components that make up a case and the guidelines to compose each one effectively.

10.7 MATCHING ALGORITHMS

To perform case matching, the system analyzes a newly input case (also called a search case), consisting of data received from the Search Panel, and develops a numerical similarity score. A similarity score is assigned to every stored case and all scores will be compared with one another to determine the most appropriate case. The search case and every stored case are characterized by a set of features. Scoring is based on the number of matching features between the search case and the stored cases. Each set of features is consisted of a case description string together with numerical data, and a number of answered questions.

The scoring of each case is performed in two separate stages. As discussed before, the initial case search is triggered after the case description is entered into the Search Panel. This is the first stage where a score (the description-score) is assigned to the case description. The second scoring stage is a repetitive process that assigns a score (question-score) by comparing every question answered for the search case against the corresponding question of the stored cases.

The total similarity score of a stored case is the summation of the description-score and all question-scores. A total of 100 points may be distributed among the case descriptions and all necessary questions. ShootDem-Ks allocated 35% of the total similarity score for the case description. In other words if the user can give a perfect case description on how the product fails, a maximum of 35 points can be scored based solely on this description. The remaining 65 points are allocated as a composite of the matches of the remaining questions.

For questions that demand a "Yes/No" answer or an answer that must be selected from a predefined list, CBR Express performs a simple string match of the answer against the corresponding answers of the stored cases in the case base. These answers are treated as features of the search case. In general, if a search feature matches exactly a stored feature the raw question-score of the stored case is increased by the "match-weight" of the answered question. This weight is an integer scaling factor set by the system designer on the Question Panel. Questions that can help drive the system to a single solution are assigned with a larger match-weight. For features that may result in partial matches, the raw question-score of the stored case will be increased by some fraction of the feature's defined match-weight. This is the case for questions requiring numerical or text answers.

ShootDem-Ks also assigns mismatch weights (where a mismatch will decrease the stored case's raw question-score) to features that can quickly eliminate irrelevant cases and questions from the Search Panel when these features fail to match. For example, if the user enters the following description for a new problem:

Modem self-test ok but will not dial properly.

the system will then ask:

Any dial-tone signals come out from the modem speaker?

If the user responds with a "No" to this question, the system immediately knows that the feature "dial-tone presence" is mismatched. From this point onward, all questions that assume the dial-tone is present are irrelevant. The system may conclude that the user should check whether a telephone wire is properly inserted between the modem and the phone outlet. On another suggestion, the system may ask the user to check on whether the phone is operating at all.

The raw question-score is summed up for each stored case, and is then normalized over the 65 points that are assigned for the total question-score. The normalization process guarantees that the total similarity score lies in the range between 0 and 100. A total score of 100 points indicates a perfect match. A threshold level of 90 points is set for ShootDem-Ks in order to retrieve highly accurate matches. This behavior is consistent with the system's design philosophy of being an exclusive case base. Cases with scores lower than 90 points may be made up of combinations of match and mismatch weights, making it more difficult for the system to describe the significance of these cases.

10.7.1 String Feature Matching

The default algorithm for scoring "Yes/No" questions and "List-Selection" questions in CBR Express is string matching. The text value of a presented feature is compared to the text value of a feature in a stored case. An exact match between the two compared features will increase the raw question-score. No partial matching is performed here.

10.7.2 Word Feature Matching

The default algorithm for scoring text questions in CBR Express is word matching. Word matching is used for text features to better handle the kind of "fuzziness" that may be present in natural-language descriptions. All the words of the presented case's text feature is compared to all the words of every stored case's corresponding text feature . The raw question-score depends on the number of words both cases have in common and the total number of words in text feature of the presented case. For example, consider that a presented case has the following text feature value:

> "MODEM FAILED WITHOUT CARRIER SIGNAL AFTER SENDING THE FIRST FAX PAGE WHILE THREE MORE PAGES ARE NEEDED."

Also consider that a stored case has the following text feature value:

> "MODEM RETURNS NO CARRIER MESSAGE AND DROPS THE PHONE LINE AFTER SENDING A SINGLE PAGE FROM A FAX DOCUMENT THAT CONTAINS MULTIPLE PAGES."

The eight matching words are: MODEM, CARRIER, AFTER, SENDING, THE, FAX, PAGE and PAGES. This matching algorithm totally disregards the order of the words. The degree of partial matching is calculated based on the proportion of matching words and the total number of words in the presented case's feature. Using the above example, eight out of seventeen words are matched in the presented case. Consequently, $^{8}/_{17}$ times the match-weight of the feature is added to the raw question-score of the stored case.

10.7.3 Character Feature Matching

The default algorithm for scoring case descriptions in CBR Express is character matching. Character matching is similar to word matching, but trigrams are used as the basic matching unit rather than individual words. A trigram is a sequence of three consecutive characters excluding spaces and punctuation marks. The purpose of using trigrams in this matching process is to reduce the impact of misspellings in determining the description-score.

To ensure a more accurate match, character feature matching preprocessed the description string to remove as much "noise" text as possible. Commonly used suffixes such as -AL, -ING and -ED are being removed from the remaining words. The algorithm is intelligent enough to prevent excessively truncation from words such as DIAL, RING and RED. The preprocessor continues to substitute synonyms for predefined words, e.g., MALFUNCTION, PROBLEM and BREAKDOWN are all replaced by FAILURE. Next, punctuation marks are removed from the description string. To prevent case-sensitive distinctions, the entire string is converted into uppercase. Finally the preprocessed string is broken into consecutive trigrams. The trigrams of a stored case description are matched against the trigrams of the presented case description. Similar to the word matching algorithm, the order of the trigrams is immaterial. The degree of partial matching is based on the proportion of the matching trigrams and the total number of trigrams in the presented case. Using the modem example, a stored case description may look like:

"MODEM FAILURE."

The sixteen consecutive trigrams for this description are: '--M', '-MO', 'MOD', 'ODE', 'DEM', 'EM-', 'M--', '--F', '-FA', 'FAI', 'AIL', 'ILU', 'LUR', 'URE', 'RE-', and 'E--'. Suppose the user misspelled modem as "madem" and entered the search description as:

"MADEM FAILURE."

In this case, thirteen out of sixteen trigrams in the search description will be matched with those of the stored case. Nevertheless, the three mismatched trigrams: '-MA', 'MAD', and 'ADE' merely decrease $^{3}/_{16}$ of the total match-weight. The description-score will still be increased by $^{13}/_{16}$ times the match-weight of the case description. On the other hand, if the word matching algorithm is applied to match these two case descriptions, the description-score will only be increased by $^{1}/_{2}$ times the match-weight of the case description. This example demonstrates how character matching can reduce the adverse effect of misspelled words in case descriptions. In ShootDEM-Ks, the character matching algorithm is only applied on matching features in case descriptions. This is because character matching is computationally intensive, uses more system memory and it is more time consuming as compared to the other matching algorithms.

10.7.4 Number Feature Matching

The default algorithm for scoring numeric questions in CBR Express is number matching. Number matching is not simply utilized to evaluate whether the search numeric feature and the case numeric feature are equal or not. If such a strict interpretation is needed, a text question should be used instead of a numeric one. Number matching actually measures the similarity of two numeric features. The raw question-score is increased or decreased based on how near the stored case's numeric feature is to the presented case's numeric feature.

Number matching utilizes the distance between two numbers to determine the score of a feature. To perform this matching process, the range of values the numeric feature may take must be programmed into the system. This range is defined by a minimum allowable value and a maximum allowable value for the feature to be matched. The values must be either integer or floating-point numbers. The system then takes an interval of 10% of this range and uses it as an acceptance window and executes one of the three actions described below:

1) If the search numeric feature is identical to the corresponding stored case's numeric feature, the raw question-score is incremented by the full amount of the match-weight of the question.

2) If the net distance between the search numeric feature and the corresponding stored case's numeric feature is more than 10% apart, the raw question-score is decremented by the full amount of the mismatch-weight of the question.

3) If the search numeric feature is within 10% of the corresponding stored case's numeric feature, the raw question-score is increased in linear proportion to the proximity of the two numeric features.

The ability to use a linear proportion scoring in this kind of matching gives great flexibility to the system developer in balancing the score contribution of numeric features in a case. For instance, by increasing the mismatch-weight one can get a narrower acceptance window.

10.8 PERFORMANCE AND BENCHMARK

The performance of the case-based ShootDem-Ks system and its rule-based counterpart, ShootDem, is discussed in this section. Performance in this discussion refers to two separate sets of issues: 1) issues relating to implementing and maintaining the systems; and 2) issues relating to operating and executing the systems.

Time and cost are the factors used to measure the cost-effectiveness of the implementation of each system. Figure 10-10 compares ShootDem-Ks and ShootDem, depicting a breakdown of overall development cost and time into four disjoint components: software development tools, knowledge acquisition, knowledge encoding, and knowledge maintenance. The comparison is

	ShootDem	ShootDem-Ks
METHOD	Rule-based 632 rules	Case-Based 385 cases
DOMAIN	PC peripherals	PC peripherals
SUBDOMAIN	7 modems and 3 video cards	7 modems and 3 video cards
SUCCESS RATE:		
CONSISTENCY	100%	85%
ACCURACY	78%	89%
SOFTWARE TOOLS *(relative pdays)*	50	60
KNOWLEDGE *ACQUISITION TIME (pdays)*	385	50
ENCODING TIME (pdays)	985	95
MAINTENANCE TIME (pdays)	550	10
TOTAL COST (estimated pdays)	1970	215

Figure 10-10 Benchmark results of the two knowledge systems: the rule-based ShootDem and the case-based ShootDem-Ks.

based on the number of person-days (pdays) required for the entire job. Pday is the unit used to measure the amount of time a person needs to spend on a job. This unit can be translated into a dollar amount provided that the level of expertise of the person(s) working on the job is known. It is assumed that the average pay of a typical engineering staff is $23 per hour. In other words one pday for a typical engineer is worth $184, using 8 hours per day for a standard work day. There is no intention here to present an accurate dollar amount of the cost measurements. All numerical values are shown as relative comparisons between the two different systems.

The results indicated that the cost of developing ShootDem was 9.2 times higher than developing ShootDem-Ks. ShootDem-Ks is more cost-effective to implement, in terms of time spent during knowledge acquisition and knowledge encoding, when compared with its rule-based counterpart. It is obvious that system maintenance is only necessary when products are being introduced or modified and when new problems concerning the products are reported. This comparison consolidates all the time spent on updating the systems from the initial pilot version to the most current version into a single measurement, the maintenance time. It is also realized that since ShootDem-Ks is developed after the rule-based ShootDem, all relevant knowledge accumulated for ShootDem is carried over at the time ShootDem-Ks was developed. This reduces the actual development time for ShootDem-Ks by a considerable amount. However, in order to fairly compare the two systems, it is assumed that ShootDem-Ks was developed from the ground up and that this system was not benefiting from any time savings due to any previous work.

In terms of operational efficiency, the comparison uses "ease-of-use" and "success rate" as measurements of merit. "Ease-of-use" measures how effective a system interfaces with the operator, to extract enough information for deriving a solution to a new query. It is quite unfair to compare the text-based/character-based interface provided by ShootDem to the graphics-based interface provided by ShootDem-Ks. Since ShootDem-Ks is implemented with CBR Express, it enjoys all the facilities that the system's graphical front-end provides. By using CBR Express, a built-in customer tracking database, special browsers, and the Toolbox interface for creating animated graphical actions are all available to ShootDem-Ks. However, the main objective of this comparison is to measure how easy it is to operate the system, as it is being reported by different operators. Focus is placed on the amount of keystrokes needed to answer a question, and the average number of questions needed to be answered before a reasonable solution is generated by the system. Operators are discouraged to excessively rate fancy features such as animated graphics that only ShootDem-Ks can provide.

Success rate measures the accuracy and consistency of the solutions suggested by each system in solving new problems. The rule-based system suffers from accuracy that may be contributed to poorly translated rules or insufficient knowledge to encode the necessary rules. However, ShootDem has a 100% consistency. Given the same set of information, ShootDem is capable of deriving the same solution every time. On the other hand, ShootDem-Ks is very capable of providing accurate solutions but the solutions may not always be the same ones for the same type of questions. This is related to how ShootDem-Ks utilizes natural-language processing in handling case descriptions . The same product failure conditions can be described in many ways by the user. The choice of words and style for writing the case description may affect the matching and scoring processes when comparing features of similar cases. It is not surprising that ShootDem-Ks is less consistent than the rigidly structured rule-based ShootDem.

There is no attempt to directly measure the execution times of each system due to the complexity of such a measurement. The first concern here is that the systems execute on a different platform. ShootDem is a text-based application running under MS-DOS while ShootDem-Ks is a graphic-based application running under Windows 3.1. The execution speed of ShootDem-Ks can be impaired by house-keeping jobs executed in the background by Windows 3.1. Operators using both systems reported that it takes an average time of eight to twelve seconds from the time the last question required by a query is answered to the instance when an appropriate solution is generated. Typically, ShootDem is reported to be faster than ShootDem-Ks by a couple of seconds. This is mainly due to the fact that there is no graphical screen to be redrawn in the text-based ShootDem. Furthermore, the Windows 3.1 GUI tends to slow down ShootDem-Ks when the GUI is conducting house-keeping jobs in the background.

10.9 CONCLUSIONS

The presented work has demonstrated how case-based reasoning is employed to build systems that can effectively assist manufacturers in providing technical support to customers. Inspired by the human thought process, CBR "remembers" previous problems and either adapts their solutions or uses their outcomes to evaluate new cases. Case-based systems eliminate the need for encoding any acquired knowledge into rules. This leads to a substantial amount of time savings in implementing a case-based system as compared to the traditional rule-based expert system. Maintenance time for a case-based system is minimum, since new cases can be added to the system during normal query operations. Similar to ShootDem-Ks, other case-based systems such as CASCADE [12], for the diagnosis of VMS operating system crashes, and COAST, for the diagnosis of problems with the WPS-PLUS wordprocessor [13] have demonstrated great efficiencies in assisting the help desk to answer end-users' queries.

Results from this work shows that, for help desk applications, rule-based systems are less cost-effective, in terms of knowledge acquisition, knowledge encoding, and knowledge maintenance, than case-based systems. Moreover, in rule based systems, knowledge is rigidly structured (in the form of rules). In this situation, interactions between the technical support staff and the customer are totally restricted by the terms that the rule-based expert system is programmed to handle. Customers may not always know the exact technical terms to describe their problems. It becomes the burden of the technical support staff to translate non-technical information provided by customers, into terms that are recognizable by the rule-based expert system. Consequently, the rule-based expert system will not perform satisfactorily if the translations of the customer's descriptions are inaccurate. On the other hand, the case-based system ShootDem-Ks is equipped with a free-form natural language interface. The technical staff simply enters the problem description in the customer's own words. The character feature matching capabilities, provided by CBR Express, allows ShootDem-Ks to perform adequately even when the words in a case description are misspelled.

The knowledge of the case-based ShootDem-Ks can be readily extended by adding new cases and their solutions to the case base. When the operator encounters a novel case that the system cannot solve, this case can be registered into the system as an unresolved case search. The unresolved cases are passed on to the product engineers who can examine these cases in detail and determine the information required to solve them. The unresolved cases are then modified accordingly by the product engineers, and become part of the working knowledge of the system. The whole process for modifying and converting an unresolved case into one that can be used for solving similar cases in the future could be completed in a few minutes.

Further studies are needed to investigate on how case-based knowledge can be tied into other manufacturing systems to improve the overall production efficiency. For instance, ShootDem-Ks maintains an ample amount of knowledge for troubleshooting product failures. Some of this knowledge can be utilized by other CAD/CAM (Computer-Aided Design/Computer-Aided Manufacturing) systems to assist design engineers in developing better products and to improve the production process. CBR technology can also be applied directly to CAD/CAM systems for solving product design[14] and production sequence planning[15] problems.

REFERENCES

1. C. K. Riesbeck and R. C. Schank, *Inside Case-Based Reasoning*, Lawrence Erlbaum Associates, Inc., Hillsdale, New Jersey, 1989.

2. A. Newell and H. Simon, *Human Problem Solving*, Prentice Hall, Englewood Cliffs, New Jersey, 1972.

3. E. A. Feigenbaum, B. G. Buchanan, and J. Lederberg, "On Generality and Problem Solving: A Case Study Involving the DENDRAL Program," Machine Intelligence, American Elsevier, New York, 1971, pp. 165-190.

4. C. L. Forgy, "RETE: A Fast Algorithm for the Many Pattern/Many Object Pattern Match Problem," Artificial Intelligence, Vol. 19, 1980.

5. P. Haley, "Data-Driven Backward Chaining," Proceedings of the Second Annual CLIPS Conference, NASA Johnson Space Center, Houston TX, September 1991.

6. M. Rychener, "Production Systems as a Programming Language for AI," Ph.D. Thesis, Carnegie-Mellon University, 1976.

7. K. J. Hammond, *Case-Based Planning: Viewing Planning as a Memory Task*, Academic Press, San Diego, California, 1989.

8. R. O. Duda and P. E. Hart, *Pattern Classification and Scene Analysis*, John Wiley and Sons, New York, 1973.

9. J. L. Kolonder, R.L. Simpson, Jr. and K. Scara-Cryanski, "A Process Model of Case-Based Reasoning in Problem Solving," Proceedings: International Joint Conference in Artificial Intelligence, Morgan Kaufmann, San Mateo, California, 1985, pp.284-290.

10. J. L. Kolodner, "Improving Human Decision Making Through Case-Based Decision Aiding," AI Magazine, Vol. 12, No. 2., Summer 1991, pp.52-68.

11. Inference Corporation, *Case-Based Reasoning in ART-IM*, Inference Corporation, El Segundo, California, 1992.

12. E. Simoudis, "Using Case-Based Retrieval for Customer Technical Support," IEEE Expert, October 1992, pp. 7-12.

13. E. Simoudis and J. S. Miller, "The Application of CBR to Help Desk Applications," Proceedings: Case-Based Reasoning Workshop, Morgan Kaufmann Publishers, May 1991, pp. 25-36.

14. A. Goel, J. Kolodner, M. Pearce, R. Billington and C. Zimring, "Towards a Case-Based Tool for Aiding Conceptual Design Problem Solving," Proceedings: Case-Based Reasoning Workshop, May 1991, Morgan Kaufmann Publishers, pp. 109-120.

15. P. Pu and M. Reschberger, "Assembly Sequence Planning Using Case-Based Reasoning Techniques," Knowledge-Based Systems, Vol. 4, No. 3, September 1991, Butterwoth-Heinemann Ltd., pp. 123-130.

11

PARTIALLY OVERLAPPED SYSTEMS: THE SCHEDULING PROBLEM

M. Kamel and H. Ghenniwa
Department of Systems Design Engineering
University of Waterloo
Waterloo, Canada.

11.1 INTRODUCTION

Many environments, such as manufacturing, involve multiple machines, multiple robot systems, and(or) multiple computer systems. Each of these systems is capable of performing specific operations which might overlap with other systems' capabilities. All of these systems, even though they belong to different environments, are related to one common class, **multi-machine systems**. Multi-machine systems, in general, act in some form of parallelism to achieve specific goals. The performance of such systems is measured by certain criteria such as the system throughput. However, several problems, such as job partitioning and scheduling, must be solved in order to achieve this objective. A great deal of research has been done to solve these problems for systems in different environments like in computer engineering (parallel processors and distributed systems) [2, 17], and in manufacturing (job-shop, flow-shop, open-shop, and parallel machine systems) [15, 18-20].

Scheduling problem, in general, is defined as the allocation of jobs to the appropriate machines over time to perform a collection of jobs [1]. Solving a scheduling problem is the same as answering the following two questions: (1) which machines will be selected to perform each job? (2) when will each job be performed? In other words, the essence of scheduling problems gives rise to two intertwined sub-problems, assignment and sequencing. The assignment problem arises whenever there are many choices of allocating a job to machines. The sequencing problem arises whenever there are many choices of distributing jobs and (or) their operations over the time scale. The sequencing problem has to be addressed in order to solve problems due to assigning more than one job (and (or) operation) to one machine, job dependency, job ready time, and job due date.

A considerable amount of work has been devoted to solve the scheduling problem by operations research practitioners, management scientists, and mathematicians since the early 1950's. Most of this work is based on the so called classical scheduling assumptions. These assumptions ignore many important characteristics of many environments. The classical scheduling, for example, assumes that the environments, in general, are consists of l processing stages, $l \geq 1$. At each processing stage there are either a (1) k-type parallel processors (machines with identical capabilities), $k \geq 1$, and each job can be processed by any of these processors; or (2) shop machines (machines with disparate capabilities) in which there are k_i identical machines i = $1,\ldots,m$ and any job requiring that stage needs to be processed on one and only one of these machines. Although this assumption restricts the machines capabilities to be either identical or disparate, some special cases of partially overlapped systems, in which the jobs are consisting of a single operation, can be modeled as unrelated parallel machines [14]. Nevertheless, a large number of partially-overlapped systems cannot be modeled under these assumptions. Flexible manufacturing systems is an example of these systems.

In this chapter, first in Section 11.2, we introduce a general model of multi-machine systems which is able to represent different multi-machine environments including partially-overlapped systems. This model is based on the system's structure and its processing characteristics. Three classes of multi-machine systems, based on their structures, are identified: **general-purpose**, **special-purpose**, and **partially-overlapped** systems. Then in the rest of this chapter we focus on solving three cases of scheduling problems for partially-overlapped systems.

The first case, in Section 11.3, is to schedule a set of simple single operation jobs, each job requires a unit-time, on partially-overlapped systems. For this case, in Section 11.3.1.1, we propose an algorithm which finds an optimal schedule that minimizes the maximum completion time (makespan). The algorithm has a worst case time complexity of $O\left(\min\left\{n^{.5},m\right\}nm\right)$, which is an improvement over the best existing algorithms [5, 6]. The second case, in Section 11.4, is to schedule a set of simple, arbitrary-time single operation jobs on partially-overlapped systems, such that the makespan is minimized. First we prove that this problem is NP-hard. Then, in Section 11.4.1, we develop a heuristic algorithm for this scheduling problem. Finally, in Section 11.5, we examine the case of scheduling a set of simple, multiple operation jobs, each operation requires a single time-unit, on partially-overlapped systems, such that the number of machines assigned to process each job and the difference between the total time (time-load) assigned to each machine for a pair of machines are minimized. This problem is proved to be NP-hard. To solve this problem we explored two techniques. The first is an optimization algorithm with exponential worst case performance, while the second is a heuristic technique. In the first, the problem, in Section 11.5.1, is modeled using a mathematical representation. This representation is then utilized to solve the scheduling problem using multiple objectives formulation. The second technique, in Section 11.5.2, is a decomposition heuristic algorithm which utilizes a graphical representation of the problem.

11.2 MULTI-MACHINE SYSTEMS: A GENERAL MODEL

In this section we introduce a model of multi-machine systems which is general enough to represent shop machines, parallel machines, and partially-overlapped systems. A multi-machine environment consists of a multi-machine system (**M**) and a set of jobs (**J**) which need to be processed by **M**. Formally,

Definition 1 a multi-machine system is a system consisting of a set of m machines denoted by **M** $=\{M_i | M_i \underline{\underline{\Delta}} (C_i, T_i), 1 \le i \le m\}$, where

- machine $M_i, 1 \le i \le m$, is an entity capable of performing some sort of operations;
- capability set C_i is a set of operations which M_i can perform; where $C_i = \{o_{i,1}, \ldots, o_{i,\alpha_i}\}$, and α_i is the number of operations belonging to M_i;
- the processing time set, T_i, of the operations given in C_i; where $T_i = \{t_{i,1}, \ldots, t_{i,\alpha_i}\}$, t_{i,ρ_i} is
- the time required to process o_{i,ρ_i} on M_i, and $1 \le \rho_i \le \alpha_i$.

A set of jobs, each job is a set of operations, may have constraints such as (1) ready time: time at which a job is ready to be processed; (2) due date: time in which a job should be completed; (3) operation precedence: an order in which the operations need to be processed. In this chapter we focus our attention on simple jobs which are independent jobs with zero ready time and no due date. Formally,

Definition 2 a set of n simple jobs is denoted by **J** $=\{J_j | J_j \underline{\underline{\Delta}} (O_j), 1 \le j \le n\}$, where O_i is a set of the operations required by job J_i.

Based on the above definitions machine M_i is an appropriate machine for J_i, iff $C_i \cap O_j \ne \phi$. In this context we define a multi-machine structure as the pattern of machine-job relationships which represent the possible ways of processing a given set of jobs by the machines of the system. Therefore, multi-machine systems can be classified, based on their structure, into the following:

1. **General-Purpose** (G.P), which consist of machines that have identical capability sets, or $C_i = C_l$, $\forall i,l$, $1 \le i,l \le m$.

2. **Special-Purpose** (S.P), which consist of machines with a distinct capability set for each machine, or $C_i \cap C_l = \phi, 1 \le i,l \le m.$, and $i \ne l$.

3. **Partially-Overlapped** (P.O), which consist of machines with capability sets that are partially overlapped. Formally, for a subset of machines (group) in a P.O system (**M**); say group k denoted by $\mathbf{M}_k = \{M_{l_1}, \ldots, M_{l_s}\} \subset \mathbf{M}$, where $l_r \in \{1,\ldots,m\}, r = 1,\ldots,s$, and $1 \le s \le m$; the capability set of machine $M_i \in \mathbf{M}_k$ can be represented as $C_i = \varsigma_i^k \cup \vartheta_i^k$, where ς_i^k is the special capability set which does not belong to any machine in group k except to M_i; and ϑ_i^k is the capability set that overlap between the capability of machine M_i and that of some other machines in group k. Note, this definition of partially-overlapped systems allows the partitioning of the machine's capability set (into special and overlapping sets) is not fixed, but they differ according to the capability sets of the group members.

Furthermore each class, based on structure, can be classified based on the system's processing characteristics (the processing speed, jobs, and operations precedence order). Here we provide two examples:

1. G.P systems can be classified based on the type of machines' processing speed into three sub-classes. A G.P system is called

(a) **Identical** (I.G.P) if the machines' processing speed is characterized by $t_{i,k} = t_{1,k}, 1 \le i \le m, 1 \le k \le \alpha$.

(b) **Uniform** (U.G.P) if the machines' processing speed is characterized by $t_{i,k} = I_i\, t_{1,k}, 1 \le i \le m, 1 \le k \le \alpha$, and $I_i \in \Re^+$ (set of positive real numbers) is the speed factor.

(c) **Unrelated** (Un.G.P) if the machines' processing speed is characterized by $t_{i,k} = R_{i,k}\, t_{1,k}, 1 \le i \le m, 1 \le k \le \alpha$, and $R_{i,k}$ is the speed factor.

2. S.P systems can be classified based on the precedence constraints between the operations of each job into three sub-classes. An S.P system is called

(a) **Identical** (S.P.I), if it processes a set of jobs that have the same initial precedence constraint between their operations. This is commonly known as a flow-shop.

(b) **Arbitrary** (S.P.A), if it processes a set of jobs such that each job is characterized by a set of operations that may have a different initial precedence constraints from those of the other jobs. This is commonly known as a job-shop.

(c) **Free** (S.P.F), if it processes jobs that have no initial precedence constraints between their operations. This is commonly known as an open-shop.

11.2.1 MULTI-MACHINE STRUCTURE: GRAPH REPRESENTATION

In this section we focus on modeling multi-machine environments based on the structure only. Recall that a multi-machine structure represents the relationships between machines and jobs. To relate between the different structures of multi-machine systems we assume that these systems are processing a set of simple jobs. Thus, the structure of multi-machine systems can be modeled using a bipartite graph as shown in Figure 11.1.

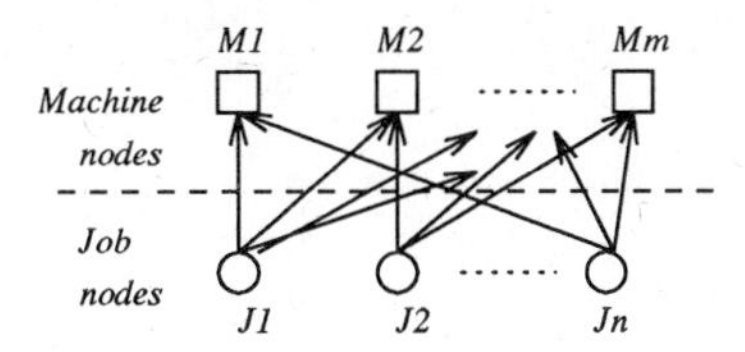

Figure 11.1: A graph model for multi-machine systems

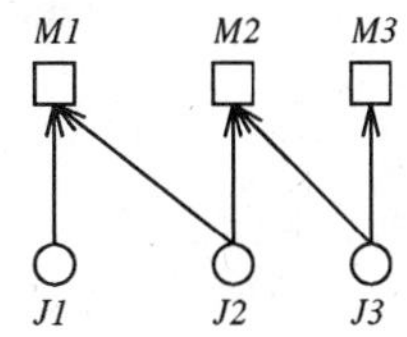

Figure 11.2: The logical OR structure

In this model there are two types of nodes, one to represent the machines, and the other to represent the jobs. (J_j, M_i) is the arc between job node J_j and the appropriate machine node M_i. This arc exists iff $C_i \cap O_j \neq \phi$. The graph representation of the system structure is called logical structure, and can be classified into the following three classes:

1. **OR structure**, which represents the logical structure of G.P systems, an example is shown in Figure 11.2 (e.g. job J_2 can be processed by either M_1 or M_2).

2. **AND structure**, which represents the logical structure of S.P systems, an example is shown in Figure 11.3 (e.g. job J_1 has to be processed by both machines M_1 and M_2, but not at the same time).

3. **AND-OR structure**, which represents the logical structure of P.0 systems, an example is shown in Figure 11.4 (e.g. J_4 can be processed either by M_3 or by M_1 and M_2 together but not at the same time).

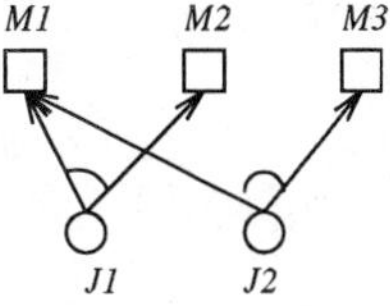

Figure 11.3: The logical AND structure

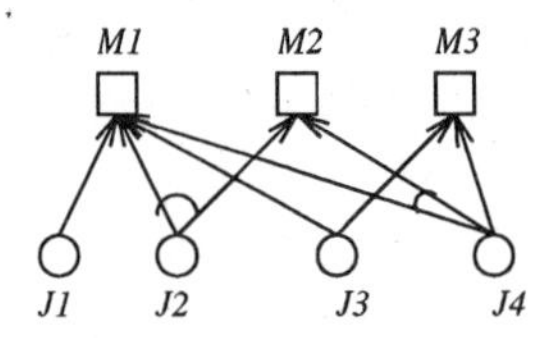

Figure 11.4: The logical AND-OR structure

It follows from the above that the logical structure of P.O systems can represent both G.P and S.P systems as special cases of its structure. In this sense,

- a G.P system is a P.O system with the property of $\vartheta_i^k = C_1$, and $\varsigma_i^k = \phi, \forall i, k$;
- an S.P system is a P.O system with the property of $\vartheta_i^k = \phi, \forall i, k$.

Therefore any multi-machine system can be classified as a special class of P.O systems, which makes P.O systems the most general class among all other classes of multi-machine systems.

In the rest of this chapter we focus on three different scheduling problems of partially-overlapped systems processing a set of simple jobs.

11.3 UNIT-TIME SINGLE OPERATION JOBS

In this section, we examine the scheduling of n simple jobs, each requires a single time-unit, on m partially overlapped machines such that the makespan is minimized. Although the single time-unit assumption simplifies the general case of the scheduling problem, solving this problem is not obvious. In addition, there are some applications where the problem can be adequately represented under this assumption. One example is to schedule jobs on machine-centers (or work-cells) in flexible manufacturing such that the maximum number of jobs assigned to each machine-center is minimized. Each machine-center is characterized by a capability of performing specific types of jobs which might partially overlap with other machine-centers; this may be due to the different tool wearing of the machine-centers. For the above problem we introduce the following definitions

Definition 3 a schedule is called **feasible** if each job is allocated to an appropriate machine.

Definition 4 a feasible schedule is called **optimal** if it minimizes the makespan.

Definition 5 an optimal schedule is called **balanced** if each machine has been assigned to a total processing time (time-load) equal to T, where $(T - \lceil n/m \rceil) \in \{0,1\}$, for m machines and n jobs.

11.3.1 SOLUTION METHODOLOGY

In this section we propose a method which utilizes a bipartite network representation of the scheduling problem in order to find an optimal solution that minimizes the makespan. The main strategy of this method is to find a balanced schedule, if it exists. Otherwise, the method finds an optimal schedule for some of the jobs, under the constraint that each machine cannot be assigned to a time-load greater than that of the load balanced case. Consequently, some jobs will be left unassigned. Then, the method tries to schedule the unassigned jobs optimally, by constructing a sub-network involving these jobs and the appropriate machines, and applying the above strategy for the sub-network. This mechanism is repeated until no job is left unassigned.

Each iteration of this method consists of two main steps, constructing a bipartite network to model the scheduling problem, and a maximum flow algorithm [11] to solve the scheduling problem. In constructing bipartite network N_κ, for n_κ jobs and m_κ machines at iteration κ, jobs and machines are represented by different nodes, as shown in Figure 11.5. Then, for each j and i, $1 \le j \le n$, $1 \le i \le m$;

- job node J_j is connected to a start node s by arc (s; J_j) which has a unit value capacity; meaning that J_j requires a unit processing time.
- job node J_j is connected to machine node M_i, iff $J_j \in C_i$, by arc (J_j; M_i) which has a unit value capacity; meaning that M_i is an appropriate machine for J_j..
- machine node M_i is connected to a terminal node t by arc (M_i; t) which has a capacity value of c_κ; meaning that the upper bound of the time-load of M_i is c_κ.

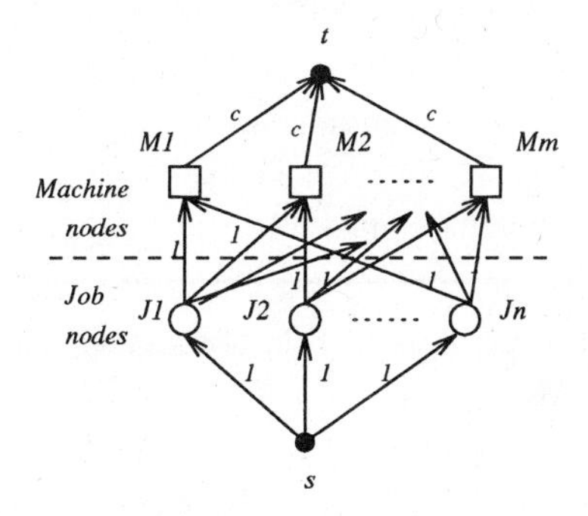

Figure 11.5: A bipartite network model for the scheduling problem.

The maximum flow algorithm [11], then, utilizes N_κ to solve the scheduling problem. A maximum flow algorithm is designed to determine a flow pattern, in N_κ, from node s to node t, with a flow value equal to f_{max} (maximum flow) such that for all possible flow patterns in N_κ from s to t there is no other flow pattern with flow value of f that is greater than f_{max}. Therefore, a flow pattern from s to t in N_κ with f_{max} represents an optimal schedule of jobs to machines such that the time-load of each machine is bounded by c_κ time-units [6]. In this sense, the balanced schedule can be represented in N_κ by making $c_\kappa = \lceil n_\kappa / m_\kappa \rceil$. Clearly, the flow pattern with $f_{max} = n_\kappa$ represents a balanced schedule. If such a schedule does not exist, then the generated flow pattern, which has $f_{max} < n_\kappa$, represents an optimal schedule for some of the jobs, while the rest of the jobs will be left unassigned. To explain how these jobs can be assigned such that the optimal solution is obtained, consider

Definition 7 let, at iteration κ, f_i be the value of the flow through (M_i; t) which represents the time-load assigned to M_i. Then M_i is called `saturated' at iteration κ iff $f_i = c_\kappa$.

Theorem 1 At iteration κ in the above method, to schedule n_κ simple jobs on m_κ partially-overlapped machines, if a balanced schedule does not exist then an appropriate machine for an unassigned job must be one of the saturated machines that are generated at the same iteration.

Proof:
Let f_κ be a flow in N_κ, and for each path there is a non-negative integer $\iota_\kappa(P) = \min_{a\in A(P)} c_\kappa(a) - f_\kappa(a)$; where, $A(P)$ is the set of all arcs in path P; $c_\kappa(a)$ is the capacity of arc a, and $f_\kappa(a)$ is the flow through arc a. A path is not being used to its full capacity if $\iota_\kappa(P) > 0$; and if the path is from s to t, then it is called *f*-incrementing. Furthermore, a flow f_κ in N_κ is a maximum flow ($f_{\max}$) iff N_κ contains no f-incrementing paths [3].

Now, suppose, at iteration κ, a bipartite network N_κ is used to model the scheduling problem, and the capacity of (M_i; t), $\forall i$, is assigned to a value of $\lceil n_\kappa / m_\kappa \rceil$. By applying a maximum flow algorithm (optimization algorithm) on N_κ, we get the flow pattern from s to t with $f_{\max}$. This implies that there will be no f-incrementing path in N_κ which can make the flow value from s to t greater than $f_{\max}$. Now, assume there is no balanced schedule, for which $f_{\max} < n_\kappa$. Therefore, there should be some paths from s to $\underline{\mathbf{M}}$, $\underline{\mathbf{M}} \subset \{M_1,\ldots,M_m\}$, not being used to their full capacity. In other terms, there should be some jobs left unassigned. However, it has been assumed that a feasible assignment does exist. To find such an assignment, assume that M_l, an appropriate machine for one of the unassigned jobs ($M_l \in \underline{\mathbf{M}}$), is an unsaturated machine (i.e. $f_l < \lceil n_\kappa / m_\kappa \rceil$). Hence, it is possible to allocate this job to M_l, which will increase $f_{\max}$ by one. Thus, an *f*-incrementing path has been found. This contradicts the fact that $f_{\max}$ is optimal. Therefore, M_l should be saturated, or $f_{\max} = \lceil n_\kappa / m_\kappa \rceil$.

Based on the above theorem, the method then iterates to consider a sub-network of the problem model which only consists of the unassigned job nodes and the appropriate saturated machine nodes. This mechanism is then repeated until all jobs are assigned and the optimal allocation is obtained.

11.3.1.1 THE ALGORITHM

Based on Theorem 1, we formulate algorithm P.O.S.U (Partially Overlapped system with Single Unit-time operation jobs) as follows:

P.O.S.U Algorithm

1. Set $\mathbf{M} = \{M_i | M_i \underline{\underline{\Delta}}(C_i), 1 \le i \le m\}$; $\mathbf{J} = \{J_1,\ldots,J_n\}$;
2. Construct a bipartite network from jobs and machines, such that:
   2.1 each job node J_j; $1 \le j \le n$, is connected to the source node s by an arc of a unit value capacity;
   2.2 each job node such that $J_j \in C_i$; $1 \le j \le n$, and $1 \le i \le m$, is connected to a machine

node M_i by an arc of a unit value capacity;

2.3 Set $m = m'$, where m' is the number of the appropriate machines (Definition 1);

2.4 each machine node is connected to the terminal node t by an arc with a capacity value of $\lceil n/m \rceil$;

3. Use maximum flow algorithm to assign jobs to machines and return f_{max};
4. If $(f_{max} = n)$ Then terminate with optimal solution;
5. Else determine the elements of the sub-network as follows:

5.1 Find the set of all saturated machines **Ms**, and the set of the unassigned jobs **Ju**, where:

$$\mathbf{Ms}=\{M_{\alpha_1},\ldots,M_{\alpha_y}\}, 1 \le y \le m, \alpha_d \in \{1,\ldots,m\}, 1 \le d \le y, M_{\alpha_d} \in \mathbf{M};$$

$$\mathbf{Ju}=\{J_{\beta_1},\ldots,J_{\beta_x}\}, 1 \le x \le n, \beta_e \in \{1,\ldots,n\}, 1 \le e \le x, J_{\beta_e} \in \mathbf{J};$$

5.2 Set $m = y$; $n = x$; $\mathbf{M} = \mathbf{Ms}$; and $\mathbf{J} = \mathbf{Ju}$;

5.3 Goto step 2;

Theorem 2 Algorithm P.O.S.U correctly determines an optimal allocation.

Proof:

It has been proven that applying a maximum flow algorithm on a bipartite network N_κ is characterized by generating a flow pattern from source node s to terminal node t with f_{max}. And for all possible flow patterns in N_κ from s to t, there is no other flow pattern with a flow value of $f > f_{max}$ (pp. 197 in [3]).

In constructing the network model of the scheduling problem, step 2 in P.O.S.U, firstly, the capacity of $(s; J_j)$ and $(J_j; M_i)$ are assigned a value of one, and $(J_j; M_i)$ exists iff $J_j \in C_i$, in steps 2.1 and 2.2 respectively. This guarantees that by applying a maximum flow algorithm, step 3, each job will be assigned only to one appropriate machine. Thus each generated allocation is feasible. Secondly, at step 2.4, the capacity of $(M_i; t)$ is assigned a value of $\lceil n_\kappa / m_\kappa \rceil$. This guarantees that by applying a maximum flow algorithm, at each iteration, each machine would not be assigned to a time-load more than that of the balanced scheduling case, for n_κ jobs and m_κ machines. Thus, if a balanced schedule exists then it would be found. Otherwise, the generated solution is an optimal allocation for some of the jobs, because there is no job which is assigned to a saturated machine that can be reassigned to an unsaturated machine such that a better solution can be found. This follows from the characteristic of the maximum flow algorithm mentioned above.

At the next iteration, the algorithm considers a sub-network of the problem which only consists of the unassigned job nodes and the appropriate saturated machine nodes that are generated in the previous iteration. Since each of these machines is saturated, then all of them should have been assigned to the same time-load from the previous iteration. Therefore, by repeating the same procedure to schedule the unassigned jobs to the appropriate saturated machines, the P.O.S.U terminates with an optimal solution.

11.3.2 COMPLEXITY ANALYSIS FOR P.O.S.U ALGORITHM

In this section we prove that the computational complexity of P.O.S.U algorithm is $O(\min\{n^{.5},m\}nm)$ which is better than the best available algorithm [6] by $\log_2 n$. It has been proven [6] that for a bipartite network N with $n \times m$ nodes, $n \geq m$, and the characteristics given in Section 11.3.1, the maximum flow of N can be found with an algorithm having a computational complexity given by $O(\min\{n^{.5},m\}nm)$.

Lemma 1 In allocating n simple jobs to m partially-overlapped machines, if a balanced allocation does not exist then the number of unassigned jobs is less than n and the number of saturated machines is less than m.

Proof: trivial.

Lemma 2 Assume that the balanced allocation does not exist, and the P.O.S.U algorithm will iterate for k times to find the optimal allocation. Let, at the κth iteration of P.O.S.U, n_κ and $n_{\kappa+1}$ be the total number of jobs to be assigned and the total number of the unassigned jobs, respectively, also, m_κ and $m_{\kappa+1}$ be the total number of the appropriate machines and the total number of the saturated machines, respectively. Then the relationship between these quantities is given by

$$n_{\kappa+1} \leq n_\kappa(1-1/r_\kappa), \text{ where } r_\kappa = m_\kappa / m_{\kappa+1}. \tag{11.1}$$

Proof:

It follows directly from Lemma 1 that for unbalanced allocation

$$\left.\begin{matrix} n_{\kappa+1} < n_\kappa \\ m_{\kappa+1} < m_\kappa \end{matrix}\right\}, \ 0 \leq \kappa \leq k-1,$$

where, k is the number of iterations required to reach optimality. Thus,

$$r_\kappa = m_\kappa / m_{\kappa+1} > 1 \tag{11.2}$$

Suppose that the set of unsaturated machines is denoted by **Mu** $=\{M_{\eta 1},\ldots,M_{\eta s}\}$, $1 \leq s \leq m_\kappa$, $\eta_\rho \in \{1,\ldots,m\}, \rho = 1,\ldots,s$, and $M_{\eta_\rho} \in$ **M**. For unbalanced allocation

$$n_{\kappa+1} = n_\kappa - f_{\max}. \tag{11.3}$$

From the definition of maximum flow

$$f_{\max} = m_{\kappa+1}\lceil n_\kappa / m_\kappa \rceil + \sum_{\rho=1}^{s} f_{\eta_\rho}. \tag{11.4}$$

Thus, from Equations (11.3) and (11.4) we get

$$n_{\kappa+1} = n_\kappa - m_{\kappa+1}\lceil n_\kappa / m_\kappa \rceil - \sum_{\rho=1}^{s} f_{\eta_\rho} \le n_\kappa - \left(m_\kappa / r_\kappa \lceil n_\kappa / m_\kappa \rceil\right).$$

Therefore,

$$n_{\kappa+1} \le n_\kappa (1 - 1/r_\kappa). \tag{11.5}$$

Theorem 3 The time complexity of algorithm P.O.S.U is $O(\min\{n^{.5}, m\} nm)$.

Proof:

At each iteration of algorithm P.O.S.U solving the maximum flow problem is the key factor of determining its computational complexity. The computational complexity of the maximum flow algorithm used is $O(\min\{n_\kappa{}^{.5}, m_\kappa\} n_\kappa m_\kappa), 0 \le \kappa \le k-1$. Thus, the computational complexity of P.O.S.U with k iterations can be calculated as follows. First, consider the case where $n_\kappa^{.5} > m_\kappa$. Therefore, the time complexity of the algorithm is $O\left(\sum_{\kappa=0}^{k-1} n_\kappa m_\kappa^2\right)$. Let r_κ equal to some value r represents the worst case. From Equation (11.2) and Inequality (11.5) we get:

$$\sum_{\kappa=0}^{k-1} n_\kappa m_\kappa^2 \le \sum_{\kappa=0}^{k-1} n_o \left(\frac{r-1}{r^3}\right)^\kappa m_o^2 (1/r^{2\kappa}).$$

The right hand side can be rewritten as:

$$\sum_{\kappa=0}^{k-1} n_o m_o^2 \left(\frac{r-1}{r^3}\right)^\kappa = \frac{1-\left(\frac{r-1}{r^3}\right)^\kappa}{1-\left(\frac{r-1}{r^3}\right)} n_o m_o^2 .$$

Therefore,

$$O\left(\sum_{\kappa=0}^{k-1} n_\kappa m_\kappa^2\right) \le O\left(\frac{1-\left(\frac{r-1}{r^3}\right)^\kappa}{1-\left(\frac{r-1}{r^3}\right)} n_o m_o^2\right) = O(C n_o m_o^2),$$

where, $C = (1-\beta^k)/(1-\beta)$, and $\beta = (r-1)/r^3$. Next, we show that $C < 2$. First assume that $C \ge 2$ Thus, $(1-\beta^k)/(1-\beta) \ge 2$. Consequently, $\beta = (r-1)/r^3 > 1/2$, or

$$r^3 - 2r + 2 < 0. \tag{11.6}$$

Inequality (11.6) implies $r < 1$ (contradiction with the Inequality at (11.2)), therefore, $C \ge 2$ cannot be true, and accordingly $C < 2$. From the above, it follows that the time complexity of

P.O.S.U algorithm is $O(n_o m_o^2) = O(nm^2)$, when $n^{.5} > m$. Similarly, it can be shown that the time complexity of the algorithm is $O(n^{1.5}m))$ for the case where $n^{.5} < m$. Therefore the time complexity of P.O.S.U algorithm is $O(\min\{n^{.5}, m\}nm)$.

11.3.3 EXAMPLE

Consider a fully automated work cell, *AWC-1*, consisting of five multiple-operation machines, denoted by M_1, M_2, M_3, M_4, and M_5. Each of them is characterized by its own capability set. M_1 is capable of drilling with a diameter range of (*5mm-15mm*), and milling with an angle range of (0° ,45°). M_2 is capable of drilling with a diameter range of (*2mm-12mm*), reaming with a diameter range of (*2mm-7mm*), grinding with wheels of diameter range of (*5mm-20mm*), and milling with an angle range of (0° ,45°). M_3 is capable of tapping with a diameter range of (*5mm-20mm*). M_4 is capable of tapping with a diameter range of (*2mm-10mm*), and reaming with a diameter range of (*8mm-10mm*). Finally, M_5 is capable of reaming with a diameter range of (*2mm-10mm*), grinding with wheels of diameter range of (*2mm-10mm*), and tapping with a diameter range of (*5mm-15mm*).

A set of nine independent objects need to be processed by *AWC-1*. Each object requires a single operation. The first object requires drilling with diameter of *9mm.* The second requires drilling with diameter of *12mm.* The third requires milling with angle of 30° . The forth requires milling with angle of 45° . The fifth requires tapping with diameter of *10mm.* The sixth requires reaming with diameter of *8mm.* The seventh requires reaming with diameter of *7mm.* The eighth requires reaming with diameter of *5mm.* Finally the ninth requires grinding with *8mm* wheel.

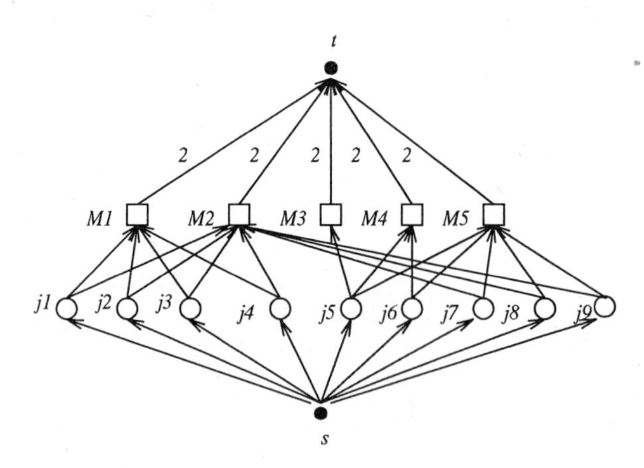

Figure 11.6: The initial network model of *ACW-1*.

The main focus of this example is to solve the scheduling problem on the basis of the machines' capability of performing specific types of operations, regardless to their processing speed. Hence, it is assumed that all the operations have processing speed of unit-time. Thus the makespan criterion, here, is equivalent to minimize the maximum number of jobs that are allocated to each machine. *AWC-1* is a multi-machine system consisting of five machines, further, it is a partially-overlapped system. For example, M_1 and M_2 have a partially overlapped range of drilling which is (*5mm-12mm*). Formally, *AWC-1*$=\{M_i | M_i \underline{\underline{\Delta}}(C_i), 1 \le i \le 5\}$. The set of objects (jobs) is denoted by $\mathbf{J}=\{J_1, J_2, J_3, J_4, J_5, J_6, J_7, J_8, J_9\}$. Thus, the capability set of each machine is formulated as $C_1=\{J_1, J_2, J_3, J_4\}$, $C_2=\{J_1, J_2, J_3, J_4, J_7, J_8, J_9\}$, $C_3=\{J_5\}$, $C_4=\{J_5, J_6\}$, and $C_5=\{J_5, J_6, J_7, J_8, J_9\}$.

The initial network which represents the scheduling problem of *AWC-1* is shown in Figure 11.6. At the first iteration of P.O.S.U algorithm we get:

steps 1-3:$\{J_1, J_2\}$ is allocated to M_1 ;$\{J_3, J_9\}$ is allocated to M_2; $\{J_5\}$ is allocated to M_3; $\{J_6\}$ is allocated to M_4; and $\{J_7, J_8\}$ is allocated to M_5. $f_{\max}=8$;

step 4: $f_{\max} \neq 9$;

step 5: **Ms**$=\{M_1, M_2, M_5\}$; **Ju** $=\{J_4\}$; *m = 3; n = 1.*

At the second iteration a 1×2 network (M_5 is inappropriate machine for J_4) is constructed as shown in Figure 11.7. Finally, the generated schedule is: $\{J_1, J_2, J_4\}$ is allocated to M_1 ;$\{J_3, J_9\}$ is allocated to M_2; $\{J_5\}$ is allocated to M_3; $\{J_6\}$ is allocated to M_4; and $\{J_7, J_8\}$ is allocated to M_5. This solution, as obtained by the algorithm, is an optimal solution for this case, which has a makespan of *3*.

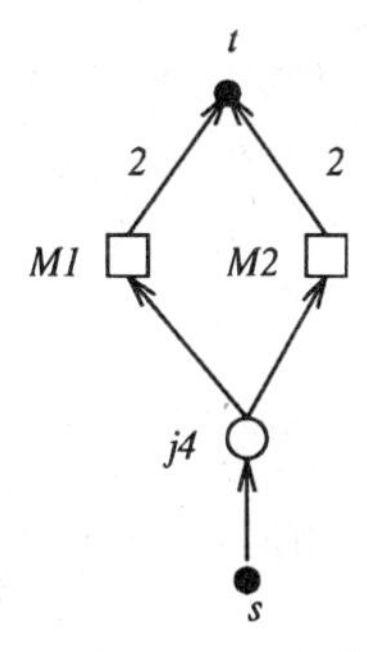

Figure 11.7: The second iteration: a 1×2 network.

Note, in using this method the size of the network is reduced, in the second iteration, to 1×2 nodes. This reflects the fact that the computational complexity of P.O.S.U algorithm is dominated by that of the first iteration.

11.4 SINGLE ARBITRARY-TIME OPERATION JOBS

In this section we examine an extended version of the scheduling problem in Section 11.3 by allowing each job to require an arbitrary processing time; or to schedule a set of n simple single operation jobs denoted by $\mathbf{J}=\{J_1, \ldots, J_n\}$, each job requires an arbitrary processing time, on a partially-overlapped system denoted by $\mathbf{M} = \{M_i | M_i \underline{\underline{\Delta}}(C_i, T_i), t_{i,\rho_i} \in \{t_1, \ldots, t_n\}, 1 \leq i \leq m\}$, where t_j is the processing time of J_j on any appropriate machine in **M**, such that the makespan is minimized. To guarantee that a feasible assignment does exist it is assumed that: $\mathbf{J} \subset \bigcup_{i=1}^{m} C_i$.

Theorem 4 Scheduling a set of single arbitrary-time operation jobs on P.O systems is NP-hard.

Proof:

Assume a P.O system denoted by $\mathbf{M} = \{M_i | M_i \underline{\Delta}(C_i, T_i | t_{i,\rho_i} \geq 1), 1 \leq i \leq m\}$ is used to process a set of jobs denoted by, $\mathbf{J} = \{J_1, \ldots, J_n\}$. Let $\mathbf{J}_k$ be a subset of **J**, and each job in $\mathbf{J}_k$ requires the same group of machines $\mathbf{M}_k$. Further, assume that $\mathbf{J}_k$ can be partitioned into two subsets $\mathbf{J}_{k1}$ and $\mathbf{J}_{k2}$, such that, $\mathbf{J}_{k1} \subseteq \bigcup_{M_i \in \mathbf{M}_k} \varsigma_i^k$, and $\mathbf{J}_{k2} \subseteq \bigcup_{M_i \in \mathbf{M}_k} \vartheta_i^k$ i. Since each job in $\mathbf{J}_{k1}$ requires a specific machine, therefore, scheduling these jobs is solvable in polynomial time. Whereas, each job in $\mathbf{J}_{k2}$ can be processed by any machine $\mathbf{M}_k$. Therefore, there exists a schedule for jobs in $\mathbf{J}_{k2}$ which minimizes the makespan iff there is a schedule to simple single arbitrary-time operation jobs on G.P systems, which is known to be *NP-hard* [9].

11.4.1 SOLUTION METHODOLOGY

An optimal solution for the above problem is unlikely to be found (Theorem 4). Thus, a heuristic algorithm is proposed to generate solutions that are near to the optimal and better than the random solutions. In this section we develop a strategy for an algorithm which extends the Multifit algorithm [10], which is designed for G.P systems. The Multifit algorithm combines binary search and *first fit decreasing* order (FFD) bin-packing algorithms. The idea of this algorithm is to identify an initial solution space by determining the initial lower and upper bounds for the makespan of the schedule. Then the binary search algorithm utilizes these boundaries to suggest a value for the makespan. This value is, then, used by the FFD algorithm to determine if there is a schedule for the given set of jobs on the given number of machines satisfies this value of makespan. Accordingly, the binary search updates the values of the upper and the lower bounds. If the updated boundaries are equal then a solution with a makespan close to the optimal is found. Otherwise, the process is repeated.

For P.O systems the solution space, or the values of the initial lower and upper bounds of the schedule's makespan can be determined as follows:

Lower-bound (Cl): for a P.O system with m machines processing n jobs, each requires an arbitrary processing time, there is no schedule can have a makespan better than **Cl** [11], where

$$\mathbf{Cl} = \max\left\{\left(\sum_{j=1}^{n} t_j\right) / m, \max\{t_j\}_{1 \leq j \leq n}\right\}.$$

Upper-bound (Cu): assume a P.O system with m machines (**M**) processing a set of n jobs (**J**), each job requires an arbitrary processing time. Further, suppose that **J** is divided into k subsets, $\bigcup_{l=1}^{k} \mathbf{J}_l = \mathbf{J}$, such that each job in $\mathbf{J}_l$ can be processed by any machine in $\mathbf{M}_l$, which is a set of

m_l machines and $\mathbf{M}_l \subset \mathbf{M}$. Then the upper bound for the makespan of scheduling $\mathbf{J}_l$ on $\mathbf{M}_l$ is equal to

$$Cu_l = \max\left\{(2/m_l)\sum_{J_j \in \mathrm{J}_l} t_j, \max\{t_j | J_j \in \mathrm{J}_l\}\right\}$$

Thus, the upper bound of the completion time of each machine can be calculated as $\mathrm{T}_i = \sum_{l=1}^{k} \tau_l$, where $\tau_l = Cu_l$ if $M_i \in \mathbf{M}_l$, otherwise $\tau_l = 0$. Hence, it is obvious to conclude that there is no need to consider schedules with makespan greater than **Cu**, where **Cu** $= \max_i \mathrm{T}_i$. Then these boundaries, **Cl** and **Cu**, are utilized to determine a near-optimal solution. The algorithm starts by determining some value **C** (a suggested value for the makespan) which is the midway value between **Cl** and **Cu**. Next a *directed first fit decreasing* order (DFFD) algorithm is invoked to check how close **C** is to the optimal. This can be achieved by transferring the scheduling problem to a bin-packing problem. Each job is considered as an item with a size equal to the processing time of the job. These items are then required to be placed in bins of capacity **C**. DFFD is named as a directed algorithm because of the structure of P.O systems in which a job can not be processed by any machine (as the case with G.P systems) but only by the appropriate machines. This restricts the number of possible choices of the appropriate machines for each job. For example, assume a P.O system with two machines, M_1 with $C_1 = \{J_1, J_2\}$ and M_2 with $C_2 = \{J_1\}$. Hence, there are two possible choices of appropriate machines for J_1, namely, M_1 and M_2, but one choice for J_2 which is M_1. Therefore, in order to deal with this restriction the set of items is represented as a linked list structure such that each item is pointed to its appropriate bin. Then the items list is divided into sub-lists. A sub-list is a collection of items that have the same order, or the same number of possible choices. The items in a sub-list is put in non-decreasing order from the perspective of the size. Finally, the sub-lists are put in an increasing order. Then DFFD algorithm utilizes this list to produce a number m', which is assumed to be close to the minimum number of bins that are required to pack the items. Then, if $m' > m$, **C** becomes the new lower bound, otherwise, **C** becomes the new upper bound. This procedure repeated until the solution space cannot be divided further, or **C**= **Cu** = **Cl**. And hence, the approximate solution is found.

P.O.S.A Algorithm

1. Calculate the initial values for **Cl** and **Cu**;
2. If (**Cl**= **Cu**) Then terminate with a near-optimal solution.
3. **C** = (**Cl** + **Cu**)=2;
4. **DFFD(C,**m'**)**;
5. If ($m' > m$)
   5.1 Then **Cl**= **C**;
   5.2 Else **Cu** = **C**;
6. Goto 2;

Function: DFFD(C,m'**)**

1. $m' = m$;

2. For $j = 1; n$
   2.1 Stop = False;
   2.1 Do
     2.1.1 $i = J_j$.machine→No;
     2.1.2 If (Load(M_i) + $t_j \geq$ **C**)
     2.1.3 Then
       2.1.3.1 Load (M_i) = Load(M_i) + t_j
       2.1.3.2 Stop = True;
     2.1.4 Else J_j..machine = J_j..machine→next;
   2.2 While((J_j..machine $\neq$Null) and (Not Stop));
   2.3 If (Not Stop) Then $m = m + 1$;
3. End For;
4. Return (**End of DFFD**).

11.4.2 EXAMPLE

Consider a fully automated work cell, *AWC-2*, consisting of four multiple-operation machines, denoted by M_1, M_2, M_3, and M_4. Each of them is characterized by its own capability set. M_1 is capable of drilling with a diameter range of (*5mm-15mm*), and milling with an angle range of (0^0, 35^0), and grinding with wheels of diameter range of (*5mm-20mm*); M_2 is capable of drilling with a diameter range of (*2mm-10mm*), reaming with a diameter range of (*2mm-7mm*), and milling with an angle range of (40^0 ,45^0); M_3 is capable of tapping with a diameter range of (5mm-*20mm*), drilling with a diameter range of (*10mm-15mm*), and reaming with diameters of (*5mm, 6mm, 8mm*); M_3 is capable of tapping with a diameter range of (*2mm-10mm*), and reaming with diameters of (*7mm, 10mm*), drilling with a diameter range of (*5mm-10mm*), and milling with an angle range of (0^0 , 45^0).

A set of nine independent objects need to be processed by *AWC-2*. Each object requires a single operation. The first object requires drilling with diameter of *9mm*. The second requires drilling with a diameter of 12*mm*. The third requires milling with an angle of 30^0 . The forth requires milling with an angle of 45^0 . The fifth requires tapping with a diameter of *10mm*. The sixth requires reaming with a diameter of *8mm*. The seventh requires reaming with a diameter of *7mm*. The eighth requires reaming with a diameter of *5mm*. Finally the ninth requires grinding with an *8mm* wheel.

AWC-2 is a multi-machine system consisting of four machines, further, it is a P.O system because the machines' capabilities are partially overlapped. For example, M1 and M2, each is capable of drilling with a diameter range of (*5mm-10mm*). Therefore, *AWC-2* $=\{M_i | M_i \underline{\underline{\Delta}}(C_i, T_i), 1 \leq i \leq 4\}$. The set of objects (jobs) is denoted by **J**$=\{J_1, J_2, J_3, J_4, J_5, J_6, J_7, J_8, J_9\}$. Thus, the capability set of each machine is formulated as $C_1 = \{J_1, J_2, J_3, J_9\}$, $C_2 = \{J_1, J_4, J_6, J_8\}$, $C_3 = \{J_2, J_5, J_7, J_8\}$, and $C_4 = \{J_1, J_3, J_4, J_5, J_6\}$; where the processing time of these jobs is given by t_1 = *7* time-units, t_2 = *6* time-units, $t_3 = t_4$ = *5* time-units, t_5 = *4* time-units, t_6 = *3* time-units, and $t_7 = t_8 = t_9$ = *2* time-units. The jobs list is divided into three sub-lists: *sub1*= $\{J_7, J_9\}$, *sub2*= $\{J_2, J_3, J_4, J_5, J_6, J_8\}$, and *sub3*= $\{J_1\}$. After four iterations the algorithm generates the following schedule with a makespan of *12* time-units: $\{J_2, J_9\}$ is assigned to M_1; $\{J_4, J_6\}$ is assigned to M_2; $\{J_5, J_7, J_8\}$ is assigned to M_3; and $\{J_1, J_3\}$ is assigned to M_4. This solution, as obtained by the algorithm, is an optimal solution for this case. In general, however, this will not be the case.

11.5 SCHEDULING MULTIPLE UNIT-TIME OPERATION JOBS

In this section we examine an extended version of single time unit scheduling problem in which a job is a set of multiple operations, and each operation requires a single time-unit. Formally, the problem is to schedule a set of n simple multiple operations jobs denoted by **J** $= \{J_j | J_j \underline{\underline{\Delta}} (O_j | \ |O_j| \geq 1), 1 \leq j \leq n\}$, on a partially-overlapped system denoted by **M** $= \{M_i | M_i \underline{\underline{\Delta}} (C_i, T_i | t_{i,p_i} = 1), 1 \leq i \leq m\}$, such that the **goodness objectives** (independence and load-balance) are satisfied, where the

1. **Independence objective** is to minimize the number of machines that are required to process each job.
2. **Load-balance objective** is to minimize the difference between the time-load of each machine for a pair of machines.

To guarantee that a feasible assignment does exist it is assumed that: $\bigcup_{j=1}^{n} O_j = \bigcup_{i=1}^{m} C_i$.

First we justify the selection of the performance criteria (goodness objectives). A job in P.O systems might need to be processed by more than one machine. Assigning a job to more than one machine adds problems due to the travel of the job between machines in order to be processed. We call these problems interaction overhead. These problems are similar to the ``inter-process-communication" that are being addressed in computer engineering field in the context of G.P systems [16]; and to the intercellular part movements between cells in flexible manufacturing systems [8]. Here, such as in FMS, we deal with environments in which the time cost of interaction overhead can become comparable to the processing time. From this perspective, the independence objective is considered as a performance measure because

Observation 1 The interaction overhead between machines can be minimized by minimizing the number of machines that are required to process each job.

In addition to the independence objective, other important objectives must also be considered to achieve a `good' scheduling strategy for P.O systems. Here we consider minimizing the makespan, and maximizing the utilization of the machines. We deal with these objectives through the load-balance because

Observation 2 An assignment of jobs to machines, in multi-machine systems, in which the load-balance is obtained guarantees the minimum makespan and the maximum utilization of the machines.

A closer look at the relationship between these objectives (independence and load- balance) shows that they are conflicting with each other. While the independence objective tends to assign the whole of a job to a single machine, the load-balance tries to make the time-load distribution evenly among the machines. Thus, satisfying both objectives together seems to be a hard, if not impossible, problem to address.

Theorem 5 Scheduling a set of simple multiple operations jobs, each operation requires a single time-unit, on P.O machines; in a way it minimizes the makespan and the interaction overhead, is *NP-hard.*

Proof:
Consider a P.O system with a unit time processing speed denoted by **M** $=\{M_i | M_i \underline{\underline{\Delta}}(C_i, T_i | t_{i,\rho_i} = 1), 1 \le i \le m\}$, is used to process a set of jobs denoted by **J** $=\{J_j | J_j \underline{\underline{\Delta}}(O_j \| O_j | \ge 1), 1 \le j \le n\}$. Assume that each job in **J** can be partitioned into l_j sub-jobs $\forall j$, and denoted by $J_j = \{J_{jk} | J_{jk} \underline{\underline{\Delta}}(O_{jk}, |O_{j_k}| \ge 1), 1 \le k \le l_j\}$, where J_{jk} is a subjob k of J_j. Each of these subjobs requires one machine to process its operations. Then, for minimum interaction overhead the objective becomes

$$\begin{aligned} &\min\ l_j,\ \forall j\,; \\ &\text{s.t} \\ &1 \le l_j \le m\,; \\ &O_j = \bigcup_{\kappa=1}^{l_j} O_{j_\kappa}\,; \\ &|O_{j_\kappa}| \ge |O_{j_{\kappa+1}}|,\ 1 \le \kappa \le l_j - 1. \end{aligned}$$

$|O_{j_\kappa}|$ is equivalent to the processing time required by sub-job κ of J_j. Thus, for each possible l_j, $\forall j$, there exists a schedule to this problem which minimizes the makespan iff there is a schedule to simple single arbitrary-time operation jobs on P.O systems, which is *NP-hard* (Theorem *4*).

To solve this problem, in the following subsection, we formulate the above objectives mathematically.

11.5.1 PROBLEM FORMULATION

In this section we develop a mathematical formulation for the assignment problem of a P.O system. First, we determine the necessary and sufficient information that is required to model this problem. Consider a P.O system, **M**, with three machines, M_1, M_2, and M_3, each characterized by a capability set given by $C_1 = \{o_1, o_2, o_3, o_4\}$ for the first machine, and $C_2 = \{o_3, o_4, o_5\}$ for the second, and $C_3 = \{o_4, o_5\}$ for the third; each operation requires a single time-unit. Now, consider a single job J_1 consists of four operations, $O_1 = \{o_1, o_2, o_3, o_5\}$. Two processing choices are possible for J_1 on **M**, the first is to process J_1 on M_1 and M_2, and the second is to process J_1 on M_1 and M_3. For the first $\varsigma_1^1 = \{o_1, o_2\}$, $\varsigma_2^1 = \{o_5\}$, and $\vartheta_1^1 = \vartheta_2^1 = \{o_3, o_4\}$. For the second processing choice, $\varsigma_1^2 = \{o_1, o_2, o_3\}$; $\varsigma_3^2 = \{o_5\}$, and $\vartheta_1^2 = \vartheta_3^2 = \{o_4\}$. In other words, in the first choice J_1 requires two operations o_1 and o_2 from M_1's special capability set, and o_5 from M_2's special capability set. We call the set $\{o_1, o_2, o_5\}$ the necessary operations required by J_1 on M_1 and M_2. Whereas, the necessary operations of J_1 on M_1 and M_3 are $\{o_1, o_2, o_3, o_5\}$. Clearly that both of these choices require J_1 to be processed by the same number (two) of machines, however, the load-balance in the first choice is better than that in the second. This can be achieved by assigning o_3 to M_2. In this sense, a selection of an assignment from those which satisfy the independence objective, such that the load-balance is achieved, is determined by the number of necessary operations. For this, we introduce the **critical heuristic**: for a set of solutions which satisfy the independence objective select the one that minimizes the number of necessary operations required by each job.

Proposition 1 For a set of assignments of simple multi-operation jobs to a P.O system which satisfy the independence objective, the load-balance can be achieved by minimizing the number of necessary operation that are required from the machines.

Proof:
Assume a P.O system, which consists of three machines M_1, M_2, and M_3, processing a job, *J*, which consists of *n* operations. Suppose that *J* has two processing choices. In the first *J* can be processed by M_1 and M_2, in which n_1^1 of *J*'s operations can only be processed by M_1, and n_2^1 of them can only be processed by M_2, where $n_1^1 + n_2^1 < n$. The second choice represents *J* processed by M_1 and M_3, in which n_1^2 of its operations can only be processed by M_1 and n_3^2 of them can only be processed by M_3, such that $n_1^2 + n_3^2 < n$.

Consider the first choice in which it is a must that M_1 processes n_1^1 operations, and M_2 processes n_2^1 operations from *J*, due to the necessary operations. Thus, at this point the time-load difference between the time-load of M_1 and M_2 can be given by $ld_1 = |n_1^1 - n_2^1|$. This difference, ld_1, can be reduced (or minimized) only by an appropriate distribution for the set of the overlapped operations, which consists of $\delta_1 = n - (n_1^1 + n_2^1) = n - nc_1$ operations. Similarly, for the second processing choice, $ld_2 = |n_1^2 - n_3^2|$, and $\delta_2 = n - (n_1^2 + n_3^2) = n - nc_2$. Thus, to balance the load among the machines the objective becomes $\max_k \delta_k = \max_k (n - nc_k)$, such that $nc_k \leq n$, for a processing choice *k*, or equivalently

$$\min_k ld_k = \min_k nc_k \ .$$

Since nc_k for processing choice *k*, represents the necessary operations that are required from all the machines that are involved in processing the job; thus this formulation can be generalized for any number of machines.

Therefore, for a P.O system, the assignment problem can be formulated in terms of the processing choices and the necessary operations required by each job:

$$\mathrm{A}_j = \left[p_{ji\kappa_j}, nc_{ji\kappa_j}\right],$$

where, $nc_{ji\kappa_j}$ is the number of necessary operations needed by J_j from M_i in processing choice κ_j,

$$p_{ji\kappa_j} = \begin{cases} 1 & \text{if } J_j \text{ requires } M_i \text{ in processing choice } \kappa_j \\ 0 & \text{if } J_j \text{ does not require } M_i \text{ in processing choice } \kappa_j \end{cases},$$

where, $1 \le \kappa_j \le k_j$, and k_j is the number of processing choices of job J_j. In terms of the systems' structure a processing choice refers to a possible, non-redundant arc between job J_j and the appropriate machines. By non-redundant we mean that for an AND arc in the structure which connects a job node, having a set of operations denoted by O, to a group of machine nodes denoted by $\mathbf{M}$, there should be no other group of machines $\underline{\mathbf{M}}$ such that $\underline{\mathbf{M}} \subset \mathbf{M}$ and $O \subset \bigcup_{M_i \in \underline{\mathbf{M}}} C_i$. For example, if machine M_1 is capable of processing all the operations of J_1, than the AND arc connecting J_1 with M_1 and M_2 is redundant. Therefore, the `goodness' objectives can be formulated as follows:

*** Independence objective**

introducing the critical heuristic leads to a *bi-criterion* formulation for the independence objective, where

- the primary criterion is $\mathrm{I}_j = \min_{\kappa_j} \sum_{i=1}^{m} p_{ji\kappa_j}, \quad \forall j;$

s.t

$$p_{ji\kappa_j} \in \{0,1\}, \quad \forall j, i, \kappa_j;$$

- the secondary-criterion is $\Omega_j = \min_{\kappa_j} \sum_{i=1}^{m} p_{ji\kappa_j} nc_{ji\kappa_j}, \quad \forall j;$

s.t

$$p_{ji\kappa_j} \in \{0,1\}, \quad \forall j, i, \kappa_j;$$
$$nc_{ji\kappa_j} \ge 1, \quad \forall j, i, \kappa_j;$$

*** Load-balance objective**

$$L_d = \min_{\kappa_j} \sum_{i}^{m} \sum_{l=(i+1)}^{m} \sum_{j=1}^{n} \left| p_{ji\kappa_j} t_{ji\kappa_j} - p_{jl\kappa_j} t_{jl\kappa_j} \right|;$$

s.t

$$p_{ji\kappa_j} \in \{0,1\}, \quad \forall j, i, \kappa_j;$$
$$t_{ji\kappa_j} \ge 0, \quad \forall j, i, \kappa_j$$

where $t_{ji\kappa_j}$ is the time load assigned to machine i due to the processing of job j at processing choice κ_j.

The above formulation can be utilized as a multiple objectives formulation to solve the assignment problem. First, and hence forth, we assign a priority levels to the objectives as follows. The load-balance has the highest level. For this objective we assume that a decision maker has to decide on certain tolerance of the theoretical lower limit without making difference to the perceived quality of the solution. The independence objective is considered at the second priority level. The formulation of the independence objective is also reduced to a bi-criterion 0-1 integer programming problem, as shown above.

To solve this problem, a secondary criterion problem [7], firstly, the problem is solved from the perspective of the primary objective (the load-balance). Then the problem is considered for the secondary objective under the constraint that the primary objective remains optimum. This can be achieved by solving the primary objective problem as a simple integer programming problem, then solve for the secondary objective while keeping the solution for the first objective. Similarly we can solve for the bi-criterion independence objective. Nonetheless, it is well known that solving *integer programming* problems has a worst case time complexity of an exponential order [13]. Therefore, it is more practical to seek an approximate solution to the problem rather than insisting on the optimal. In the following sections we investigate finding such a solution in a polynomial time using the graphical representation.

11.5.2 HEURISTIC SOLUTION USING GRAPHICAL REPRESENTATION

In this section we develop a heuristic algorithm to solve the above problem which utilizes the graphical representation of the system's structure. First we modify the logical structure of the P.O systems in a way it can be used to model the assignment problem. It is clear that the AND-OR structure of P.O systems makes the assignment problem hard to deal with. Consequently, it becomes necessary to simplify the structure. One approach is to reduce the topology of the AND-OR structure to that of the OR structure, while preserving the characteristics of the original structure, we call it OR-like structure. To explain this, consider a P.O system processing a set of jobs, *JS-1*, its logical structure shown in Figure 11.8. *JS-1* is given such that $J_1 \Rightarrow (M_1 \text{ AND } M_2)$, $J_2 \Rightarrow (M_2)$, and $J_3 \Rightarrow (M_1 \text{ OR } M_2)$, where `$\Rightarrow$' means can be processed by. For each distinct set of machine nodes, that are connected to a job node by an AND arc, we construct a virtual machine node. For the above example, $M_{12}$(read as `machine one two" and characterized by a capability set $C_{12} = (C_1 \cup C_2)$) is created and added to the system's graph. By introducing the virtual machine, (M_{12}), the logical structure of *JS-1* becomes free of AND arcs, as shown in Figure 11.9; where $J_1 \Rightarrow$ (M_{12}), $J_2 \Rightarrow$ (M_1), and $J_3 \Rightarrow$ (M_1 OR M_2). Although the OR-like structure provides P.O systems with a simplified graph representation, an extra level of nodes is introduced over the AND-OR structure. An Or-like graph consists of

- Real machine nodes: nodes to represent all actual machines given in the system.
- Virtual machine nodes: nodes to represent virtual machines; each of these machines is characterized by a capability set equal to the union of the capability sets of some real machines (called members).
- Job nodes: nodes to represent all jobs given in the processed job set.

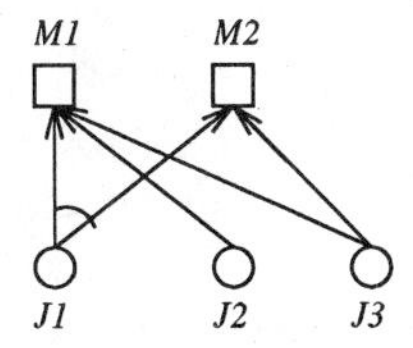

Figure 11.8: A simple graph representation for a P.O system

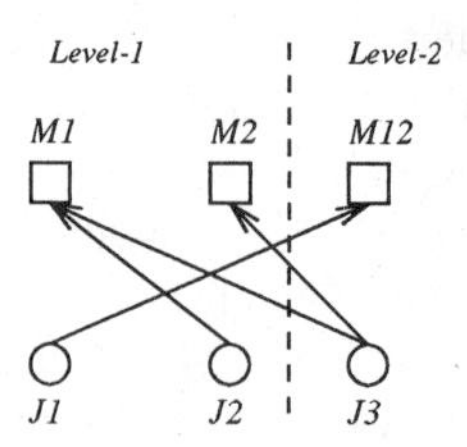

Figure 11.9: An OR-like representation for *JS-1*

To distinguish between real-machine and virtual-machine nodes in the graph, real-machines are classified as *level-1* machines, and virtual-machines are classified as *level-k* machines, where *k*, $k > 1$, is equal to the number of members that are involved in the virtual machine, as shown in Figure 11.9. This, in turn, introduces extra complication for the independence objective. Because, a *level-2* machine node represents a virtual machine which is a combination of two real machines, whereas, a *level-4* machine node represents a combination of four real machines. To deal with this problem the graph is divided into sub-graphs, each consists of all machine nodes of the same level, their arcs, and job nodes that are connected to these machine nodes. A sub-graph which contains machine nodes of *level-k,* for simplicity, is called *level-k.* Finally, the number of necessary operations that are required by a job on an appropriate machine (real or virtual) can be represented as weights on the arcs between them. In the following, we propose a method which utilizes this representation to solve the scheduling problem.

11.5.2.1 THE STRATEGY

In this section we propose a decomposition strategy based on two heuristics. The first is to satisfy the goodness objectives for each sub-graph separately. The second heuristic is applied when the satisfaction of the load balance fails. In this case, the current sub-graph is merged with the next level sub-graph to construct a new sub-graph. Then the first heuristic is applied for the new sub-graph, and so on until no job is left unassigned. Both the decomposition strategy and the graph representation reduces the independence objective to finding a feasible assignment at the lowest possible level. To show this, consider an assignment problem of a P.O system which is represented by an OR-like graph *G*. Let $\Gamma = \{\ell_1, \ell_2, \ldots, \ell_v | \ell_x$ is a subgraph of level-*x*, and $1 \le x \le v\}$ be the set of all sub-graphs of *G*. From the definition of the nodes' level and their corresponding sub-graphs, then the independence objective can be rewritten as

$$I_j = \min_{1 \le x \le v} x, \ \forall j \ ;$$

s.t

$$J_j \in l_x \ .$$

Or, for each job node we try to find the assignment at the lowest possible level. Then, the critical objective is to minimize the total weight for each sub-graph. Now we reach a point where we can introduce our method of obtaining an approximate solution for the assignment problem.

The main strategy of the proposed method is to solve the assignment problem by decomposing the system graph into sub-graph levels of increasing order. Then, the assignment problem for each sub-graph is solved locally using the linear-sum assignment algorithm [4]. Linear-sum assignments algorithms are designed to utilize Balanced-Complete (BC) bipartite graphs. A Balanced-Complete bipartite graph is defined as a simple bipartite graph with a bipartition (**X**;**Y**), **X** and **Y** are sets of nodes, having the following properties: (1) $|\mathbf{X}| = |\mathbf{Y}|$, (2) each node of **X** is connected to each node of **Y** . However, these algorithms can be used to accommodate systems with non-BC graphs such as the P.O systems. In some cases of P.O systems, the number of machines are not equal to the number of jobs, thus $|\mathbf{X}| \neq |\mathbf{Y}|$. In addition, some of the machines, in P.O systems, are not capable of performing some operations, and hence, the second property is violated. The problem, here, is how to convert a non-BC graph model of a P.O system to a BC graph in a way it represents the assignment problem. Then how to extend the linear sum assignment algorithm to produce a feasible assignment for the P.O systems which satisfies the goodness objectives.

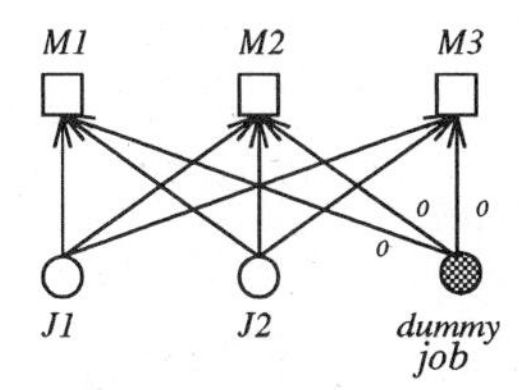

Figure 11.10: Converting a non-BC graph into a BC , $m > n$

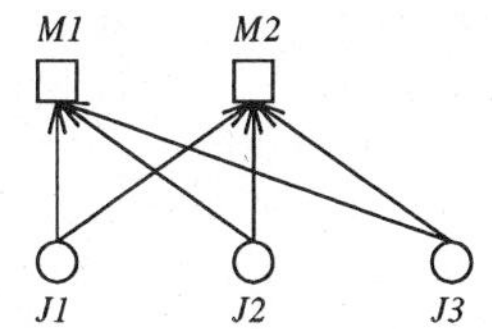

Figure 11.11: A BC graph, $m < n$

First consider the case when there are more machine nodes than job nodes, $m > n$, in the sub-graph. The perfect load-balance, in this case, is to assign at most one job to each machine ($\lceil n/m \rceil = 1$). Thus, to convert such a sub-graph to a BC graph, in a way it represents the perfect load-balance, the number of job nodes is increased, to match the number of machine nodes, by adding dummy job nodes. A dummy job is characterized by an empty operation set, and hence, each dummy job node is connected to each machine node by an arc with a zero weight, as shown in Figure 11.10.

Now, consider the other case of $|\mathbf{X}| \neq |\mathbf{Y}|$ in which $m < n$, as shown in Figure 11.11. The perfect load-balance, in this case, is to assign at most $\lceil n/m \rceil > 1$ jobs to each machine. To convert such a sub-graph to a BC graph that is able to represent the perfect load-balance, the number of machine nodes is increased, to match the job nodes, by creating ($\lceil n/m \rceil - 1$) copies of each existing machine node (original node). The set of original and its copy nodes represents one machine. The new sub-graph contains a total of $m\lceil n/m \rceil$ machine nodes, originals and copies, however, n is not necessarily an integer multiple of m. This means $m\lceil n/m \rceil \geq n$, and hence, its possible that the new graph is still a BC graph. Thus, the number of job nodes, in case of $m\lceil n/m \rceil > n$, should also be increased by ($m\lceil n/m \rceil - n$) dummy job nodes to match the new number of the machine nodes. To preserve the load-balance objective each machine copy node is connected to the same job nodes of its original by the same arcs except for the arcs that are connected to the dummy job nodes. The perfect load-balance for assigning the dummy-job nodes, with a number of nodes less than m, is to assign at most one dummy job to each machine (original and copies). This can be represented by assigning zero weights to the arcs between dummy job nodes and original nodes, and infinity, ∞, weights to the arcs between copy nodes and dummy job nodes, as shown in Figure 11.12.

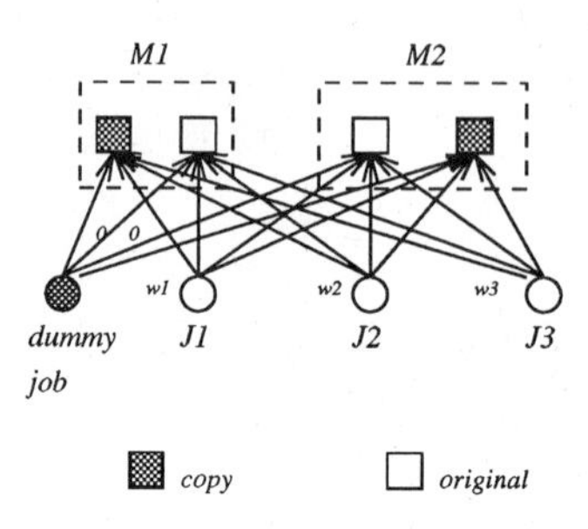

Figure 11.12: Converting a non-BC graph into a BC graph, $m > n$.

Finally, consider the case of a non-BC graph which violates the second property of a BC bipartite graph. This graph can be converted to a BC graph by adding extra arcs with ∞weights between the unconnected machine (original and copies) and job nodes. Note, these arcs graphically represent infeasible assignment between these nodes.

At this stage, the linear sum assignment algorithms can be used to solve the assignment problem for P.O systems, although the generated solution may be infeasible. It is easy though, to check if the assignment is feasible or not. Simply, by checking the total number of the necessary operations (*total necessary-load*) of the assignment. If it is less than ∞ then a balanced-load assignment is found. If it is not, then there are some jobs not appropriately assigned yet . These jobs are assigned to machines through arcs with ∞ weights (*unprocessed jobs*). Now, we extend the linear sum assignment algorithms to produce a feasible assignment for P.O systems. Let ω_ι^k be the total necessary-load of machine ι (real or virtual) at *level-k*. Also assume that the algorithm generates an assignment for *level-k* sub-graph with a summation of total necessary-loads given by

$$W_{\min} = \omega_1^k + \omega_2^k + \omega_3^k + \ldots + \omega_\kappa^k.$$

Definition 7 A machine is called **not-needed** if it has been assigned to a total necessary-load of ω such that $\omega = \infty$; but if $\omega < \infty$ and at most one dummy job is in the set of jobs that are assigned to that machine, then it is called **needed.**

Assume, without loss of generality, that the generated solution is infeasible assignment with two unprocessed jobs, J_{i_2} and J_{i_3}, which are assigned to machine 2, and machine 3 respectively (with arcs of ∞ weights). According to Definition 8, machine 2 and machine 3 are **not-needed** machines, and each of the other machines is a **needed** machine. To produce a feasible assignment these unprocessed jobs need to be reassigned to the appropriate machines.

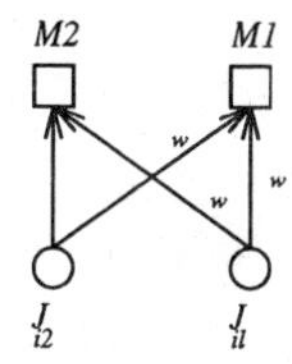

Figure 11.13: Re-assigning an unprocessed job to an appropriate **needed** machine.

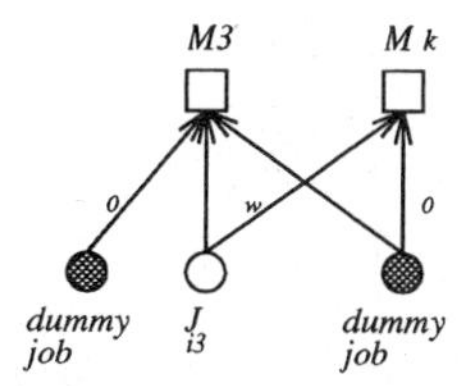

Figure 11.14: Re-assigning an unprocessed job to a **needed** machine.

Lemma 3 An unprocessed job can only be reassigned to an appropriate **needed** machine.

Proof:

Note by definition a **needed** machine can be assigned only to at most one dummy job, this leaves us with three cases to examine.

Case-1, assume that machine 1 is a **needed** machine with no dummy job and connected to J_{i_2} by an arc with some weight $w<\infty$. Assume job J_{i_1} is connected to machine 1 with an arc of weight $w''<\infty$, and to machine 2 with an arc of weight w', as shown in Figure 11.13. Now, suppose that the linear-sum algorithm generates a solution in which J_{i_2} is assigned to machine 2, and J_{i_1} is assigned to machine 1. However, it is possible to generate another solution in which J_{i_1} is assigned to machine 2 and J_{i_2} is assigned to machine 1. This solution is characterized with a total necessary-load less than $W_{\min}$ (which contradicts the fact that the linear-sum algorithm is an optimization algorithm), unless $w'=\infty$ and $w''<w$ which is possible. Since $w<\infty$, therefore, machine 1 (**needed**) is appropriate for J_{i_2} (unprocessed).

Case-2, assume that machine κ is **needed** with one dummy job and connected to J_{i_3} with some weight w such that $0<w<\infty$, as shown in Figure 11.14. Suppose that the linear-sum algorithm generates an assignment in which J_{i_3} is assigned to machine 3. Since the algorithm is an optimization algorithm, then the only job node that machine κ can exchange with another machine (say machine 3) is the dummy job. Now, if machine 3 has not been assigned to a dummy job, then there is another possible solution in which J_{i_3} is assigned to machine κ and the dummy job that was assigned to machine κ is assigned to machine 3. This solution would have a total necessary-load less than $W_{\min}$ (which contradicts the fact that the linear-sum algorithm is an optimization algorithm). This leaves the fact that machine 3 must have been assigned to a dummy job. Therefore, if the dummy job of machine κ and J_{i_3} are switched between machines κ and 3, we get:

$$W = \omega_1^k + \omega_2^k + \dot{\omega}_3^k + \ldots + \dot{\omega}_\kappa^k > W_{min}$$

where $\omega_3^k = \dot{\omega}_3^k = \infty$ because no machine can be assigned to more than one dummy job; and $\dot{\omega}_\kappa^k = \omega_\kappa^k + w > \omega_\kappa^k$. Since, this is a feasible assignment which is assumed to be ignored by the algorithm, and $w < \infty$; therefore, machine κ (**needed**) with one dummy job is appropriate machine for J_{i_3} (unprocessed job).

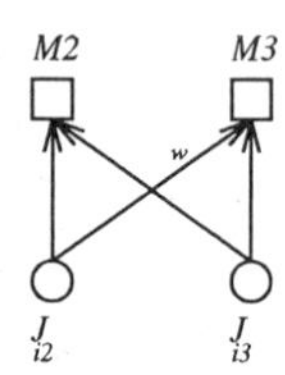

Figure 11.15: Re-assigning an unprocessed job to a **not-needed** machine.

Case-3, suppose that J_{i_2} is connected to machine 3 with some weight w, as shown in Figure 11.15. Assume that the linear-sum algorithm generated a solution in which J_{i_2} is assigned to machine 2 and J_{i_3} is assigned to machine 3. However, there exists another assignment in which J_{i_2} is assigned to machine 3 and J_{i_3} is assigned to machine 2. This solution is associated with a total necessary-load less than W_{min} (this contradicts the fact that the linear-sum algorithm is an optimization algorithm), unless $w = \infty$. Thus, machine 3 (**not-needed**) is not appropriate machine for J_{i_2} (unprocessed job).

Using lemma 1, then, the linear-sum algorithm can be extended to distribute the unprocessed jobs between the appropriate **needed** machines such that the load-balance is acceptable. If such an assignment exists, say at *level-k'*, then the accumulated time-load of each machine due to this assignment and the ones before, is passed to the next level as necessary operations for the corresponding machine. But, if a balanced assignment doesn't exist at certain level (say at *level-k'*), then both *level-k'* and *level-(k'+1)* sub-graphs are merged into a new sub-graph. Then, we solve for a feasible assignment which considers the goodness objectives. In this situation the independence objective, within the merged levels, is treated as to make the chance of selecting machines of the lower levels is greater than that of the higher levels. One simple approach to achieve this is the offsetting technique. By this, the machines at the lower levels will be selected over that at the next higher levels, except when there is no alternative at the lower levels; then, machines at the next higher level are explored, and so on.

Finally, the algorithm solution excludes the dummy jobs, and treats the original and copy nodes for the same machine as one machine node. Furthermore, each virtual-machine is also unfolded and its assignment solution is distributed among its real-machines. For example, assume that job J_1 is assigned to a virtual-machine M_{12}. In the assignment solution, M_{12} is unfolded to M_1 and M_2, and operations of J_1 is distributed among M_1 and M_2 job lists. Therefore, the assignment solution includes assignment lists between real-machines and jobs only. Now we reach the point where we can introduce the algorithm steps.

11.5.2.2 THE ALGORITHM

In this section the steps of an approximation algorithm called P.O.M.U (Partially Overlapped system with Multiple Unit-time operation jobs) which utilizes the above strategy is listed.

P.O.M.U Algorithm

1) Set $\mathbf{M} = \{M_i | M_i \underline{\underline{\Delta}} (C_i, T_i), 1 \le i \le m\}$;
2) Set $\mathbf{J} = \{J_j | J_j \underline{\underline{\Delta}} (O_j), 1 \le j \le n\}$;
3) Set assignment-list = ϕ;
4) Set acceptable-assignment = True;
5) Set $k = 0$;
6) If ($\mathbf{J} \ne \phi$) then
   - Set $k = k + 1$;
   - *level-k* = Get-OR-like(k);
   - If ($\mathbf{J}_k \ne \phi$) then
     - If (acceptable-assignment = False) then
       - *level-k = level-k* $\cup$ *Offsetting(level-($k-1$))*;
       - Set $m' = |\mathbf{M}_k|$;
       - Set $n' = |\mathbf{J}_k|$;
     - perfect-level-k = Perfect-matching(*level-k*);
     - acceptable-assignment = Assign(perfect-level-k);
     - $\mathbf{J} = \mathbf{J} - \mathbf{J}_k$;
     - If ((acceptable-assignment = True) and ($\mathbf{J} = \phi$)) then
     - Update the assignment-list using temp-assignment-list;
7) Goto 6.

Function: Get-OR-like(k)

1) Set $\mathbf{J}_k = \phi$;
2) Set $\mathbf{M}_k = \phi$;
3) For $j = 1$; $|\mathbf{J}|$
   - Find all possible combinations of exactly k machines, $CM_{j,y} = \{M_{\beta_{y1}}, \ldots, M_{\beta_{yk}}\}$,

     where y is the number of the combinations, $1 \le \beta_{yi} \le m, 1 \le i \le k$,

     and $M_{\beta_{yi}} \in \mathbf{M}$, that are required to process O_j;
   - If $y \ne 0$ Then $\mathbf{J}_k = \mathbf{J}_k \cup J_j$;
4) If $\mathbf{J}_k \ne \phi$ then
   - For $j = 1$; $|\mathbf{J}_k|$

- For $y = 1; y_j$
- For $i = 1; k$
- Determine the number of necessary operations, $\left|O_j \cap \varsigma^i_{\beta_{yi}}\right|$;
- If $(k \neq 1)$ then
- Identify a corresponding virtual machine for $CM_{j,y}$, denoted by $M_{(\beta_{y1} \ldots \beta_{yk})}$;
- Determine the number of necessary operations of $M_{(\beta_{y1} \ldots \beta_{yk})}$ which is equal to the sum of the necessary operations for each associated machine ;
- Connect J_j to $M_{(\beta_{y1} \ldots \beta_{yk})}$ with an arc having weight equal to the number of necessary operations of $M_{(\beta_{y1} \ldots \beta_{yk})}$;
- Set $\mathbf{M}_k = \mathbf{M}_k \cup M_{(\beta_{y1} \ldots \beta_{yk})}$;
- Set $m' = |\mathbf{M}_k|$;
- Set $n' = |\mathbf{J}_k|$;

5) Return($\mathbf{M}_k$; $\mathbf{J}_k$).

Function: Offsetting(*level-k*)

1) Set offset = max{weights of *level-k*};
2) For each arc at *level-*$(k-1)$
 - arc-weight = arc-weight + *offset*;
3) Return(*level-*$(k-1)$.

Function: Perfect-matching(*level-k*)

1) For $y = 1; m'$
 - For $j = 1; n'$;
 - If ($M_{(\beta_{y1} \ldots \beta_{yk})}$ is not connected to J_j) then
 - Connect $M_{(\beta_{y1} \ldots \beta_{yk})}$ to J_j with arc having ∞ weight;
2) If ($m' < n'$) then
 - For $l = 1; (m' - n')$
 - Create a $dummy_l$ job node;
 - For y = 1; m'
 - Connect $dummy_l$ node to $M_{(\beta_{y1} \ldots \beta_{yk})}$ with arc having zero weight;
 - Else
 - If ($m' < n'$)
 - For $y = 1; m'$
 - For $x = 1; (\lceil n'/m' \rceil$-1)
 - Create a $copy_x$ node of the original $M_{(\beta_{y1} \ldots \beta_{yk})}$;
 - For j = 1; n';

- Connect $copy_x$ node to J_j with arc having the same weight of the arc between $M_{(\beta_{y1} \ldots \beta_{yk})}$ and J_j;
- If ($n' \ mod \ m' > 0$) then
- For $l = 1$; ($\lceil n'/m' \rceil$ - n)
  - Create a $dummy_l$ job node;
  - For $y = 1; m'$
  - Connect $dummy_l$ node to the original $M_{(\beta_{y1} \ldots \beta_{yk})}$ with arc having zero weight;
  - For $x = 1$; ($\lceil n'/m' \rceil$ - 1)
  - Connect $dummy_l$ node to $copy_x$ of $M_{(\beta_{y1} \ldots \beta_{yk})}$ with arc having ∞weight;

3) Return($\mathbf{M}_k$; $\mathbf{J}_k$, *copy-nodes*, *dummy-nodes*)

Function: Assign(perfect-level-k)

1) Set temp-assignment-list =ϕ;
2) Set cost = minimal weighted perfect matching in the edge weighted bipartite graph of perfect-level-k;
3) If (cost = ∞) then
   - Determine a list for each of the **needed**-machines, and the unprocessed-jobs;
   - Reassign the unprocessed-jobs to **needed** machines;
4) For each arc satisfied in Steps (2) and (3), add its job and machine nodes into temp-assignment-list;
5) For $y = 1;m'$
   - Unfold $M_{(\beta_{y1} \ldots \beta_{yk})}$ into its associated machines in **M** ;
   - Adjust the time-load for each associated machine;
   - Distribute the overlapped operations among the associated machines in a way it reduces the time-load difference between them;
6) Set total-load = *0*;
7) Set tolerance = *value*;
8) Set feasible = True;
9) For $i = 1; m$
   - total-load = total-load+ time-load-of-M_i;
10) Set average-load = total-load / m;
11) Set upper-bound = average-load(1 + tolerance);
12) Set lower-bound = average-load(1 - tolerance);
13) For i = $1; m$
    - If (upper-bound < time-load-of-M_i < lower-bound) then
    - Set feasible = False;
14) Return(feasible).

In the following the performance of the algorithm is illustrated by a detailed example.

11.5.2.3 EXAMPLE

Consider a fully automated work cell, *AWC-3*, consisting of three multiple operation machines. Each of them is characterized by its own capability set. The first machine is capable of drilling with a diameter range of (*5mm-15mm*); tapping with a diameter range of (*5mm-20mm*); milling with an angle range of (0° ,45°). The second is capable of drilling with a diameter range of (*2mm-10mm*); reaming with a diameter range of (*2mm-10mm*); grinding with wheels diameter range of (*5mm- 20mm*). The third is capable of drilling with a diameter range of (*8mm-20mm*); reaming with a diameter range of (*6mm-20mm*); milling with an angle range of (30° ,90°). All the operations have processing time of a single time-unit. A set of five independent objects need to be processed by *AWC-3*. Each object requires a set of operations. The first object requires drilling with diameters of *9mm* and *12mm*; tapping with a diameter of *10mm*; reaming with a diameter of *5mm*. The second requires drilling with a diameter of *9mm*; milling with an angle of 30° ; grinding with an *8mm* wheel; reaming with a diameter of *7mm*. The third requires a drilling with diameters of *9mm* and 12*mm*; reaming with diameters of *7mm* and *8mm*. The forth requires milling with angles of 30° and 45° ; grinding with an *8mm* wheel; drilling with a diameter of *9mm*. Finally, the fifth requires drilling with diameters of *9mm* and *12mm*; milling with angles of 30° and 45° .

AWC-3 is a multi-machine system consisting of three machines, M_1, M_2, and M_3. *AWC-3* is classified as a P.O system because its machines' capabilities are partially overlapped. For example, $M_1$and M_2, each is capable of performing drilling with a diameter range of (*5mm-10mm*). Therefore, *AWC-3* $=\{M_i | M_i \underline{\underline{\Delta}}(C_i, T_i | t_{i,\rho_i}=1), 1 \leq i \leq 3\}$. Each of the objects can be considered as a job, J_j, for $1 \leq j \leq 5$.

Only a subset of the machines' capabilities is considered here, because not all of them are required by the given set of jobs. Hence, the following is a list of all operations that are required by the jobs.

- Drilling with diameters of *9mm* and *12mm* we call them o_1 and o_2 respectively.
- Milling with angles of 30° and 45° we call them o_3 and o_4 respectively.
- Tapping with a diameter of *10mm* we call it o_5.
- Reaming with diameters of *8mm, 7mm,* and 5mm we call them o_6, o_7, and o_8 respectively.
- Grinding with a wheel of 8mm diameter we call it o_9.

Accordingly, the capability sets of the machines are $C_1 = \{o_1, o_2, o_3, o_4, o_5\}$, $C_2 = \{o_1, o_6, o_7, o_8, o_9\}$and $C_3 = \{o_1, o_2, o_3, o_4, o_6, o_7\}$; $t_k = 1, 1 \leq k \leq 9$. The operation sets required by the jobs are $O_1 = \{o_1, o_2, o_5, o_8\}$, $O_2 = \{o_1, o_3, o_7, o_9\}$, $O_3 = \{o_1, o_2, o_6, o_7\}$, $O_4 = \{o_1, o_3, o_4, o_9\}$, and $O5 = \{o_1, o_2, o_3, o_4\}$.

The OR-like bipartite graph model of the system is shown in Figure 11.16. This graph model consists of two sub-graphs, called *level-1* and *level-2*. *Level-1* sub-graph is not a BC bipartite graph which is then converted into a BC graph, as shown in Figure 11.17. The algorithm disconnects J_3 from *level-2* sub-graph, and then finds the best assignment solution in *level-1*. The assignment solution of *level-1* is shown in Figure 11.18.

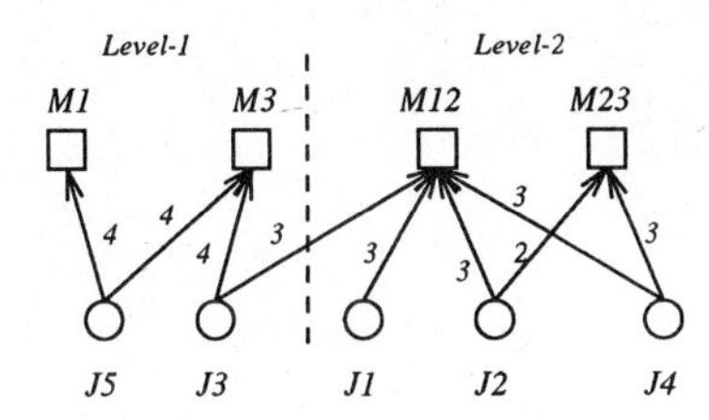

Figure 11.16: An OR-like bipartite graph model of *AWC-3*.

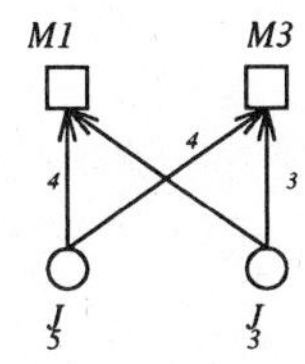

Figure 11.17: *Level-1* as a BC sub-graph

Next, the algorithm considers *level-2* sub-graph, where J_3 is excluded. This sub-graph is then converted to a BC as shown in Figure 11.19. The assignment solution is shown in Figure 11.20. Each assignment at this level is unfolded to the corresponding real machines. For example, the assignment of J_1 to M_{12} is unfolded, and the operations of J_1 is distributed among M_1 and M_2, as shown in the machines' load condition, in Figure 11.20.

This solution, as obtained by the algorithm, is an optimal solution for this case. The sequencing algorithm given in [12] is then used to obtain the `complete' schedule shown in Figure 11.21 with maximum completion time of 8. This schedule requires no preemption and no idle time on any of the machines. In general, however, this will not be the case.

Level-1

----

job .. J_5 .. Assigned to .. M_1 .

job .. J_3.. Assigned to .. M_3 .

The machines Load condition

Machine 1: J_5 $\{o_1, o_2, o_3, o_4\}$

Machine 3: J_3 $\{o_1, o_2, o_6, o_7\}$

==================================

Figure 11.18: The assignment solution for level-1 machines

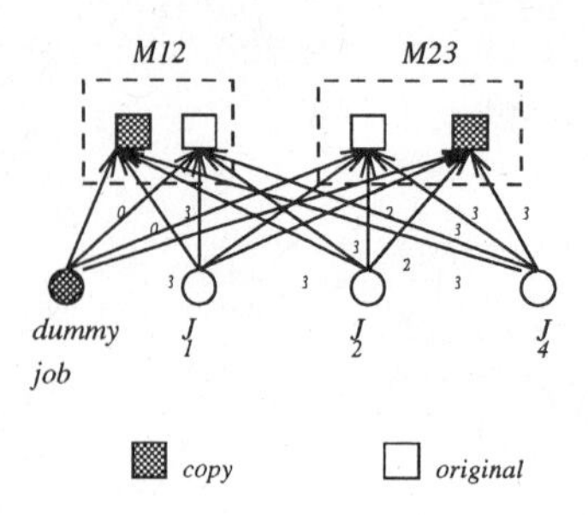

Figure 11.19: Level-2 as a BC sub-graph

Level-2

----

job .. J_1 .. Assigned to .. M_{12} .
job .. J_1 .. Assigned to .. M_{12} .
job .. J_2 .. Assigned to .. M_{23} .
The machines Load condition

Machine 1: J_1 $\{o_2,o_5\}$; J_1 $\{o_3,o_4\}$; J_5 $\{o_1,o_2,o_3,o_4\}$
Machine 2: J_1 $\{o_1,o_8\}$; J_2 $\{o_1,o_9\}$; J_4 $\{o_1,o_9\}$
Machine 3: J_2 $\{o_3,o_7\}$; J_3 $\{o_1,o_2,o_6,o_7\}$

==

Figure 11.20: The assignment solution for level-2 machines

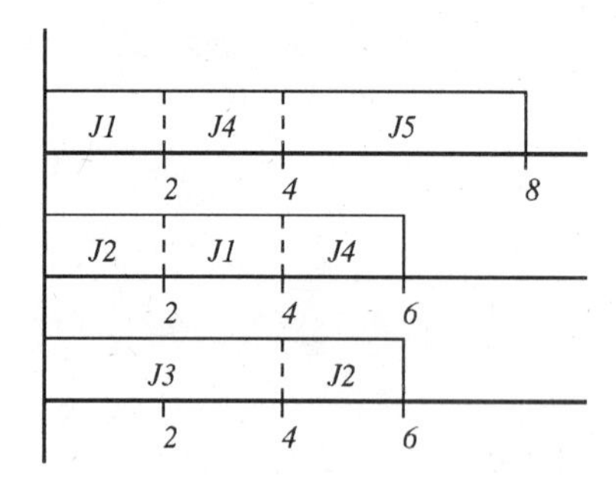

Figure 11.21: The `complete' schedule with makespan equal to 8 time-units.

11.6 CONCLUSION

Most of the research done in classical scheduling theory has been devoted either to scheduling single operation jobs on G.P systems, or to scheduling multiple operation jobs on S.P systems. Meanwhile a large number of environments are more efficient and applicable if they are designed as P.O systems. Solving the scheduling problems in this case is challenging.

In this chapter, a general model for multi-machine systems, based on the structure and the processing characteristics, has been proposed. Three main classes of multi-machine systems, based on structure, have been identified, general purpose (G.P), special purpose (S.P), and partially-overlapped (P.O) systems. P.O systems are shown to be the most general class of multi-machine systems.

Three scheduling problems of partially-overlapped systems have been examined: (1) scheduling a set of unit-time single operation jobs such that the makespan is minimized; (2) scheduling a set of arbitrary-time single operation jobs such that the makespan is minimized; (3) scheduling a set of multiple operations jobs, each operation requires a single time-unit, such that the load balance and the independence objectives are satisfied.

P.O.S.U algorithm, an optimization algorithm, with a time complexity of $O(\min\{n^{.5}, m\}nm)$ has been developed for the first problem. The second and the third problems have been proven to be in NP-hard class. P.O.S.A heuristic algorithm has been developed for the second problem, and P.O.M.U heuristic algorithm for the third problem. The performance of each algorithm has been demonstrated on an example of a manufacturing system.

ACKNOWLEDGEMENT

This work has been partially supported by a grant from Manufacturing Research Corporation of Ontario, Canada (a Government of Ontario Center of Excellence).

REFERENCES

1. Baker, K. Introduction to Sequencing and Scheduling, John Wiley & Sons Inc., New York, 1974.

2. Bokhari, S. Assignment Problems in Parallel and Distributed Computing, Kluwer Academic, Boston, 1987.

3. Bondy, J. and V. Murthy, Graph Theory with Application, Macmillan Press, 1976.

4. Burkard, R. and U. Derigs, Assignment and Matching Problems: Solution Methods with Fortran Program, Springer, Berlin, 1980.

5. Chang, R. and R. Lee, ``On a Scheduling Problem where a Job can be Executed only by a Limited Number of Processors," Comput. Opns. Res., vol. 15, no. 5, pp. 471-478, 1988.

6. Chen, Y. and Y. Chin, ``Scheduling Unit Time Jobs on Processors with Different Capabilities," Comput. Opns. Res., vol. 16, no. 5, pp. 409-417, 1989.

7. Chen, C. and R. Bulfin, `` Scheduling Unit Processing Time Jobs on a Single Machine with Multiple Criteria," Comput. Opns. Res., vol. 17, no. 1, pp. 1-7, 1990.

8. Chow, W. S. and O. Hawaleshka, ``Minimizing intercellular part movements in manufacturing cell formation", Int. J. Prod. Res., vol. 31, no. 9, pp. 2161-2170, 1993.

9. Coffman, E. G. Jr. (Ed), Computer and Job-Shop Scheduling Theory, John Wiley, New York, 1976.

10. Coffman, E. Jr., M. Garey, and D. Johnson, ``An Application of Bin-Packing to Multiprocessor Scheduling," SIAM J., vol. 7, no. 1, pp. 1-17, 1978.

11. Even, S. and R. Tarjan, ``Network Flow and Testing Graph Connectivity", SIAM. Comput., vol. 4, pp. 507-518, 1975.

12. Gonzalez, T. and S. Sahni, `` Open-Shop Scheduling to Minimize Finish Time,"J. ACM, vol. 23, no. 4, pp. 665-679, Oct. 1976.

13. Karp, R. M., ``Reducibility among Combinatorial Problems," Complexity of Computer Computations, R.E. Miller and J.W. Thatcher, (Eds) Plenm Press NY, pp 85-103, 1972.

14. Lenstra, J., D. Shmoys, and E. Tardos, ``Approximation algorithms for Scheduling unrelated Parallel Machines," *Math. Programming,* 1990.

15. Lawler, E., J. Lenstra, A. Rinnooy Kan, and D. Shmoys, `` Sequencing and Scheduling: Algorithms and Complexity," Tech. Report BS-R8909, Center for Mathematics and Computer Science, Amsterdam, The Netherlands, June 1989.

16. Lee, C., J. Hwang, Y. Chuw, F. and Anger, `` Multiprocessor Scheduling with Interprocessor Communication Delays," Opns. Res. Letters, vol. 7, no. 3, pp. 141-147, June 1988.

17. Lo, V. and J. W. S. Liu, `` Task Assignment in Distributed Multiprocessor Systems," in Proc. of the 1981 International Conf. On Parallel Processing, pp. 358-360.

18. Luh, P., D. Hoitomt, E.max, and K. Pattipapti, `` Schedule Generation and Recognition for Parallel Machines," IEEE Trans. on Robotics and Automat., vol. 6, no. 6, pp. 687-696, 1990.

19. Rinnooy Kan, A. H. G., Machine Scheduling Problems: Classification, Complexity and Computations, Martinus Nijhoff, The Hague, Holland, 1976.

20. Shin K. and Q. Zheng, ``Scheduling Job Operations in Automatic Assembly Line," IEEE Trans. Robotics Automat., vol. 7, no. 3, pp. 333-341, June 1991.

12

OBJECT ORIENTED APPROACH TO FEATURE-BASED PROCESS PLANNING

John M. Usher
Mississippi State University

The design and development of an object-oriented system for process planning is presented. The design demonstrates the use of the object-oriented paradigm not only as an efficient means for the representation of planning knowledge, but also as a means of organizing and encapsulating the functionality of the system with the data it manipulates. This modularity results in a design that can be easily extended to include additional functionality and address other processes. The design also expands on traditional piece-part planning by extending the part model to support planning for end-products composed of multiple parts and sub-assemblies. This support is included to permit the incorporation of an assembly planning component. Planning for all components of an end-product can now be performed in one run of the system. A prototype planning system has been developed and implemented for end-products composed of rotational parts.

12.1 INTRODUCTION

Process planning forms the link between design and manufacturing, providing a means for the translation of design requirements and specifications into a set of instructions describing how to manufacture a part. This planning function includes the determination of the routing for a part, the processes involved in its production, process parameters, machines, and tooling.

Today, the tools of design include the use of CAD systems for generating design data. Likewise, manufacturing engineers employ CAM systems which make use of the design data. It is the computer-aided process planning (CAPP) system that promotes the integration of CAD and CAM systems. Although not a simple task, this level of integration is needed to support the future of autonomous manufacturing and the requirements of computer integrated manufacturing.

The majority of commercially available process planning systems make use of the variant approach to planning. Although variant systems provide improvements over manual planning, there is a need to continue research efforts toward the development of a robust generative system that can support process planning for a broad range of part types and assemblies, provide the highest-level of integration, and support autonomous manufacturing.

Development of a truly generative process planning system has been stifled for several reasons. These include:

- the need for a means of representing and storing design (dimensions, geometry, etc.) and manufacturing information (materials, tolerances, surface finish, etc.) about the product in a form useful for programmed software systems [1],
- the complexity of system design and implementation that arises due to the large number of functions involved in the planning process,
- the difficulty in acquiring, organizing and representing the knowledge required to implement these functions [1,2], and
- the need to allow the planning system to evolve as the user's and developer's understanding of the planning process, operations, and parts, matures, and as additional knowledge is acquired and verified.

This research proposes the use of an object-oriented paradigm as a means for the design and implementation of an intelligent process planning system that addresses these obstacles.

In addition to the obstacles listed above, the current generative planning systems reported in the literature focus on process planning for individual machined parts. However, what is needed within manufacturing is a planning system that can generate all the process plans for an end-product composed of multiple parts and sub-assemblies. This type of system represents a higher-level planning system. A first attempt at implementing such a system could be structured around multiple calls to a piece part planner. However, such an approach results in a static design with limited expandability.

The work reported in this paper focuses on applying the object-oriented paradigm to the design of an intelligent process planning system capable of planning for end-products composed of multiple sub-assemblies and piece-parts. Such a design is not limited to process planning of machined parts. This same system design can also be extended to encompass both process and assembly planning for the end-product providing a more complete planning system for manufacturing use. Also, due to the nature of the object-oriented programming environment, the resulting system design:

- provides the functionality to simplify and improve the efficiency of the planning functions (specifically operation, tool, and machine selection),
- is easier to change and expand than designs based on conventional programming, and
- can evolve as our understanding of planning, processes, parts, and tooling evolves.

The paper will begin with an overview of process planning and a review of object-oriented design. This will be followed by a presentation of the object-oriented design approach used in this study and include an illustration of this design using an example application involving process planning for an end-product composed of multiple parts.

12.2 BACKGROUND ON PROCESS PLANNING

Process planning is an activity that determines the procedure necessary to transform raw materials into finished goods taking into consideration the available production resources. This task is performed based on an interpretation of the part design from an engineering drawing. Depending on the application, process planning may involve several or all of the activities listed in Table 12-1 [2,3,4]. The results of these activities are documented in the form of a process route sheet for use in production.

The purpose of a CAPP system is to automate and standardize the process planning function. However, the automation of the planning activities is complicated by the highly complex and ill-structured nature of the problem and the interrelationships that exist between many of these activities. Two recognized approaches to the automation of the process planning function that have emerged over the years are the variant and generative approaches. Earlier systems favored the variant approach to planning with the latest research efforts focusing on generative techniques employing artificial intelligence to enhance their capabilities.

TABLE 12-1: Process planning activities.

Interpretation of product design data
Selection of machining processes
Selection of cutting tools
Selection of machine tools
Determination of setup requirements
Sequencing of operations
Determination of production tolerances
Determination of cutting conditions
Design of jigs and fixtures
Calculation of process times
Tool path planning & NC program generation
Generation of process route sheets

12.2.1 Variant Approach

The variant approach to planning follows the principal that similar parts require similar plans. Therefore, planning for a new part can be simplified by first retrieving an existing plan for a similar part and editing that plan to suit the design differences introduced by the new part. Implementation of this approach has made use of the concepts of group technology involving the identification and categorization of parts into families based on similarities in their design and manufacturing attributes. The use of a classification and coding system provides a means for using a computer to identify to which part family a new part belongs such that a standard family plan can be retrieved and edited. This approach to automated planning is used by such systems as: CAPP [5], MIPLAN [6, 7], and MULTIPLAN [8].

The variant approach to process planning represents the favored approach for commercial system implementations since it is simpler to construct than a generative system [2]. Compared with manual planning, use of a variant system reduces the time required to generate new plans and improves the consistency of those plans, but it still requires the input of an experienced process planner for handling the plan modifications and can not perform planning for parts that can't be identified as belonging to an existing part family.

12.2.2 Generative Approach

The second method, generative planning, requires that the system generate a new process plan from scratch for each new part. This method relies on the use of decision logic and algorithms to determine the process plan based on an interpretation of the geometrical and technological data specified in a CAD drawing. Due to the complexity of this approach a generative CAPP system is more difficult to design and implement than a system based on the variant approach. Likewise, due to the vast knowledge requirements of the generative approach, the number of different parts that the system can handle is limited.

Representing the planning logic in a generative system is a difficult task. Several common methods that have been used include decision trees and tables [9, 10], analytical models [11, 12, 13], and production rules [14, 15, 16, 17]. Reviews of these systems and many others can be found in one of the numerous survey articles [3, 4, 18, 19, 20, 21]. In addition to the benefits provided in the variant approach, a generative CAPP system does not require the aid of a human planner, and can produce plans for parts not belonging to an existing part family.

12.3 FEATURE-BASED GENERATIVE CAPP

A main difference between generative systems and variant systems is that the knowledge of a generative system is organized on the basis of part features not complete parts [16]. Each feature represents a subset of the part geometry which has significance in manufacturing (i.e., hole, groove, chamfer, etc.). The planning system then views each part as an assembly of manufacturing form-features and tries to determine the sequence of operations that can generate those features. This approach permits the planning system to make use of predefined (default) relationships between specific features and manufacturing operations and resources thereby, simplifying the planning process. For example, low tolerance round holes can be produced by using drilling or boring, while high-tolerance holes require reaming and grinding [17].

Planning based on form-features requires that the CAPP system first be able to transform the geometrical and technological data of a part model from a CAD system into a set of manufacturing form-features. This mapping of design features into manufacturing features is commonly achieved through the use of feature recognition and extraction algorithms [22, 23, 24, 25]. However, current systems are constrained in terms of the number of different feature types they can recognize, the types of parts they can consider (rotational or prismatic), and the complexity of the geometry they can handle (i.e., 2D, 3D, solids, free-form surfaces, etc.). Also, these systems are unable to directly extract technological data concerning such attributes as surface finish and tolerances since this information is not a part of a CAD model. Research to overcome this limitation involves investigating new methods of binding the technological data to the part model representation [26, 27].

Some argue that the need for feature recognition could be circumvented if the design system itself made use of standardized features for the construction, representation, and storage of the part [17, 28]. This use of explicit manufacturing features at the modeling level will relieve the designer from having to manipulate low-level graphic entities in order to convey design intent [29]. One approach to design at this level involves a user selecting a blank, and specifying what operations (i.e., make-hole, make-slot, etc.) should be performed on the workpiece to produce the final part [30]. This results in the simultaneous generation of a part design and an initial process plan. One problem with this approach is that it places the burden of process selection and sequencing on the designers, meaning that they must assume the role of both the design and manufacturing engineer.

A second common approach favors a more general mechanism where the designer selects machine-formable features from those stored in a library and builds a model of the part as an assembly of these features. This model is then used directly to reason about what sequence of operations and resources are needed to generate the defined part [28, 31, 32]. However, problems associated with this approach include determining a means for associating technological data with a feature and providing sufficient flexibility in the definition of a feature such that the creativity of the designer is not constrained.

Independent of the approach taken to arrive at a feature-based model of the part, a feature-based generative CAPP system performs its reasoning based on these features to arrive at a plan. In this approach, the capabilities of the manufacturing resources are defined in terms of what feature they can produce and how well they can produce them. This association simplifies the knowledge representation and inferencing mechanism used to implement the planning strategy.

12.4 OBJECT-ORIENTED PROCESS PLANNING

12.4.1 The Object-Oriented Paradigm

In recent years, there has been a growing interest in object-oriented design and programming [33]. This interest has carried over to the design of manufacturing systems with special emphasis on its application to the design of complex software systems [34,35]. The attractiveness of object-oriented software design (OOD) and programming (OOP) is in the advantages it provides in the design and development of complex software systems. These advantages include [34]:

- object-oriented design results in smaller systems,
- object-oriented systems are more flexible in terms of making changes and handling the evolution of the system over time,
- object-oriented systems are "designed to evolve incrementally from smaller systems" making it easier to create and maintain larger and more complex systems, and
- object-oriented decomposition results in reusable modules that can be applied to the development of other systems reducing future development efforts.

These advantages are important when considering the number of functions that take place in planning and the overlap between these functions. A search of the literature reveals that most

CAPP systems are incomplete and do not provide a full implementation of the process planning activities. It may be that many of these systems become too complex to manage in terms of further development and are abandoned. Compared with traditional designs, the use of OOD and OOP for the development of a process planning system provides the tools for addressing the complexity of process planning and the capability to incrementally add functionality as the system matures, thereby permitting the developer to work toward the creation of a complete manufacturing planning system.

12.4.2 The Object-Oriented Approach

Object-oriented software design (OOD) is an alternative approach to traditional "top-down" software design. This design approach requires that the developer identify a set of system objects from the problem domain and express the operation of the system as the interaction between these objects. Instead of a system comprised of algorithms that carry out the individual steps of the process, the program is built around a set of objects. Each of these objects models the behavior of an object in the real world. This behavior is defined by what an object is capable of doing. Asking an object to perform a function requires that it be sent a message. Therefore, the software system operates by allowing the objects of the system to pass messages to one another requesting that they perform certain functions (Figure 12-1).

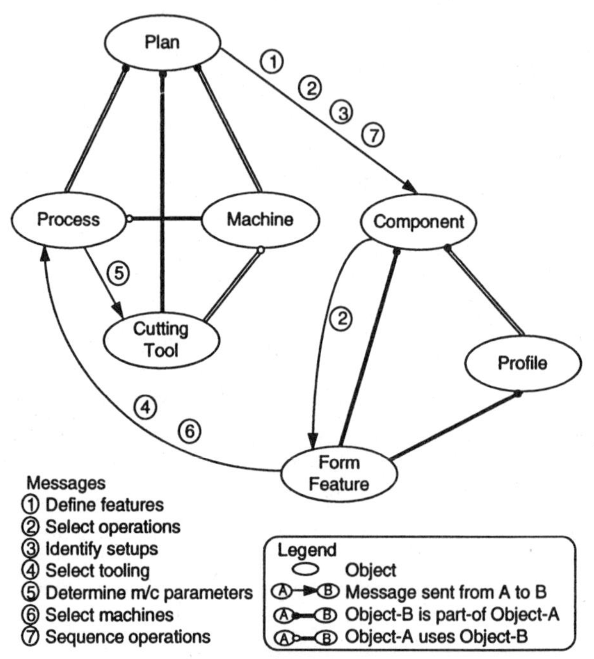

Figure 12-1: Object hierarchy for planning with example messages.

The main elements of object-oriented design include abstraction, encapsulation, hierarchies, and polymorphism. Each of these is briefly outlined below.

Abstraction: For process planning, the objects of the system include entities such as those shown in Figure 12-1. The implementation of these objects within the system is in the form of object classes, each of which represents an abstraction of a set of real world objects that share a common structure and behavior [34]. Through abstraction, a system object is defined by the things it knows about itself (its internal state or characteristics) and the things it can do (its operations or behaviors). Therefore, in programming terms, the objects of the system must encompass both the data concerning the characteristics of the object and the functions which describe its behavior. This binding of data with functions is known as encapsulation.

As an example, consider the definition of a object representing a machined part. From the planning perspective, this definition would include the part's features, the geometry of those features, their relationship, and characteristics (surface finish, tolerances, etc.). To model this view of a part, a developer would create an object class representing these characteristics and the operations that express the behavior of the part. For rotational parts this view might look like that given in Figure 12-2. Note that the behavior of this object includes the capability to define its features, select the operations for production, and sequence those operations.

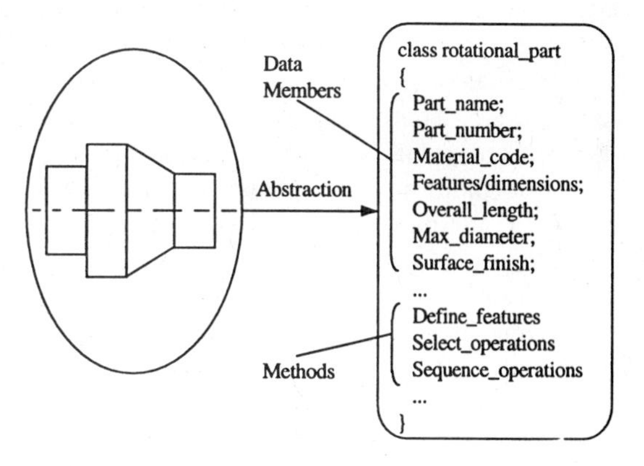

Figure 12-2: Abstraction of a rotational part.

Encapsulation: In addition to encapsulation as the binding of data with functions, it also addresses object visibility. This pertains to "the process of hiding all of the details of an object that do not contribute to its essential characteristics" [34]. The idea is to create a system composed of objects that are available for use by client objects without knowledge of the details of the implementation of each object's data members and functions (called "methods" in OOP parlance). This concept of limited visibility permits the development of a system where no part of the system is dependent on the internal details of any other part. The ability to change the representation of an abstraction without disturbing any of its client objects is a major benefit of encapsulation. This independence is a fundamental requirement for the development of stable and maintainable complex software systems, a desirable feature for a CAPP system [36].

Hierarchy: Another important element of an object-oriented design deals with the hierarchy that exists between the real world objects. This hierarchy is represented within the ordering of the object classes. The two types of hierarchies that can be represented include the composition ("part-of") hierarchy and the inheritance ("kind-of") hierarchy.

A composition hierarchy defines a structural relationship between object classes. For example, consider the *plan* object shown in Figure 12-1. This diagram defines a *plan* as being composed of *processes*, *machines*, and *cutting tools*. A composition hierarchy is implemented by embedding objects within other objects. This method defines the relationship that the embedded objects are "part-of" that parent object.

When one object is related to others by the fact that they share common attributes and behavior, a hierarchy can be created with the base class representing those common class elements and the derived classes defining those elements distinct to each object type. This results in an inheritance hierarchy where one or more derived classes inherit the characteristics and behavior of a base class while maintaining distinct characteristics and behaviors of itself within its own class. The derived classes are said to be objects that are a "kind-of" the base class object.

For example, consider the representation of a machined part as an assembly of form-features that comprise its volume. These features might include cylinders, tapers, fillets, chamfers, grooves, threads, etc. Each of these distinct features are a "kind-of" form-feature. Therefore, by identifying all the attributes these features have in common, an inheritance hierarchy can be constructed consisting of the base class *form-feature* and the derived classes *cone, cylinder, fillet,* etc. (see Figure 12-3). The base class defines those data members and methods shared by all the different features. As shown, the *cylinder* class is defined as a derived class of the base class *form-feature*. Therefore, *cylinder* inherits the data attributes and behaviors of *form-feature*. These include attributes such as: location (external/ internal), position, basic tolerances, and finishes. In addition, the *cylinder* class can modify and complement the base class definition by declaring additional data members, and methods that distinguish a cylinder from the other types of form-features.

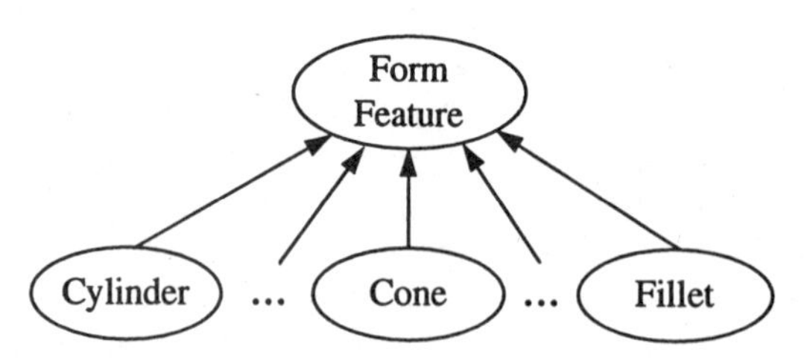

Figure 12-3: Inheritance object hierarchy.

Polymorphism: A fundamental step within process planning is that of selecting the operations to produce the part features. Therefore, in the definition of the *form-feature* class, the function *select_operation()* was identified as a method common to all form features. However, since different knowledge is applied in the selection for the different features, the behavior, and therefore the implementation of this functionality, will not be the same for each feature type. Thus, it is necessary that each of the derived classes have their own function for operation selection.

Now the *select_operation* function is defined as a public method within the interface for the base class *form_feature*. Therefore, due to the inheritance of the base class methods by the derived classes, a user (client object) knows that it can send the "select operation" message to any type of form-feature and expect it to determine the operations appropriate for its production based on the current attribute values of that feature. When the system receives a message from a client object to select the operation for a form feature, it determines at run-time the type of form feature the client object is referring to and then calls the appropriate method (for that type of feature). This functionality of object-oriented programming is called polymorphism, meaning that a method of the same name is shared within an inheritance hierarchy, but the behavior is different within each class. Therefore, each object is able to respond to the same message in their own unique way.

Polymorphism is "perhaps the most powerful feature of object-oriented programming languages next to their support for abstraction, and it is what distinguishes object-oriented programming from more traditional programming with abstract data types" [1]. With polymorphism it is possible for a client object to use the methods of the base type to access the derived types without knowing the particulars of the derived types. This results in a system that can be easily extended to add new derived types. Since the client objects make use of the base class methods, adding a new derived class is simply a matter of implementing the methods of the base class within the new derived class based on the desired behavior of the derived class. Since the client already knows how to use the base class methods, it can immediately make use of the new types. So, to add a new form feature (e.g., radial groove) to our process planning system, we would only need to create a new derived class (i.e., groove) under the form-feature class and implement those behaviors for the new class.

The object-oriented approach to software design represents a different way of thinking about the problem domain. However, this concept parallels the natural way in which we perceive the operation of systems around us.

12.4.3 Object-Oriented Process Planning Systems

As evidenced in the discussion above, the application of the object-oriented paradigm to process planning is a logical means for representing the real-world components within a manufacturing system. Although a relatively new paradigm, the use of this approach in the development of CAPP systems is reported in the literature for several systems of varying application.

In terms of specific applications, Lee, et al. [17] develop a system that combines OOP and rule-base techniques for planning the manufacture of progressive dies. Their use of objects seems to focus mainly on its benefits as an improved data structure with no discussion of the functions (methods) defined within the object classes. Srinivasan, et al. [37] discusses the use of an object-oriented approach for the design and development of a process planning system for electronic assemblies. They also developed two other non-OOP systems for the same application each making use of a different knowledge-based approach to inferencing: forward chaining and backward chaining. Based on their comparison, they found that objects are a more suitable means of representing the diversified knowledge in a CAPP system, and that the object-oriented system solves the problem more efficiently.

At an abstract level, Shue and Kashyap [38, 39] propose an object-oriented data model as the basis for the development of a framework of object-oriented knowledge bases to represent parts and process operators for use in planning. Likewise, Krause, et al. [40] discusses the proposed use of objects for representing features in part design and using a rule-based approach to transform these design features into manufacturing features for use in planning. The authors briefly mention the use of objects for representing processes and operations, but provides no examples or discussion of other objects or the relationships between objects.

Sanii and Davis [41] present an object-oriented planner (DiCAPP) whose planning functions are distributed in a hierarchical fashion based on the time-horizon of their applicability. The DiCAPP system focuses on prismatic parts and makes use of objects in a similar fashion to the work presented in this paper. However, no details of the implementation are provided. Also, when describing a part for planning, tolerancing appears to be absent from consideration.

Feghhi, et al. [42], employ an object-oriented approach to the design and implementation of a system for part design and process planning. Using the Smalltalk language, objects are used in the definition of class hierarchies for part features and manufacturing processes. A process plan is generated through the examination of rules using backward chaining to reason from the finished part state to the raw material state. This same system was then extended to integrate the design, planning, and inspection functions for prismatic parts [43]. However, in place of the rule-based approach, case-based reasoning is used for planning. This approach makes use of past planning experiences to plan for new parts.

The major thrust of these reported studies appears to be on their use of objects as an improved data structure. This approach falls short in terms of the potential available using the object-oriented paradigm. Therefore, compared to the previous work in this area, the research addressed in this paper offers an object-oriented design and implementation that more fully utilizes the capabilities of the OOP paradigm focusing on both the definition and interaction of the system objects. Also, the methods used for implementing the planning functions differs from other strategies making better use of encapsulation and polymorphism. This results in more efficient use of the process knowledge in planning.

12.5 EXPERIMENTAL SYSTEM

The prototype system, IOOPP (Intelligent Object-Oriented Process Planner), was developed on a microcomputer using C++ as the implementation language. The system is able to perform planning for rotational parts incorporating the functionality shown in Figure 12-4. The following sections examine the design and implementation of these planning functions.

12.5.1 Part Description

A common approach used by CAPP systems to describe a part for planning purposes is to view that part as an assembly of features. The need to represent this feature information in a construct useful for planning has resulted in the use of frame-based approaches [1,44,45,46], specialized data structures or databases [16,47,48,49], and object hierarchies [17,43].

In each approach, the objective is to represent a part in terms of its features. This requires that the system be able to represent and store the geometric and technological information of each feature. This study makes use of object hierarchies to describe a part since they permit the use of inheritance in the representation of the feature attributes, and polymorphism in the definition of the methods associated with the manipulation and query of each feature. Although inheritance can be obtained through the use of data structures and frame-based approaches, neither of these techniques provides a robust means for encapsulating methods with the feature object and making use of polymorphism to simplify use and manipulation of these feature objects.

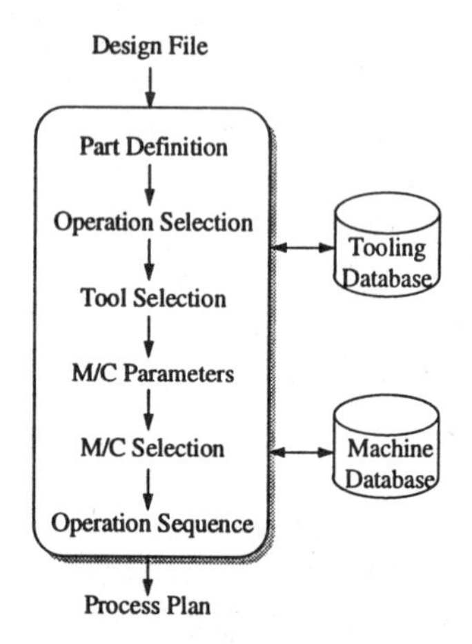

Figure 12-4: Prototype planning system.

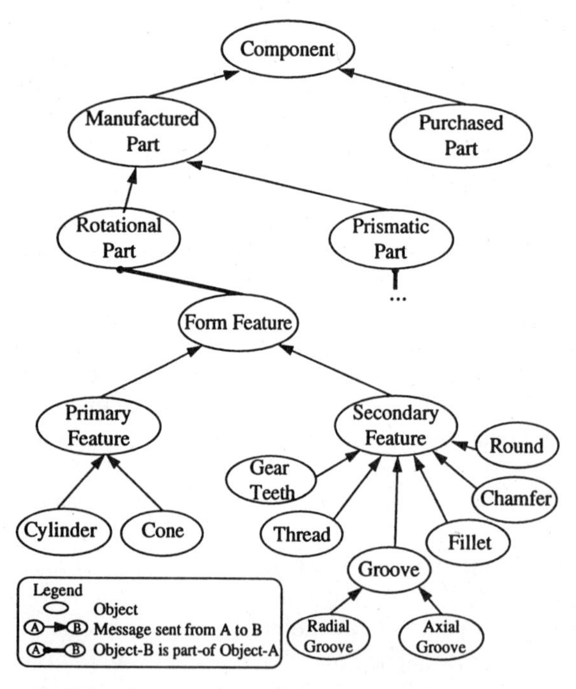

Figure 12-5: Part definition object hierarchy.

A part is modeled at two separate levels of abstraction: the component level and form features level. As shown in Figure 12-5, the component level object hierarchy specifies that a component may be identified as either a manufactured or purchased part, and that a manufactured part be classified as being either rotational or prismatic [50]. The *component* object class at the top of the hierarchy defines such general information as the part number, name, and a brief description. At the next level in the hierarchy, the objects define either the part material, batch size, and part form (rotational or prismatic) if the part is to be *manufactured*, or the vendor and lot size if the part is *purchased*. (The need to distinguish between a part being manufactured or purchased arises when modeling multi-component products and will be explained later in the discussion of product modeling and planning.) The objects at the lowest level further define a part by declaring its overall dimensions, and the form-features it possesses. The *form-feature* class does not inherit from the *rotational-parts* class, but instead defines a relationship where the rotational-part object is said to be "composed-of" one or more form-features. This relationship is indicated by the double line with a black circle on the end with the class made up of the connected class objects.

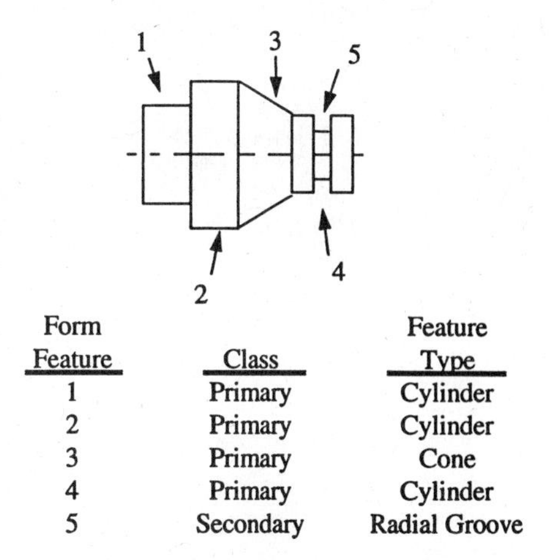

Form Feature	Class	Feature Type
1	Primary	Cylinder
2	Primary	Cylinder
3	Primary	Cone
4	Primary	Cylinder
5	Secondary	Radial Groove

Figure 12-6: Example end-product.

The object hierarchy at the form features level classifies the features of the part as either primary or secondary features [1,51]. The overall shape of the part is formed from a combination of the two types of primary features: cylinder and cone for rotational parts (or block, wedge, cylinder, etc., for prismatic parts [51]). The secondary features of the part add the detail of the part and reside on, and are defined relative to, a primary feature. For example, the external profile of the rotational part shown in Figure 12-6 would be defined as composed of four primary features (three cylinders and one cone) and one secondary feature (a groove).

Both the primary and secondary feature classes model the features parametrically rather than geometrically. For example, consider the data members within the classes: *form-feature, primary feature*, and *cylinder* (see Figure 12-7). Note that the *cylinder* primary feature class is defined in terms of its diameter, length, surface finish, tolerances, etc. The placement of attribute data within a particular object class is dependent on the relationship between the objects in the hierarchy. Beginning with the *form-feature* class and working down the hierarchy to the *cylinder* class, one can observe that the data becomes less general in nature. This class structure permits the use of inheritance between the object classes for accessing data and expressing behavior.

Use of a hierarchical approach for describing a part results in a complete, compact, and non-redundant method for defining and storing information about a part. Also, since the relationships between the features of the part is inherent in the hierarchy, defining and reasoning about these relationships becomes more efficient.

Origin of Part Definition Data: The proposed approach to be used by IOOPP to obtain the part definition data is not dependent on the use of a feature-based modeler, but is more closely associated with feature extraction techniques. The feature data for input to the system will be extracted from a part representation stored using the STEP standards [52]. The method of accessing this data will the development and implementation of an application protocol which will define the context and scope for the use of the product data and specify the method for interpretating the semantics of the stored information for use in process planning [53].

```
class form_feature {
      name;              // ID tag
      type;              // Cylinder, fillet, etc.
      process;           // Turn, mill, grind
      location;          // External or internal
      position;          // On part
      position tolerance;
      surface finish;
      operation list;
      ...
};
class primary_feature {
      length;
      length_tolerance;
      roundness;
      ...
      surface_datum_list;
      secondary_feature_pointers;
};
class cylinder {
      diameter;
      diameter_tolerance;
      concentricity;
      possible_operations;
      ...
   Methods:
      volume_removed;
      select_operation;
      ...
};
```

Figure 12-7: Object classes of the form-feature hierarchy.

However, since some of the required STEP resource models are not complete, this effort is still under study. For now, IOOPP currently obtains the part definition data from a user interpretation of the design represented in terms of the specified attributes for the objects that appear in Figure 12-5. To do this the user will examine the part drawing, identify its form features, and define the attributes of each feature (i.e., dimensions, location, etc.). This information is stored in a single ASCII file and read by the system as it instantiates the objects of the system.

Since this approach is a temporary solution, a separate menu-driven program is provided that allows the user to easily build the input file. The system prompts the user for the information about the product, each of its components, and their features. For each component, the system will query the user for the data required to explicitly define each form feature, one at a time. The result is an input file correctly formatted for use by the planning system.

12.5.2 Operation Selection

The objective of operation selection is to determine what operations possess the capability to produce the form-features that comprise the part. The selection process is driven by the type of feature, its geometry, and technological requirements, and involves mapping these specifications to the appropriate operation or series of operations for each feature. This requires that the system have access to information concerning the shape generation capabilities of the various operations and the requirements of the part features. The form-feature class hierarchy, that was described earlier, contains information regarding the shape generation requirements of each form feature. Likewise, a process class hierarchy is used to define shape generation capabilities of each operation. Therefore, operation selection is based on finding a match between the objects of these two class hierarchies.

Process Hierarchy: An inheritance hierarchy of object classes was created to represent the processes to be considered in planning. The hierarchy is composed of up to five class levels as show in Figure 12-8. Since the IOOPP system considers both process and assembly planning, the first level distinguishes between those processes for *assembly* and those for *manufacturing*. The next level then further distinguishes between the different categories of processes followed at the third level by distinct processes of that type (e.g., milling, turning, etc.). At the fourth and fifth level specific operations associated with each process are further defined to distinguish between their capabilities. As illustrated with the form-feature hierarchy example, the placement of attributes and methods within a particular class is dependent on the type of processes for which they are common. As shown, the process class hierarchy is incomplete, but since the system design is an object-oriented implementation it would not be difficult to include other processes and operations to the system. As long as the interface to these classes is left unchanged, no other changes would be required to make such additions.

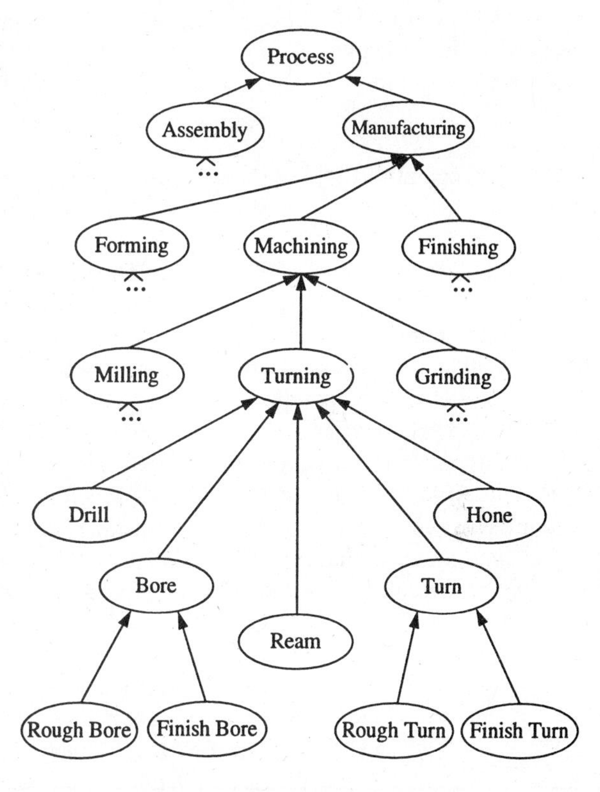

Figure 12-8: Process class hierarchy.

Selection Methodology: Methods of implementing the translation from feature specifications to required operations have included the use of direct association, [16,41,54,55,56], decision tables and trees [57,58], and production rules [59,60,29]. Direct association involves specifying for each feature what operation, or operation series, to use in its manufacture. This approach makes a selection by default without any consideration of the specific attributes of the feature. The direct approach is simple to implement and efficient, but constrains system flexibility since it is not possible to consider alternative operations for machining a feature.

The deficiency of direct association is overcome in the other methods where process knowledge relating to each operation is represented using either decision tables, decision trees, or production rules. The system can then use this knowledge to select the appropriate operation based on a feature's attributes. This permits the system to consider all operations in its selection. However the efficiency of this approach is sacrificed since the system must examine all the knowledge embodied in the tree, table, or rule-base (depending on the method used) to determine the possible alternative operations.

The approach to operation selection used by IOOPP involves a combination of direct association and production rules. First, encapsulated within each feature object class is a set of possible operations to be considered for the production of that specific feature. If the system finds only a single operation associated with a feature, then it accepts this operation as the selection and no further processing is required. However, if multiple operations are listed, selection of the appropriate alternative operations involves an examination of the rules associated with those operations listed. This approach to operation selection combines the advantages of direct association in terms of its simplicity and efficiency, with the ability to consider alternative operations. It also improves on the efficiency of consulting a rule-base by limiting the search to only those rules associated with the operations listed.

As an example, consider the *cylinder* class as shown in Figure 12-7. The *possible_operations* data member of this class lists the alternative operations that the system should consider in selection.

possible_operations[][] = {{"rough turn", "finish turn"}, {"rough face", "finish face"},
{"drill", "rough bore", "finish bore", "ream"}}

As given, there exist three alternative operation series. Those members of the list that denote an operation series define the operation precedence that may be required for the production of the associated feature.

Within the process hierarchy, each operation is represented by an object which expresses the capabilities of the operation, both geometric and technological. Similarly, the data members of the feature class objects define the manufacturing requirements for each feature type (i.e., dimensions, surface finish, tolerances, etc.). The task of the system is to check those alternative operations from the list and determine which operation's capabilities match the feature's requirements. This match is determined using a general rule defined for each type of feature within that feature's class. This rule defines what capabilities and requirements to consider in selecting an operation. For the *cylinder* feature class, this rule has the form:

IF (location = *op_location*)
AND (diameter > *min_diameter*)
AND (diameter < *max_diameter*)
AND (surface_finish > *min_surface_finish*)
AND (pos_diameter_tolerance > *min_pos_diameter_tol*)
AND (neg_diameter_tolerance > *min_neg_diameter_tol*)
THEN (selected_operation = *operation*)

The italicized variables represent those attributes from the class of the operation being considered, and the non-italicized variables represent the feature requirements specified in the feature class object, *cylinder*.

This approach eliminates the need for a central knowledge base, instead it locates the process knowledge for a feature within the feature class itself. The result is the capability to customize the rules to suit each type of feature. This process of selection is also more efficient than other knowledge-based approaches and does not require the use of an inference engine in selection. Also, if one needs to add or edit rules it is much easier to associate a rule with a feature than when using a conventional rule-based system.

When an alternative operation is an operation series (such as rough turn → finish turn), the selection process begins by considering the first operation in the series. If no match is found, then the next operation in the series is considered. This procedure is continued until either a match is found, or the end of the list is encountered indicating that this alternative is not a valid operation. This order of processing ensures that the system selects the most appropriate operation and not spend time or money performing unnecessary operations. Whenever the system selects an operation from a series, all operations that comes before the selected operation are included in the list of the required operations for that feature. This strategy maintains the precedence relationship defined by the series. Therefore, if the system has selected an operation in a series that is the third operation, then by default the system knows that the first two operations must be completed prior to the selected operation and subsequent operations are not necessary.

Once all the operations for a feature have been identified, they are written to an operation array within each feature object denoting the selected operations. Next, the process types from the third and fourth levels of the process hierarchy associated with that operation: forming, machining, or finishing, and milling, turning, or grinding are stored in the feature object. These associations will be used later when the system sequences the operations.

As illustrated in Figure 12-1, operation selection is invoked by passing the message, *select_ operation*, to the feature objects of the part. Within each feature class, the *select_ operation* method defines the procedure of selecting an operation or sequence of operations needed to produce that type of feature. This use of polymorphism permits a client object (*component*) to issue the same message, *select_operation*, to any feature prompting it to perform operation selection. This is done for each feature of a component part before the system moves on to the next step, setup identification.

12.5.3 Setup Identification

At this phase in the development of the prototype, the system is only able to plan setups for rotational parts composed of turned features processed on a lathe. It begins by identifying both the internal and external profiles of a part (represented as separate *profile* objects) and the features they contain. This division of a part into profiles is based on the feature having the maximum diameter, for external profiles, and the feature with the minimum diameter, for internal profiles. The dividing feature is included in the profile having the largest volume of material to be machined.

Following identification of the profiles, the system will determine which side of the part to machine first based on the volume of the material to be removed. The side of the part with the largest volume (sum of internal and external profiles on that side) is machined first.

12.5.4 Tool Selection

The task of tool selection involves determining all the tools which can be used to machine the features of the part (a geometrical problem), and selecting the optimum set of tools from this list (a technological problem) [61]. Although a vital element in CAPP, many prototype systems do not consider tool selection [16, 29, 55]. Those systems that do include tool selection make use of either rule-bases systems [17,59,62], tabular relations with database searches [57,63], or procedural algorithms used in combination with the first two approaches [61,64].

Although it is possible to associate a tool with a feature (an its operation), it is also necessary that the geometry of adjacent features be considered when evaluating potential tools to ensure that the tool can access the feature without interference from other features of the workpiece. For this reason, IOOPP performs tool selection by considering the features within the profiles defined (as objects) during setup identification. Based on the features within a profile, selection is performed by first identifying all tools whose parameters match the requirements of the operation selected for each feature. This list is then pruned based on an analysis of interference between the tooling and the workpiece [65].

Defined for each selected operation within the object class *operation* is a default data-member which specifies a list of tool types that can be used to perform that particular operation.

For example, the class *rough_turn* specifies that a "turning tool" is one possible tool. This value makes reference to a class within the cutting tool hierarchy shown in Figure 12-9. The tool objects within this hierarchy define the characteristics of the tooling for each particular type of tool. For example, the *turning* tool class's data members include: shank height and width, feed direction, entering angle, clamping method, insert shape, nose radius, etc. These attributes define the parameters of importance for selecting a tool of that type and point to a database of persistent objects representing available tools of that type. This database can then be queried to perform a focused search for the desired tooling as defined by the methods (functions) within the tooling object. At present the tooling database contains references to about 150 toolholders and over 400 types of inserts.

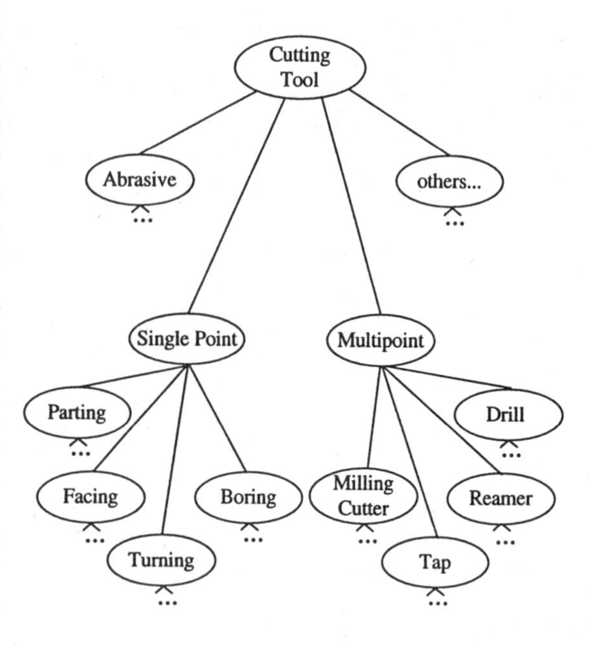

Figure 12-9: Tooling class hierarchy.

The method used by the system for selecting a tool differs based on the type of tool. This procedure is defined for each tool type within the tool's object class as the method, *select_ tool()*. Therefore, even though the method of selection is defined differently for each type of tool, client objects (an operation object such as *rough_turn*) can send the same message and expect the same response independent of the tool type.

For example, the *select_tool()* method within the *turning* tool class specifies the need to select a toolholder from a named database of toolholders, followed by an insert from an insert database. For turning tools, the procedure requires that the system first consult a small set of rules to determine suitable values for the tooling parameters: clamping method, direction of cut, entering angle, etc., based on information concerning the feature this tooling will cut. Next, these parameter values are used to extract from a database of toolholders those applicable for machining the feature. These two steps are then repeated for each feature in the profile. Once a set of toolholders have been determined for the feature of a profile, the system examines the geometry of the features in a profile and determines what toolholders to eliminate due to interference with the workpiece. Given the list of applicable toolholders, the system will then determine and select inserts for each toolholder. This selection takes into account part material, insert shape as required by the toolholder, nose radius, chipbreaker design, grade, and coating.

Once all the tooling has been selected for the feature within each of the part profiles, the system then examines the tooling to determine what set represents the minimum needed. This goal will minimize the number of tool changes required when in production.

12.5.5 Machining Parameter Determination

Once the tooling has been selected it is possible to determine the optimum cutting conditions for each tool for the purpose of optimizing either cost or production rate. Methods for calculating these values based on empirical equations have appeared in the literature [66, 67]. However, at this time the IOOPP system suggests values for speed, feed, and depth of cut as recommended by the vendor for each tool. These values are maintained in an external database. Changes to this component of the system to incorporate a more analytical approach are planned.

12.5.6 Machine Selection

Machine selection involves finding a match between the capabilities of a machine and the requirements of the part and process. The requirements of the part define the overall dimensions of the part to ensure that the machine's table size and work envelop is large enough to process the part. The process requirements ensure that the machine's processing capabilities (feed and speed range, horsepower, etc.) fit those defined by the cutting parameters specified earlier for each operation.

As with tool selection, not all prototype CAPP systems include the ability to select machines. Of those that do, production rules and database searches seem to be the most popular methods [16,17,57,62]. The approach to machine selection used by IOOPP parallels that used earlier for operation selection.

Each machine within the company is represented as an instance of the object class, *machine*. The data members of this class define the characteristics of a machine for use in comparison with the requirements. This data includes such information as: machine type (e.g., chucker lathe, vertical mill, etc.), process type (e.g., turning, milling, etc.), work table size, work envelop, available spindle horsepower, feed and speed range, tolerance capability, etc. Information concerning the dimensions of the part is available to each feature object by means of inheritance. Information regarding the machining parameters is defined within the cutting tool object class associated with the operation used to produce the feature. Using these information sources, the system goes through the database of sixteen machines stored as persistent objects looking for all possible matches that satisfy the criteria specified in the machine selection rule defined within the operation class for each operation type. Since it is possible for more than one machine to be capable of producing a part feature, all selected machines are stored for later consideration as alternatives. These alternatives will be useful at the time of scheduling for production, but the system currently prints out the first match as the selected machine. At this time no consideration is given to processing time and cost in the selection process.

The machine selection rule defines what part and process characteristics should be considered in the selection. For example, a rough turning operation would require a search based on the following comparisons.

```
IF    process_type = "turning"
AND   part_length < max_part_length
AND   part_diameter < max_part_diameter
AND   req_feed < max_feed
AND   req_feed > min_feed
AND   req_speed < max_speed
```

AND req_speed > min_speed
THEN machine = "Lathe-A03"

Due to the differences that exist between the types of machines and the operations they perform, the form and content of the selection rule will vary from operation to operation. Therefore, by defining the selection rule within an operation class it is possible to write a distinct rule for each type of operation.

As in operation selection, this approach to machine selection eliminates the need for a separate expert system with inference engine and rule-base. Since the association between a rule for selecting a machine and the operations its pertains to is implicit in its placement within an the operation object, the resulting selection procedure is more efficient, easier to maintain, and extend than the traditional rule-based approach.

12.5.7 Operation Sequencing

The essence of operation sequencing involves determining what order to perform the selected operations such that the resulting order satisfies the precedence constraints of the operations. Some CAPP systems proceed from operation selection directly to sequencing without consideration for what tooling is required and what machines will be used. This approach makes it difficult to ensure that the generated operation sequence satisfies the need to minimize the number of tool and machine changes.

Reported methods for implementing operation sequencing include the use of production rules [29,68,69], precedence graphs with procedural algorithms for evaluation and pruning [55], and predefined sequences which are refined using production rules [16,17,56,59,62]. The approach used by the IOOPP system is a variation of the third approach making use of the object-oriented paradigm to improve the sequencing process. Predefined sequences are used to provide the system with a simple means to arrive at an initial ordering of operations by process type, and rules are used to reason about the ordering of operations of each process.

At this point in the planning process, the system has grouped the part features into profiles based on their relative position and location (internal or external) on the part. Next the system performs a preliminary ordering of the features within each profile based on the machine selected and process type (turning, milling, etc.). Since it is best to perform as many operations on the same machine at one time, this association of features with machines results in a preliminary ordering of the features for each part based on the processing they require.

The strategy for this preliminary ordering of processes is based on a predefined sequence defined within the *rotational_part* class. For rotational parts this sequencing strategy states that features should be ordered such that all turning operations are performed first, followed by milling, and then grinding. Once the features are grouped by process type, the system is able to apply a different set of sequencing rules to each group dependent on the process type.

Sequencing of the operations for each process type is performed using two sets of rules. The first set, the macro-rules, perform sequencing based on a predefined order in which operations (turn, bore, groove, etc.) should be performed. This strategy is translated into a set of generic rules whose order defines the sequence and whose attributes (location, feature type, and operation) define the operation to schedule next. The system begins with the first rule and checks with each feature in a group to determine if the rule applies. This applicability is based on a match of the feature's location, type, and associated operation (shown in Table 12-2). For example, the first rule calls for the removal of material, associated with a primary feature located on the interior of the part, through drilling. If the location of the feature is internal, and it is a primary feature (cylinder or cone), and it requires the "drill" operation, then that operation should be scheduled at this time. Since the system examines the features in a group one at a time, those operations on a profile are clustered and will be machined together.

The system will process the macro-sequencing rules one by one. After it applies each rule it will access a set of micro-sequencing rules that focus on the operation of the macro step. These micro-rules take care of the fine details of sequencing for that particular operation. For example, if the macro-rule operation is to drill an internal profile, then the micro-rules would address the ordering of these multiple drilling operations such that the holes are drilled beginning with the smallest diameter first, and if needed, the inclusion of a centerdrill operation before the first drilling operation.

TABLE 12-2: Macro-sequencing rules.

Location	Feature	Operation
Internal	Primary	Drill
External	Primary	Rough turn
External	Secondary	Groove
Internal	Primary	Rough bore
Internal	Secondary	Groove
External	Primary	Finish turn
External	Secondary	Thread
Internal	Primary	Ream
Internal	Primary	Finish bore
Internal	Secondary	Thread

This approach reduces the knowledge requirements, focusing the reasoning on the knowledge that applies only to that operation. Also, if the sequencing strategy for a particular shop is different, a user could easily change the sequencing strategy and macro-rules to reflect their needs.

The end result of the sequencing function is an ordering of the operations for each defined profile of the part. Since this currently marks the last planning step, following sequencing the system will output the process plan for the part. This plan consists of a listing of each required operation along with the machine, tooling, and cutting conditions associated with each specified operation.

12.6 PART VERSUS PRODUCT PLANNING

The overall goal of this research project is to develop a planning system that couples both process and assembly planning. In order to support this endeavor, the system was designed keeping in mind this objective. Therefore, the architecture of IOOPP was conceived to permit the representation of end-products composed of various combinations of piece-parts and sub-assemblies.

12.6.1 Product Modeling

To support planning for end-products, the product model for a single part was extended to include object classes for assemblies and end-products. The composition hierarchy used to represent an end-product is shown in Figure 12-10 . This representation defines an object class structure where an end-product is composed of multiple component and assembly objects, and an assembly object is composed of some multiple of other component and assembly objects. This structure parallels the format of a common bill-of-material used to represent products for manufacture. Use of this design approach results in a means for representing all types of end-products and not just a single machined component. The hierarchical structure implicitly defines the relationship between the components of the product.

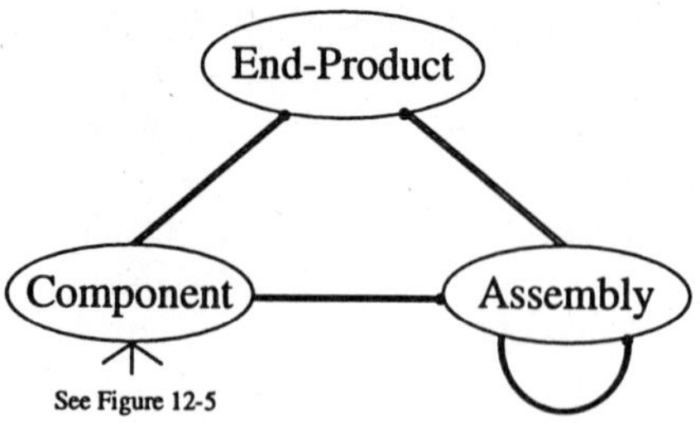

Figure 12-10: Part class hierarchy for product modeling.

12.6.2 Product Planning

Current CAPP systems are designed to plan for single parts, whereas IOOPP supports planning for all the components of an end-product. The planning begins with those components at the bottom of the product model hierarchy moving up through the tree until all components have been considered. At the completion of the planning phase, the system will present the user with a set of process plans for each manufactured part. This approach is essentially the same as multiple calls of a piece part planner.

12.6.3 Assembly Planning Extension

One other part of the product model as shown in Figure 12-10, is the definition of the assemblies within the end-product. This information is available for use in defining an assembly planning module within the overall system to provide for the development of a complete planner capable of both process and assembly planning. Due to the OOD of the system, incorporation of assembly planning can be accomplished through the definition of additional objects and object hierarchies that provide the data and functionality required. Some modification would be required to the current objects of the system. However, if the interface to these objects is not altered, then the process planning portion would be unaffected. At this time, the assembly planning function within the system is not fully operational and calls to it generate a blank process sheet.

12.7 EXAMPLE

The IOOPP system was designed and implemented using an object-oriented paradigm and the C++ programming language compiled on a microcomputer. This prototype system focuses on process planning for end-products made up of rotational parts. As an example, the system was applied to the example product shown in Figure 12-11, resulting in the product structure shown in Figure 12-12 in terms of the

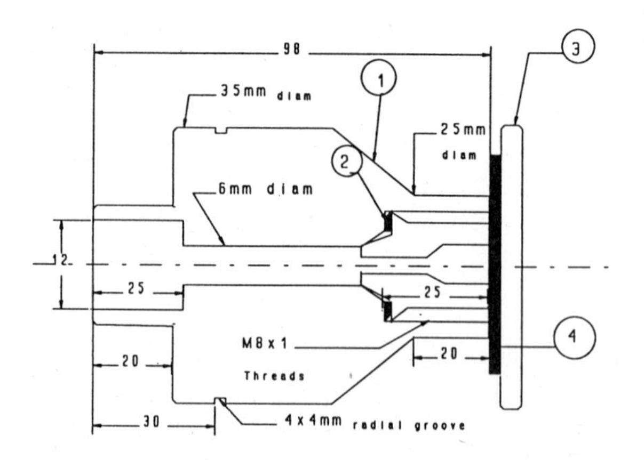

Figure 12-11: Example end-product.

object classes. As drawn, the end-product is assembled from two components and one subassembly, which itself is comprised of two components. This structure was used to demonstrate the capability of the system to handle subassemblies. Of the four total component parts, two of them are manufactured parts and two are purchased parts. The output of the system for *part-1* (the throttle body) is shown in Table 12-3.

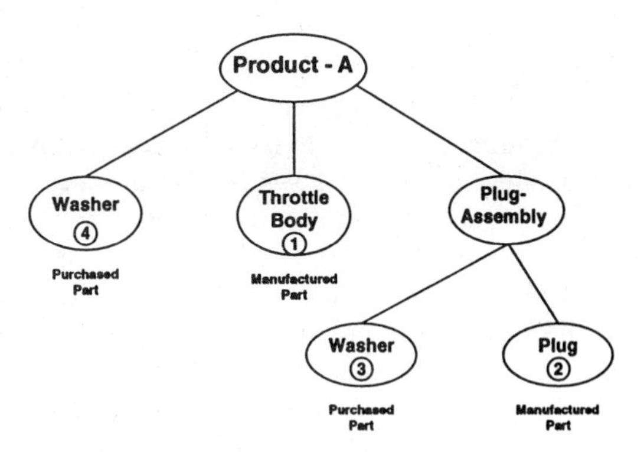

Figure 12-12: Product structure for example part.

TABLE 12-3: System output for Part 1 of the example end product.

End Product: P123
Part Number: Part-1
Raw Material: 35 mm diameter bright bar

Seq.	Operation	Description	Machine	Toolholder	Insert
1	Face	free end	Lathe-01	MCLNR2525M12	CNMG120412
2	Centerdrill	free end	Lathe-01	R210.5-XXX	-
3	Drill	6 mm diam. to 32 mm depth	Lathe-01	R410.5-0600-60-01	-
4	Rough turn	20 mm OD diam./20 mm length	Lathe-01	MCLNR2525M12	CNMG120412
5	Groove	4 mm radial groove	Lathe-01	N151.2-400-30-4G GC225	RF151.22-16-30
6	Drill	11 mm diam. to 25 mm	Lathe-01	R410.5-1100-60-01	-
7	Rough bore	12 mm diam. to 25 mm depth	Lathe-01		
8	Part-off	60 mm length	Lathe-01	N151.2-300-4E GC225	RF151.21-16-30
	Flip part				
9	Face	58 mm length	Lathe-01	MCLNR2525M12	CNMG120412
10	Rough Turn	OD taper to 25 mm diam.	Lathe-01	MCLNR2525M12	CNMG120412
11	Drill	6 mm diam. through	Lathe-01	R410.5-0600-60-01	-
12	Drill	12 mm diam. to 25 mm depth	Lathe-01	R411.5-12032D12.00	-
13	Drill	16.5 mm diam. to 25 mm depth	Lathe-01	R411.5-16532D16.50	-
14	Thread	M18 x 1.5mm pitch	Lathe-01	R166.0L-16VM01-002	R166.0KF-D10-3

12.8 CONCLUSION

This work has demonstrated the usefulness of the object-oriented paradigm in the design and construction of an automated process planning system. This application has provided a means for the development of a system that operates on the basis of message passing between objects that encapsulate the attributes and behavior of real world objects.

The resulting design of IOOPP is more flexible in terms of making changes and handling the evolution of the system over time. This makes it possible to easily continue the development of the system, adding functionality in terms of additional planning steps and expanding the current steps to consider additional part types. Process planning is a complex function, and the OOD approach is the means needed for the development of a generative system that addresses a domain large enough to be of use to industry, and simple enough to permit continual evolution while in commercial use.

The implementation of the planning functions within the object paradigm made use of the properties of inheritance and polymorphism to enhance the methods developed for selection and sequencing. These approaches eliminated the need for a separate knowledge base and inference engine instead relying on the ability to store the knowledge, and details on its use, within each object itself.

This research is continuing with the expansion of the current system to include feature recognition of part data from designs stored using the PDES/STEP standard, and consideration of prismatic parts. Also, as discussed, this same system will be expanded to incorporate assembly planning making use of the same product model to support both assembly and process planning functions. To date, modifications to the system have supported the claimed benefits of using an object-oriented design and implementation.

REFERENCES

1. Juri, A.H., Saia, A., and A. DePennington, "Reasoning About Machining Operations using Feature-Based Models," *Int. J. Prod. Res.*, 28(1), 1990, 153-171.

2. Chang, T-C., *Expert Process Planning for Manufacturing*, Addison-Wesley, 1990.

3. Alting, L., and H. Zhang, "Computer Aided Process Planning: the State-of-the-Art Survey," *Int. J. Prod. Res.*, 27(4), 1989, 553-585.

4. Ham, I., "Computer-Aided Process Planning: The Present and the Future," *Annals of CIRP*, 37(2), 1988, 1-11.

5. Link, C.H., "CAPP - CAM-I Automated Process Planning System," *Proc. 13th Numerical Control Society Annual Meeting and Technical Conf.*, March 1976, 401-408.

6. Houtzeel, A., "Computer Assisted Process Planning a First Step Towards Integration," *Proc. of Autofact*, 1980.

7. Schaffer, G., "GT via Automated Process Planning," *American Machinist*, 124(5), 1980, 119-122.

8. OIR, *MULTIPLAN*, Organization for Industrial Research, Waltham MA., 1983.

9. Wysk, R.A., "An Automated Process Planning and Selection Program: APPAS," *Ph.D. dissertation, Purdue University*, West Lafayette, IN., 1977.

10. Lee, H., and R.A.Wysk, "Formal Model of a Decision Tree for Process Planning," *IMSE Working Paper 91-143*, 1991, 1-35.

11. Bhaskaran, K., "Process Plan Selection," *Int. J. Prod. Res.*, 28(8), 1990, 1527-1539.

12. Prabhu, P. and H.P. Wang, "Algorithms for Computer-Aided Generative Process Planning," *International Journal of Advanced Manufacturing Technology*, 6(1), 1991, 3-15.

13. Lin, C-J., and H-P. Wang, "Optimal Operation Planning and Sequencing: Minimization of Tool Changeovers," *Int. J. Prod. Res.*, 31(2), 1993, 311-324.

14. Davies, B.J., and I.L. Darbyshire, "The Use of Expert Systems in Process-Planning," *Annals of the CIRP*, 33(1), 1984, 303-306.

15. Matsushima, K., Okada, N., and T. Sata, "The Integration of CAD and CAM by Application of Artificial Intelligence Techniques," *Annals of CIRP*, 31, 1982, 766-772.

16. Smith, J.S., Cohen, P.H., Davis, J.W., and S.A. Irani, "Process Plan Generation for Sheet Metal Parts using an Integrated Featured-Based Expert System Approach," *Int. J. Prod. Res.*, 30(5), 1992, 1175-1190.

17. Lee, I.B.H., Lim, B.S., and A.Y.C. Nee, "Knowledge-based Process Planning Systems for the Manufacture of Progressive Dies," *Int. J. Prod. Res.*, 31(2), 1993, 251-278.

18. Eversheim, W., and J. Schultz, "CIRP Technical Reports: Survey of Computer Aided Process Planning Systems," *Annals of the CIRP*, 34(2), 1985, 607-613.

19. Gupta, T., "An Expert System Approach in Process Planning: Current Development and Its Future," *Computers Ind. Engng*, 18(1), 1990, 69-80.

20. Peklenik, J., "Report of the 19th CIRP International Seminar on Manufacturing Systems--Computer Aided Process Planning," *Journal of Manufacturing Systems*, 7(3), 1988, 255-265.

21. Shah, J., Sreevalsan, P., and A. Mathew, "Survey of CAD/Feature-Based Process Planning and NC Programming Techniques," *Computer-Aided Engineering Journal*, Feb. 1991, 25-33.

22. Joshi, S., Vissa, N.N., and T.C. Chang, "Expert Process Planning System with Solid Model Interface," *Int. J. Prod. Res.*, 26(5), 1988, 863-885.

23. Joshi, S., and T.C. Chang, "Feature Extraction and Feature Based Design Approaches in the Development of Design Interface for Process Planning," *Journal of Intelligent Manufacturing*, Vol. 1, 1990, 1-15.

24. Lee, K., Lee, J.W., and J.M. Lee, "Pattern Recognition and Process Planning Prismatic Workpieces by Knowledge-Based Approach," *Annals of the CIRP*, 38(1), 1989, 485-488.

25. Mayer, R.J., Su, C.J. and A.K. Keen, "An Integrated Manufacturing Planning Assistant--IMPA," *Journal of Intelligent Manufacturing*, Vol. 3, 1992, 109-122.

26. Roy, U., Pollard, M.D., Mantooth, K., and C.R. Liu, "Tolerance Representation Scheme in Solid Model: Part I," *ASME 15th Design Automation Conf.*, Montreal, Canada, Sept. 1989, 1-10.

27. Roy, U., Pollard, M.D., Mantooth, K., and C.R. Liu, "Tolerance Representation Scheme in Solid Model: Part II," *ASME 15th Design Automation Conf.*, Montreal, Canada, Sept. 1989, 11-17.

28. Duan, W., Zhou, J., and K. Lai, "FSMT: a Feature Solid-Modeling Tool for Feature-Based Design and Manufacture," *Computer Aided Design*, 25(1), 1993, 29-38.

29. Wang, M., Waldron, M.B., and R.A. Miller, "Prototype Integrated Feature-Based Design and Expert Process Planning System for Turned Parts," *International Journal of Systems Automation: Research and Applications (SARA)*, Vol. 1, 1991, 7-32.

30. Jaques, M., Billingsley, J., and D. Harrison, "Generative Feature-Based Design-by-Constraints as a Means of Integration within the Manufacturing Industry," *Computer-Aided Engineering J.*, Dec. 1991, 261-267.

31. Chamberlain, M.A., Joneja, A., and T-C. Chang, "Protrusion-Features Handling in Design and Manufacturing Planning," *Computer Aided Design*, 25(1), 1993, 19-28.

32. Desai, V.S., and S.S. Pande, "GFM -- an Interactive Feature Modeller for CAPP of Rotational Components," *Computer-Aided Engineering J.*, October 1991, 217-220.

33. Cummings, S., "Gearing Up for the Object-Oriented Express," *Corporate Computing*, (1)1, 1992, 285-287.

34. Booch, G., *Object-Oriented Design with Applications*, The Benjamin/Cummings Pub. Co., 1991.

35. Usher, J.M., "An Object-Oriented Approach to the Design of Manufacturing Software Systems," to appear in *Computers in Industry*, late 1994.

36. Ingalls, D., "The Smalltalk-76 Programming System Design and Implementation," *Proc. of the Fifth Annual ACM Symposium on Principles of Programming Languages*, ACM, 1978, 9-13.

37. Srinivasan, K., and E.T. Sanii, "AI-Based Process Planning for Electronic Assembly," *IIE Transactions*, 23(2), June 1991, 127-137.

38. Sheu, P., and R.L. Kashyap, "Object-based Process Planning in Automatic Manufacturing Environments, *Proc. Int. Conf. on Robotics and Automation*, 1987, 435-440.

39. Sheu, P., and R.L. Kashyap, "Object-Based Knowledge Bases in Automatic Manufacturing Environments," *Int. J. of Advanced Manufacturing Technology*, 3(3), 1988, 39-52.

40. Krause, F-L., Ulbrich, A., and F.H. Vosgerau, "Featured Based Approach for the Integration of Design and Process Planning Systems," *Product Modeling for Computer-Aided Design and Manufacturing*, J. Turner, et al. (eds.), Elsevier Science Pub., 1991, 285-298.

41. Sanii, E., and R. Davis, "CAPP in a Hierarchically Controlled Manufacturing Environment," *Intelligent Manufacturing Structure, Control, and Integration, Vol 1 of Proceedings of Manufacturing International 90*, ASME, 1990.

42. Feghhi, S.J., Marefat, M., and R.L. Kashyap, "An Object-Oriented Kernel for an Integrated Design and Process Planning System," *27th ACM/IEEE Design Automation Conf.*, 1990, 437-443.

43. Marefat, M., "Object-Oriented Intelligent Computer-Integrated Design, Process Planning, and Inspection," *Computer*, 26(3), 1993, 54-65.

44. Kanumury, M., and T.C. Chang, "Process Planning in an Automated Manufacturing Environment," *Journal of Manufacturing Systems*, 10(1), 1991, 67-78.

45. Humm, B., Schultz, C., Radtke, M., and G. Warnecke, "IMS '91--Learning in IMS: A system for case-based process planning," *Computers in Industry*, 1991, 169-180.

46. Zust, R., and J. Taiber, "Knowledge-Based Process Planning System for Prismatic Workpieces in a CAD/CAM-Environment," *Annals of the CIRP*, 39(1), 1990, 493-496.

47. Abdou, G., and R. Cheng, "TVCAPP, Tolerance Verification in Computer-Aided Process Planning," *Int. J. Prod. Res.*, 31(2), 1993, 393-411.

48. Wang, M., and H. Walker, "Creation of an Intelligent Process Planning System within the Relational DBMS Software Environment," *Computers in Industry*, Vol. 13, 1989, 215-228.

49. Shah, J.J., and M.T. Rogers, "Expert Form Feature Modelling Shell," *Computer-Aided Design*, 20(9), 1988, 515-524.

50. Usher, J.M., "An Object-Oriented Approach to Product Modeling for Manufacturing Systems," *Computers and Industrial Eng.*, 25(1-4), 1993, 557-560.

51. Shah, J., and A. Mathew, "Experimental Investigation of the STEP Form-Feature Information Model," *Computer Aided Design*, 23(5), 1991, 282-296.

52. ISO CD 10303-1: Product Data Representation and Exchange - Part 1: Overview and Fundamental Principles, *Inter. Organization for Standardization*, TC 184/SC4 N154, 1992.

53. Kramer, T.R., Palmer, M.E., and A.B. Feeney, "Issues and Recommendations for a STEP Application Protocol Framework," National PDES Testbed, *Nat. Inst. of Stds. and Technology*, NISTIR 4755, Jan. 1992.

54. Eskicioglu, H., "The Use of Expert System Building Tools in Process Planning," *Engng. Applic. Artif. Intell.*, 5(1), 1992, 33-42.

55. Korde, U.P., Bora, B.C., Stelson, K.A., and D.R. Riley, "Computer Aided Process Planning for Turned Parts Using Fundamental and Heuristic Principles," *Trans. of ASME, J. of Engineering for Industry*, 114, Feb. 1992, 31-40.

56. Hinduja, S., and H. Huang, "OP-PLAN: An Automated Operation Planning System for Turned Component," *Proc. Instn Mech Engrs*, 203, 1989, 145-158.

57. Zhang, H., and L. Alting, "XPLAN-R: An Expert Process Planning System for Rotational Components," *Proc. of Institute of Industrial Engineers*, 1988, 54-60.

58. Prabhu, P., Elhence, S., Wang, H., and R. Wysk, "An Operations Network Generator for Computer Aided Process Planning," *J. of Manufacturing Systems*, 9(4), 283-291.

59. Yeo, S.H., Wong, Y.S., and M. Rahman, "Integrated Knowledge-based Machining System for Rotational Parts," *Int. J. Prod. Res.*, 29(7), 1991, 1325-1337.

60. Cho, K.-K., Lee, S-H., and J-H. Ahn, "Development of Integrated Process Planning and Monitoring System for Turning Operation," *Annals of the CIRP*, 40(1), 1991, 423-427.

61. Van Houten, F.J.M., "Strategy in Generative Planning of Turning Processes," *Annals of the CIRP*, 35(1), 1986, 331-335.

62. Giusti, F., Santochi, M., and G. Dini, "COATS: An Expert Module for Optimal Tool Selection," *Annals of the CIRP*, 35(1), 1986, 337-340

63. Yeh, C.-H., and G.W. Fischer, "A Structured Approach to the Automatic Planning of Machining Operations for Rotational Parts Based on Computer Integration of Standard Design and Process Data," *Int. J. Adv. Manuf. Technol.*, 6, 1991, 285-298.

64. Dhage, S., and J.M. Usher, "Computer-aided Tool Selection for Turning Operations", *Computers & Industrial Eng.*, 25(1-4), 1993.

65. Chen, S.J., and S. Hinduja, "Checking for tool collisions in turning," *Computer-Aided Design*, 20(5), June 1988, 281-288.

66. Chen, S.J., Hinduja, S., and G. Barrow, "Automatic Tool Selection for Rough Turning Operations," *Int. J. Mach. Tools Manufact.*, 29(4), 1989, 535-553.

67. Aresecularatne, J.A., Hinduja, S., and G. Barrow, "Optimum Cutting Conditions for Turned Components," *Proc. of Inst. of Mech. Engrs.*, 206(B1), 1992, 15-31.

68. Giusti, F., Santochi, M., and G. Dini, "KAPLAN: a Knowledge-Based Approach to Process Planning of Rotational Parts," *Annals of CIRP*, 38(1), 1989, 481-484.

69. Shanker, K., and K.V.S. Prasad, "Intelligent Process Planning: Rotational Parts," in *Expert Systems: Strategies and Solutions in Manufacturing Design and Planning*, Kusiak, A. (ed.), SME Pub., 1988, 187-211.

13

INTELLIGENT FEATURE EXTRACTION FOR CONCURRENT DESIGN AND MANUFACTURING

Jian(John) Dong, University of Connecticut

Hamid R. Parsaei, University of Louisville

Anup Kumar, University of Louisville

13.1 INTRODUCTION

The primary objective of Computer Integrated Manufacturing (CIM) is to develop high quality products and to bring them to the competitive global marketplace at a lower price and in significantly less time. As technological tasks in manufacturing environment have become progressively complicated, today's designers are faced not only with complexity of product design but also with a constantly increasing requirements for manufacturability, assemblability, etc.. Manufacturing decisions should be made early in the product development cycle in order design but also with a constantly increasing requirements for manufacturability, assemblability, etc.. Manufacturing decisions should be made early in the product development cycle in order to reducing cost and time to market.

Design and manufacturing, two major activities in an industry, have been performed separately for long time. This separation blocks the flow between design and manufacturing and greatly increases the product development time. To improve design and manufacturing efficiency, concurrently performing design and manufacturing is necessary.

Process planning is a linkage activity between design and manufacturing. It plays an important role in concurrent design and manufacturing. In this chapter, the research in intelligent feature extraction for concurrent design and manufacturing will be discussed. A prototype Feature-Based Automated Process Planning (FBAPP) system will be presented. FBAPP system can generate manufacturing process planning directly from a CAD system. This is an essential step towards the integration of design and manufacturing.

13.2 LITERATURE REVIEW

13.2.1 Computer Aided Process Planning(CAPP)

Process planning is defined by Chang (1991) as the function within a manufacturing facility that establishes which process and parameters are to be used (as well as those machines capable of performing these processes) to convert a piece of part from its initial form to its final form predetermined in an engineering drawing.

Traditional manual process planning involves several steps. The first step is the interpretation of the design data which is usually displayed by blue prints, or by a computer aided design (CAD) system. In this stage, batch size, geometric configuration, raw material property, dimensions, tolerances, surface roughness, heat treatment and hardness, as well as some special requirements, are studied and interpreted. The second step is the selection of manufacturing operations and suitable machines. Interpreted design information, shop capabilities, production knowledge are needed to perform tasks in the second step. The third step is the determination of operation sequences. Economic considerations and company's strategies are needed to perform this step. The forth step is the determination of clamping devices, and the orientation of the cutting tools. The appropriate cutting tools and operation parameters such as depth of the cut and feed rate are also needed to be determined. Finally, the overall machining time and non-machining time are then calculated, process sheets, operation sheets, and route sheets are prepared (Alting, 1989).

The manual process planning highly relies on the expertise of a process planner. The motivation behind most researches conducted in Computer Aided Process Planning (CAPP) was to replace experienced process planners with a computer system. The ideas of Computer Aided Process Planning (CAPP) were first presented by Niebel in 1965. Later in 1966, Schenck discussed the "Feasibility of an Automated Process Planning(APP)" in his Ph.D dissertation at Purdue University. Due to the limitation of computer hardware and software, the CAPP had not been feasible until the middle of 1970s.

Due to the rapid diminishing of experienced process planners in current industry and the importance of concurrent Computer Aided Design and Manufacturing (CAD/CAM), the research on CAPP has gained more attention than ever before. Automated Process Planning (APP) has received the highest priority among top ten areas which are most needed to be improved in the manufacturing environment(Nolen, 1989).

Three commonly-used approaches to Computer Aided Process Planning (CAPP) include: Variant Process Planning (VPP), Generative Process Planning (GPP), and Automated Process Planning (APP).

Variant Process Planning (VPP) requires a human operator to classify a part, input part information, retrieve a similar process plan from a database (which contains the previous process plans), and edit the plan to produce a new variation of the pre-existing process plan (Alting and Zhang, 1989).

The first VPP system was developed in 1976 under the direction and sponsorship of CAM-I (Computer Aided Manufacturing-International). In the same year, another system called "MIPLAN" was developed and introduced by OIR (Organization of Industrial Research).

Generative Process Planning (GPP) generates process plans utilizing decision logic, formulae, manufacturing rules, geometry-based data to determine the processes required to convert the raw materials to finished parts. The ideal GPP has no human interventions.

All present GPP systems to some extent depend on process planners (Nolen, 1989, Alting and Zhang, 1989). This is mainly due to the complexity of manufacturing processes and the geometric reasoning which consequently make it difficult to fully computerize the GPP system. However, the application of artificial intelligent (AI) and feature-based techniques to GPP seems to possess the potential of solving this problem (Alting and Zhang, 1989, Chang,1990). In early 1980s, AI technologies were applied to process planning which resulted in the development of GPP systems, such as TOM (Matsushima et. al. 1982), CMPP (Waldman 1983), EXCAP (Davis & Darbyshire 1984), and XPLAN (Lenau and Alting 1986).

Automated Process Planning (APP) generates process plans directly from an engineering design model (CAD data). In this system human involvement in decision making is not required. The major difference between an automated approach and other approaches is the use of automated CAD interface capability (Chang, 1990). APP is considered as a real bridge between design and manufacturing. At present, there is no APP system commercially available. The system discussed in this chapter is a prototype APP system.

13.2.2 Feature-based Technologies

Based on applications, a feature may have different meanings. From the manufacturing point of view, features represent shape and technological attributes associated with manufacturing operations and tools. From design point of view, features are elements used in generating, analyzing, and evaluating design. However, since features are deeply entrenched in geometric modeling, from geometric modeling point of view, features are groups of geometrical and topological entities (Shah, 1991), and used in reasoning design and manufacturing activities (Cunningham and Dixon, 1988).

Feature-based design provides designers a feature library in which a number of features are pre-defined. When designers design a part, they can select different features to form the part.

Feature-based design attempts to match the design process with the physical reality of manufacturing process, however, this may cause some barriers. First, in feature-based design, designers concentrate too much on the manufacturing process of a part. This may leave too little time to functionability which are the most important task during design stage. Secondly, since feature is application-oriented, features with the same shape may be given different names in different application areas. Some features may have the same name but carry different meanings. Thirdly, the limitation of the shape complexity and the number of features available greatly narrow the application areas.

Besides feature-based design, there is another feature-based concept for integrating design and manufacturing, which seems to match more closely with the nature of design and manufacturing processes. This technique, called feature recognition and extraction, provides much more flexibility to designers.

13.2.3 Feature Recognition/Extraction

The main task of feature recognition and extraction is to animate human geometric reasoning process to automatically extract design information for manufacturing. It is probably the most important and the most difficult task in concurrent design and manufacturing.

Up-to-date feature recognition and extraction approaches can be summarized into three categories: Graph/Pattern Matching (Choi,et al 1984, Joshi and Chang, 1988, Jakubowski, et al, 1982, Sakurai, et al, 1988), Geometry Decomposition (Tang and Woo 1991, Armstrong, et al, 1984) and Knowledge-based System (Henderson, 1988).

a. Graph/Pattern Matching

Graph/Pattern Matching involves searching a part model for primitives, such as faces and edges, and creating a graph of the entity's geometry shape. This graph is used as a template to identify features(Nitschke, D. R, 1991). The most important concept used in graph matching is edge classification. The edges can be classified into convex edge, concave edge, smooth convex edge and smooth concave edge (Shah, 1991, Chang, 1990) (Figure. 1).

Joshi and Chang (1988) used this concept in their Augmented Adjacency Graph(AAG). Sakurai and Gossard(1988) also used graph matching with a richer set of possible characteristics to describe features.

Pattern Matching refers to syntactic pattern recognition. In this system, geometric patterns are described by means of a series of straight, circular or other curved-line segments. Languages have been developed for describing these patterns and manipulating them with symbolic operators that form the grammar. A feature can be recognized by comparing with the object description in the grammar. This approach was used by Jakubowski et al (1982) and Choi (1984).

Graph/Pattern Matching is limited in its capability due to the fact that it can recognize only negative features with simple shapes such as hole, slot, etc.

b. Knowledge-based Systems (KBS)

A knowledge-based system for feature recognition was developed by Henderson and Chang (1988). Expert system rules and techniques are used to extract features from a 3D solid model using the internal boundary representation of a designed part. These recognition algorithms are also limited in their capability as they can deal with only simple negative features.

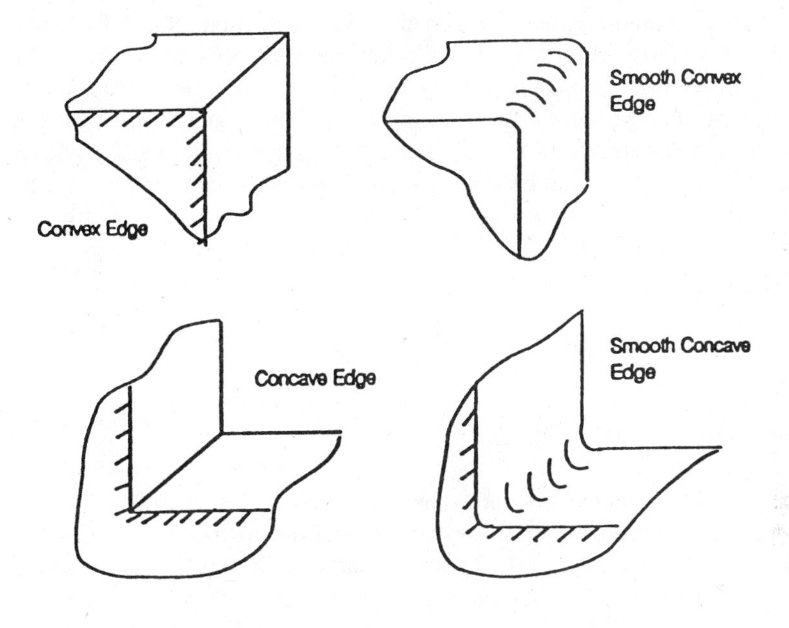

Figure 1 The Definitions of Concave Edges and Convex Edges

c. Geometry Decomposition

There are three common decomposition approaches to feature extraction/recognition. These include cell decomposition, convex-hull, and constructive solid geometry tree rearrangement.

Cell Decomposition. Cell decomposition typically uses a spatially enumerated model of a part, or a part decomposed into a number of cells. The cells that are to be removed by machining are recognized. Armstrong (1984) used this strategy to generate NC codes for a milled part. This technique is well-suited for generating NC codes for roughing cuts. However, it is not good for process planning.

Convex-hull Algorithm. This algorithm was developed by Woo(1984), and later on enhanced by Tang(1990) and Kim(1991). The technique is called "Alternative Sum of

Volumes(ASV)" or "Alternative Sum of Volumes With Partitioning(ASVP)" since it analyzes a part model and recursively computes the difference between objects and their convex hulls. As objects are identified, the part is broken down into its constituent features. The drawback of this technique is that it often produces an unusable decomposition, because it can result in removal of volumes that do not correspond to a single machining operation (usually with an odd-shaped feature). Also the stock shape can be awkward because it is the convex hull of the initial shape.

Constructive Solid Geometry (CSG) Tree Rearrangement. This algorithm, created by Lee and Fu (1987), rearranges an object's CSG tree by moving-up and shuffling features within the tree until a desired pattern of features is obtained. The primary drawback of the algorithm is that the CSG representations are not unique, therefore a feature can be defined in several ways by performing Boolean operations on primitives.(Lee and Jea,1988)

13.2.4 Feature-based Process Planning

A considerable amount of research has been conducted in geometric reasoning as well as generative process planning. Several systems have tried to integrate a CAD interface with process planning, for example, QTC (Quick Turnaround Cell) at Purdue University by Chang (1991), FRAPP (Feature Recognizer and Process Planning) at Arizona State University by Henderson (1988), and First-Cut/Next-Cut (Cutkosky,1988, 1992) at Stanford University.

In QTC and First-Cut/Next-Cut, a part is designed with the design-oriented manufacturing features. No feature recognition and extraction is involved in these two systems. QTC create a process plan when feature based design is finished. First-Cut creates a process plan while a feature is being selected. As was mentioned earlier, the feature-based design has some serious drawbacks.

In FRAPP, features are recognized only based on the appearance of a designed part. FRAPP searches all the depressions in a designed part and then attempts to recognize the features. The recognized feature information used as an input for process planning. The system does not consider the blank part shape and the intermediate specifications. The intermediate specifications are the specifications for an in-process work-piece, such as geometry, dimension, and tolerances, which are associated with each operation and are very important for manufacturing process planning (Wang and Chang 1990). In FRAPP, only very simple features such as slots and holes can be recognized.

In the prototype Feature-based Automated Process Planning(FBAPP) system, the designed part as well as the blank part are considered. Features are recognized from the removable volume point of view rather than from the designed part point of view. A Blank Surface-based Concave Edge (BS-CE) Feature Extraction approach is introduced in this study for extracting features. The extracted features are recognized with a knowledge-based approaches. Redundant entities in a feature are smoothed out and the features with various combinations of topological and geometrical information can be recognized. Feature spatial relationships are managed. The General Manufacturing Feature Information Scheme (GMFIS) is designed for transferring design information to manufacturing. Process plans are generated using expert system rules. The entire process in FBAPP is naturally closer to the thinking process of a human process planner. The intermediate specifications are also considered. A 3D graphically represented process plan can be generated in post-process planning.

13.3 PROTOTYPE FEATURE-BASED AUTOMATED PROCESS PLANNING (FBAPP) SYSTEM

Figure 2 illustrates the structure of the prototype FBAPP system. This system consists of four major components: CAD interface, production knowledge, process planning, and post-process planning.

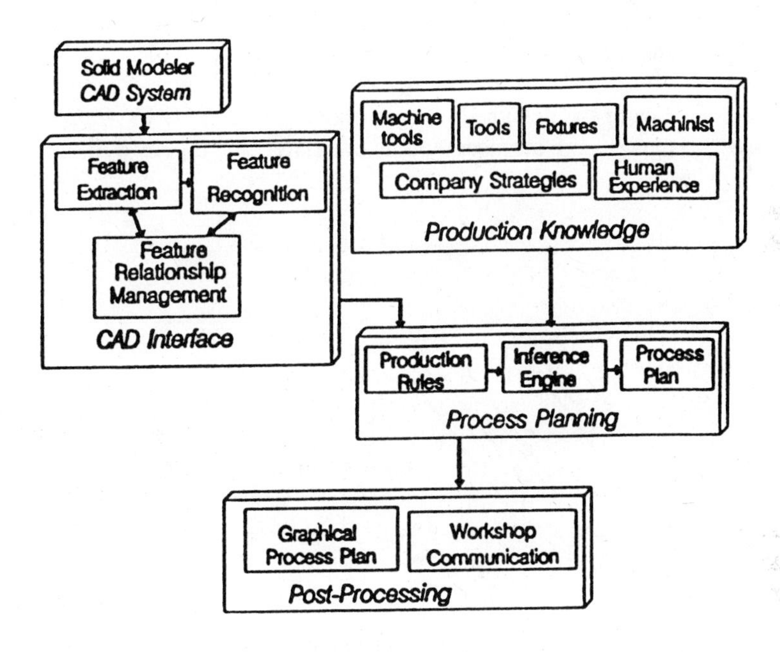

Figure 2 Structure of The Prototype FBAPP System

13.3.1 CAD Interface

To automatically generate manufacturing process plans from a CAD design, an interface is needed for transferring design information to manufacturing. Most existing CAPP systems use either a part code or a part description language as an interface to a CAD system. Since the part description is prepared manually, the quality of a process plan and the time spent to develop a process plan highly depend on a process planner's experience and knowledge. One way to automate design and manufacturing processes is to use feature extraction and recognition techniques

a. Overall Removable Volume(ORV) Generation

A process plan is the collection of operations required to transform a blank into a designed part. In order to create a process plan, both the designed part and the blank should be considered.

3D solid models of the designed part and blank can be pre-built in a solid modeling package such as I-DEAS software. Depending upon the tolerances, surface requirements and certain datum surfaces or datum axis, the ORV can be determined by subtraction of the designed part from the blank. This process can be performed by using the Boolean difference in a solid modeling package.

The specifications of the designed part are transferred to the ORV which consists of two kinds of surfaces: designed part surfaces and blank surfaces. Figure 3 illustrates the ORV determination.

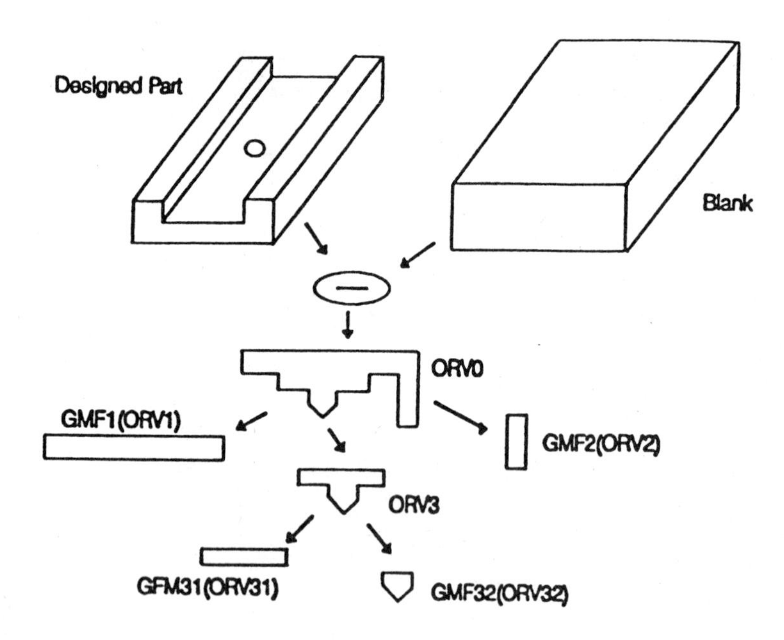

Figure 3 ORV Determination and GMF Extraction

b. Blank Surface-based Concave Edge (BS-CE) Feature Extraction Approach

The ORV can not be manufactured in one machine set up at the most time. The decomposition is, therefore, needed to extract each general manufacturing features (GMF). The BS-CE feature extraction approach is designed to decompose the ORV into GMFs (Dong 1993). This approach first identifies the largest blank surface (LBS) on the ORV, and then identifies the nearest concave edge (NCE) from the LBS. Based on the NCE, a splitting plane can be formed and the ORV can be decomposed into at least two pieces. The two pieces are continuously decomposed with the same approach until no concave edge is found on any volume. This approach allows as much material as possible to be removed in each machine setup. Figures 3 illustrates the concept and flow chart for the BS-CE approach.

c. General Manufacturing Feature (GMF) Recognition

After GMFs are extracted, the shape of each GMF must be recognized. The recognition is performed by using computer routines written in C and embedded with the CLIPS expert system rules.

In solid modeling, the shape of an object is decided by the topological and geometrical information. To recognize a feature, a series of rules must be built. The following are the rules listed for some features.

For rectangular-prisms(brick)

IF all surfaces are planes
THEN the object is a prism-like-1
IF the object is a prism-like-1
AND the object has 12 edges, 8 vertices and 6 surfaces
THEN the object is a prism-like-2
IF the object is a prism-like-2
AND each vertex is the intersection of 3 edges
THEN the object is a dovetail-like
IF the object is a dovetail-like
AND the three edges at each vertex are perpendicular to each other
THEN the object is a rectangular-prism

For dovetails

IF the object is a dovetail-like
AND the three edges at each vertex are not perpendicular to each other
AND the angular of two edges at one vertex is equal to the one at next vertex.
THEN the object is a dovetail

For tetrahedrons

IF the object is a prism-like-1
AND the object has 4 surfaces, 4 vertices and 6 edges
THEN the object is a tetrahedron

For standard cylinders

IF the object has three surfaces, one is a cylindrical surface, and the other two are plane surfaces
AND only two edges, both of them are closed circles
THEN the object is a standard cylinder

For cylinders with one circle surface and one ellipse surface(cylinder-2)

IF the object has three surfaces, one is a cylindrical surface, and the other two are plane surfaces
AND only two edges, one is a closed circle, and the other is a closed ellipse
THEN the object is a cylinder-2

For cylinders with two ellipse surfaces (cylinder-3)

IF the object has three surfaces, one is a cylindrical surface, the other two are plane surfaces
AND only two edges, they are closed ellipses.
THEN the object is a cylinder-3

For pegs(cylinder-4)

IF the object has two surfaces, one is a corn surface and the other is a plane surface
AND only one closed circle edges
THEN the object is a peg

For cylinders with one cone surfaces(Cylinder-5)

IF the object has three surfaces, one is cylindrical surface, one is a cone surface, and the other is a plane surface
AND there are only two edges in the object, and they are the closed circles
THEN the object is a cylinder-5

For cones (Cylinder-6)

IF the object has three surfaces, one is a cone surface, and the others are two plane surfaces
AND there are only two edges in the object,and they are closed circles
THEN the object is a cone

d. Handling Redundant Faces, Lines, and Points

The same shape of GMFs may have many different topological and geometrical combinations, which make the feature recognition very difficult. For instance, in Figure 4, four objects all are rectangular-prisms from both design and manufacturing points of view. However, they are very different from topological and geometric point of view. In figure 5, the component A consists of 8 topological surfaces, 12 topological vertices and 18 topological edges; component B consists of 7 topological surfaces, 8 topological vertices and 13 topological edges; component C consists of 7 topological surfaces, 14 topological vertices and 19 topological edges; component D consists of 9 topological surfaces, 12 topological vertices and 18 topological edges.

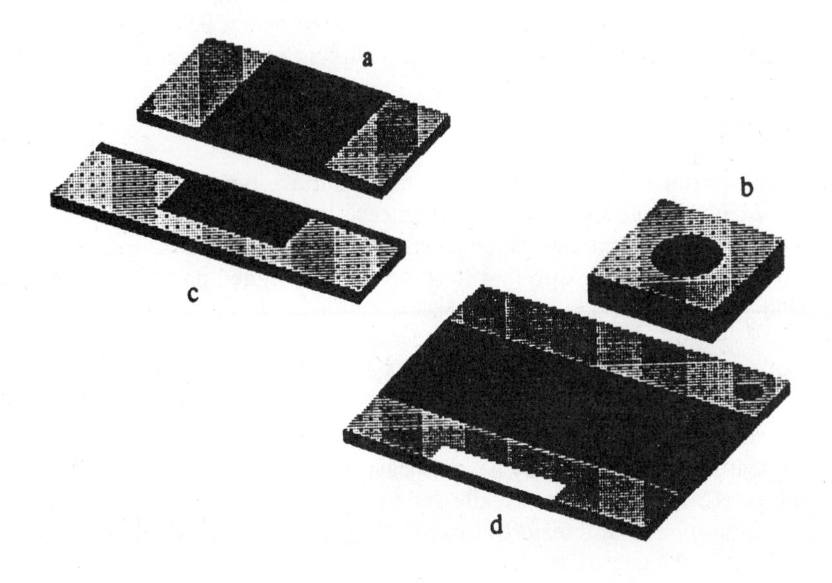

Figure 4 The Shape of a Rectangular-Prism Can be Different in Terms of Topology and Geometry

A GMF may be formed by more complicated combinations of topological surfaces, edges, and vertices. Simple expert system rules written above are not capable of recognizing a feature with various combinations of topological and geometrical information.

Before using simple expert system rules, a smoothing process is required. To recognize the shape of a rectangular prism feature, the coordinates of each point must be checked to determine the number of surfaces lying in the same plane. The surfaces in the same plane are merged into a single surface. The points, edges, and lines which are not in the boundary of the merged surface are smoothed out.

The edges which are in the same line are merged into one edge, and the points between two end vertices are smoothed out. A GMF shape can then be recognized based on the expert system rules.

e. Feature Relations

The relationship among the designed part, blank, ORV, and GMF should be maintained in order to make GMF meaningful to process planning. A GMF Information Scheme (GMFIS) is created to maintain all the information. The GMFIS includes five components: feature extraction flow information, object information, surface information, edge information, facet and point information.

The feature extraction flow information contains the information of the relationships among the designed part, blank, ORV and GMFs. The ORV and GMFs may include four types of surfaces: surfaces generated during current splitting, surfaces which belong to the designed part, surfaces which belong to blank, and the surfaces generated in previous splitting (set as blank surface).

Object information includes information such as the shape of a feature, material, the relationship between the object and surfaces.

Surface information contains the information such as surface type, facet in each surface.

The edge information includes edge types, vertices, and other information. Facet and point information contains information on the coordinates of each point and facet.

The same GMF may appear in different locations of a designed part. The information maintained in GMFIS is able to express the differences.

f. Compound Manufacturing Feature (CMF) Recognition

In a machine shop, some features such as countersink and counterbore are treated as a single manufacturing feature even though they have concave edges. These features are Compound Manufacturing Features (CMF). They are usually formed by more than one GMF and can be produced in one machine setup. These features can be recognized by identifying the position of adjacent GMFs.

For countersink:
IF adjacent two GMFs are cylinder-5 and cylinder-6
AND they share one circle plane
AND the diameter of cylinder is the same as the smaller diameter in cylinder-6
THEN the object is a countersink

For counterbore:

IF adjacent two GMFs are standard cylinder (cylinder-1) and cylinder-5
AND they share one plane
AND the center axes of two GMFs are in the same line
THEN the object is a counterbore

Some compound features meet the above rules. However, they may not be manufactured in one machine setup. This is due to size of available tools in a workshop. For example, a counterbore is usually produced with special counterbore tools. These tools decide what types of counterbore can be produced in a single machine setup.

In FBAPP system, any CMF meets above rules is considered as a counterbore or a countersink. However, during process planning, if there is no suitable tool for a CMF, manufacturing process plans are created for each GMF which forms the CMF.

13.3.2 Process Planning

Wright and Bourne (1988) conducted an experimental work to analyze the thinking process of an experienced machinist in manufacturing a designed part. Their experiment shows how the manual process planning works. During manual process planning, a process planner first reviews an engineering drawing to identify major removable features of the part. The planner then studies the design specifications (detailed information) and compares these specifications with his/her knowledge about the production. During this process, the planner may also look at some available documents to ascertain the ability of producing the designed part. Finally, based on the shop capacities, the process planner may either decide to sub-contract the work or to make the part in-house which results in the preparation of a list of manufacturing steps, which is the process plan to manufacture the part.

The CAD interface previously discussed provides the design information required for a process plan. Production knowledge is also needed to determine the required manufacturing approaches.

a. Production Knowledge

Production knowledge may include process capabilities, tool selection knowledge, fixture selection knowledge, sequencing, etc. The process capability is the only feature incorporated in this prototype system. Process capabilities can be described by the shapes, dimensions, tolerances, surface finishing, geometrical and technological constraints, and economics of the process.

The production knowledge should be properly created and stored in a computer system for an automated process planning system. Representation of the shape producing capabilities of a machine is a difficult task. This is due to the difficulty involved in mathematical modeling of the geometry of a cutting tool and possible tool motion. In manual process planning the operator often has a good understanding about the machine tool capabilities. However, in a computer system, an explicit expression is necessary. Table 1 illustrates some process capabilities, tolerances, and surface finishes using volume features.

Dimension (Size) capability of a machine tool is determined by work envelope and the tool size, which may be found in machine manuals. For example, the Cincinnati Milacron CNC Turning Center has the capacities to manufacture a part within the following dimensions: maximum outside diameter is 190 mm, maximum inside diameter is 161 mm, and maximum length is 609 mm., etc. Bridgeport RE2E3 CNC milling machine has the following work envelope: Xmax = 864 mm, Ymax = 318 mm, and Z max = 400 mm and the Brother TC-227 taping CNC machine's capacities are: Xmax = 600 mm, Ymax = 300, and Zmax = 300 mm.

Tolerance and surface finishing capabilities of a machine are decided by many factors such as tool wear, workpiece material, work environment, thermal deformation, fixture, human experience, etc. It is not possible to give a precise tolerance for each machine process, however, experimental data may be founded in machine tool handbooks or manuals Table 1 lists a few basic machine processes and their typical tolerance achieved.

Table. 1 Representation of Process Capacities, Tolerances, and Surface Finishing Using Volume Features (Units: inch) (Tien-Chien Chang, Expert Process Planning for Manufacturing, pp.124 - 127, 1990, by Addison-Wesley Publishing Company, Inc. Reprinted by permission of the publisher)

Process	Sub-Process	Cutters	Volume Capacities	Tolerances, Surface finish
Milling	Face milling	Plain Inserted-tooth	Flat bottom volume	rough / finish tol 0.002 / 0.001 flatness 0.001 / 0.001 angularity 0.001 / 0.001 paralleism 0.001 / 0.001 surface finsh(μin) 50 / 30
	Peripheral milling	Plain Slitting Saw Form Inserted-tooth Staggered-tooth Angle T-slot Woodruff keyseat cutter	Flat bottom volume Slot Formed volume T-slot Internal groove	rough / finish tol 0.002 / 0.001 flatness 0.001 / 0.001 surface finish(μin) 50 / 30
	End Milling	Plain Shell end Hollow end Ball end	Pocket, slot, flat Sculptured surface, flat	rough / finish tol 0.004 / 0.041 paralleism 0.0015 / 0.0015 surface finish(μin) 60 / 50
Drilling		Twist drill Spade drill Deep-hole drill Gun drill Trepanning cutter Center drill Combination drill Countersink Counterbore	Round hole Round hole Deep round hole Deep round hole Large round hole Shallow round hole Multi-diameter round hole Countersink hole Counterbore hole	length/dia <8 Mtl <Rc 50 dia / Tolerance 0 - 1/8 / 0.003 - 0.001 1/8 - 1/4 / 0.004 - 0.001 1/4 - 1/2 / 0.006 - 0.001 1/2 - 1 / 0.008 - 0.002 1 - 2 / 0.01 - 0.003 2 - 4 / 0.012 - 0.004 roundness 0.004 surface finish 100
		Deep-hole drill Gun drill		Dia / Tolerance <5/8 / 0.0015 >5/8 / 0.002 surface finish >100 μin straightness 0.005 in 6 in
Reaming		Shell reamer Expansion reamer Adjustable reamer Taper reamer	Thin wall of round hole Thin wall of round hole Thin wall of round hole Thin wall of round hole	Dia / Tolerance 0 - 1/2 / 0.0005 - 0.001 1/2 - 1 / 0.001 1 - 2 / 0.002 2 - 4 / 0.003 roundness 0.0005 surface finish 50 - 125

Knowledge representation of manufacturing processes in a computer system is difficult and time consuming. Artificial Intelligence (AI) and Expert Systems are considered as the most popular and practical approaches for representing the knowledge. In the FBAPP system, process knowledge is described with CLIPS expert system rules based on three machines mentioned before.

b. Process Planning

Manufacturing steps for producing each GMF or each CMF can be generated with the production rules coded with the CLIPS expert system shell. Figure 5 illustrates the generation of the process plan for each feature.

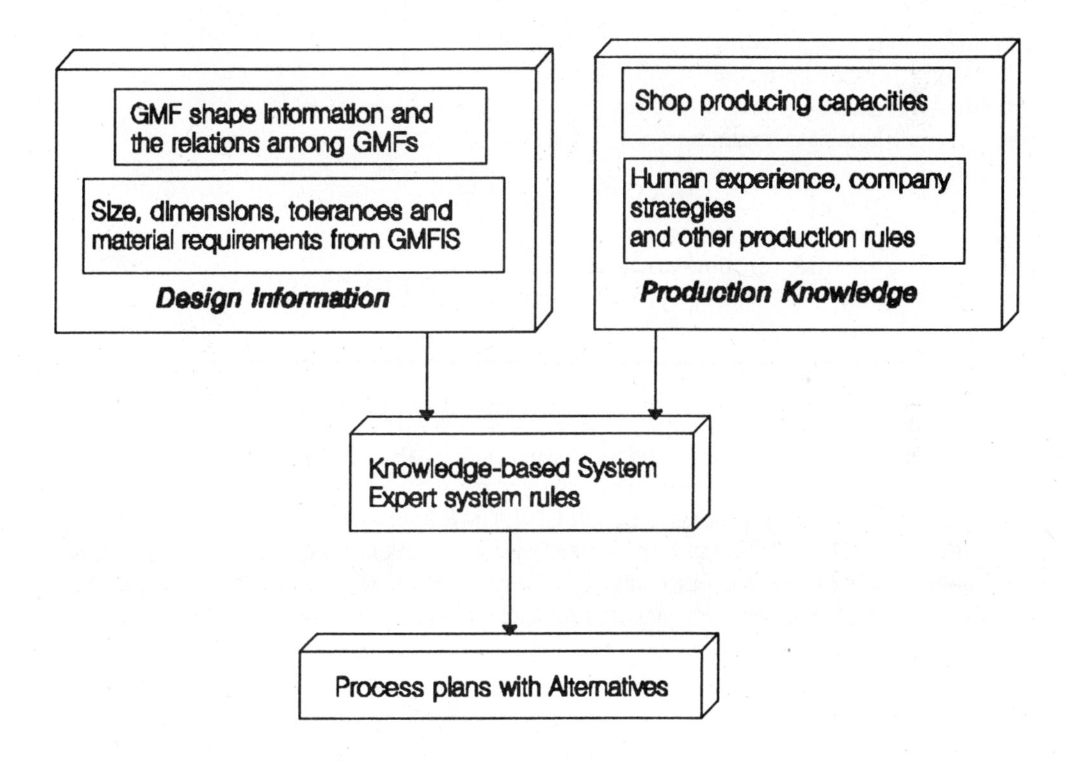

Figure 5 Generation of Process Plans for Each GMF

An example of production rules to produce a through hole using a rectangular prism blank is given below. The information about machines, tools etc. are built in the CLIPS data file.

```
(feature-shape cylinder-with-two-plane-ends)   ;          A through hole
=>
(assert (drilling or/and reaming operation))
)
;
;
(drilling or/and reaming operation)
(blank-shape rect-prism)
=>
(assert (tc-227 taping machine))
(assert (bridgeport milling machine))
)
;
;
(hole-dimensions height, radius)
(hole-radius-tolerance up-t, low-t)
(hole-high-tolerance up-t, low-t)
(hole-material)(hole-heat-treatment)
(hardness)
=>
(assert (drilling and reaming operation is needed))
(assert (drilling-feed-rate, drilling-speed, tool-material))
(assert (reaming-feet-rate, drilling-speed, tool-material))
)
;
;
```

The collection of information on processing each GMF or CMF provides the overall plan to manufacture the entire part. The process plan for the designed part will then be obtained by ordering the manufacturing process for each GMF/CMF based on its geometric relations.

Since each GMF/CMF can be processed by more than one type of operations, there may be a number of alternative process plans. The most economical process plan can be obtained through the integration of process planning and scheduling (Dong 1992).

13.3.3 Post-processing

Based on each feature, both graphic and textual representation a process plan can be generated for GMF or CMF. This representation greatly helps operators in understanding required manufacturing processes. Figure 9 illustrates a process plan and its graphic representation for cutting a slot.

13.4 THE IMPLEMENTATION

I-DEAS software was used as a testbed for the prototype FBAPP system. 3D models of a designed part and a blank were created in the I-DEAS software. Computer routines for feature extraction, feature recognition, process planning, and post-process planning were developed and integrated with the I-DEAS software.

13.4.1 I-DEAS Solid Modeling Software and FBAPP

I-DEAS (Integrated Design Engineering Analysis Software) is a product of SDRC (Structural Dynamics Research Corporation, Ohio) and is widely used as engineering design and analysis tool. Figures 6 illustrates the relationship between the FBAPP and I-DEAS.

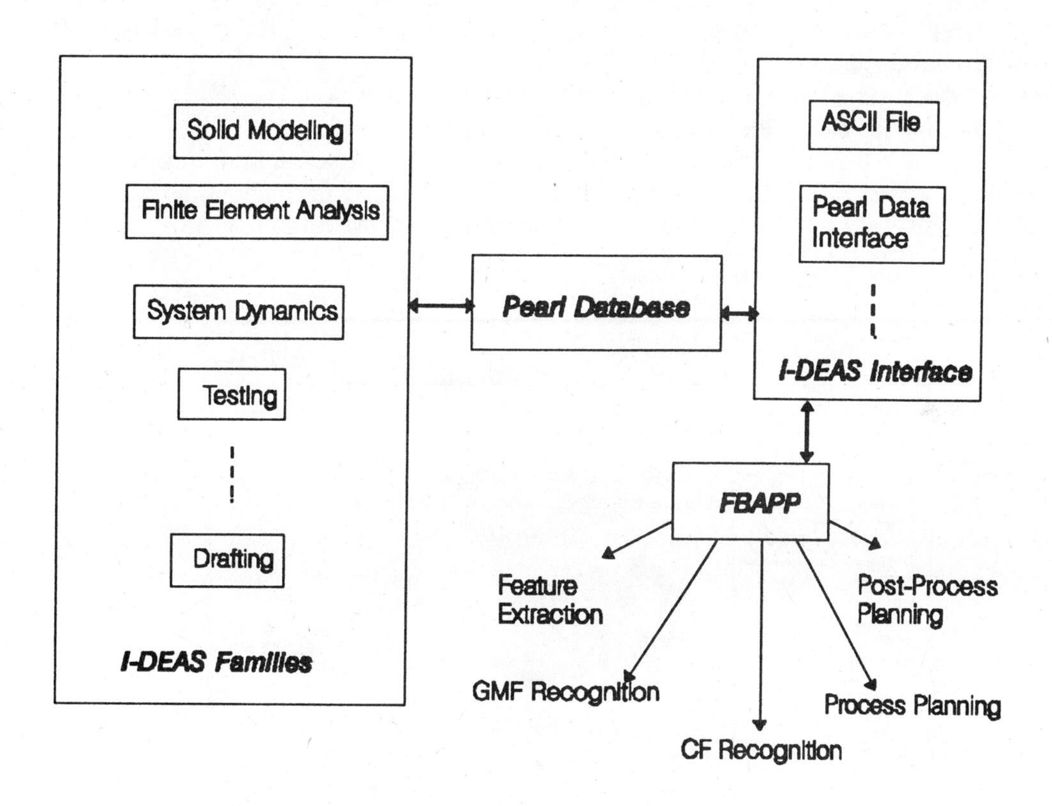

Figure 6 I-DEAS and FBAPP System

13.4.2 FBAPP and CLIPS Expert System Shell

CLIPS is an expert system shell developed by NASA. CLIPS expert system rules embedded with C programs are used for feature recognition and process plan generation in the prototype FBAPP system.

13.4.3 The Operation of Prototype FBAPP System

To run the prototype FBAPP system, the following steps are required.

1. Start I-DEAS, properly arrange four windows on a screen and load a designed part and a blank. These four windows include: I-DEAS Graphics, I-DEAS List, I-DEAS Prompt and Console windows in HP/Unix-based Terminal.
2. Run program file "fe.prg". Following the instruction in the I-DEAS List and Prompt windows to perform feature extraction step by step until all GMFs are obtained.
3. Click FBAPP in I-DEAS main menu, and then click sub-menu "GMF Recognition" to run GMF recognition.
4. Click sub-menu "CF Recognition" to run CMF recognition.
5. Click sub-menu "Process Planning" to create process plans.
6. Click sub-menu "Graphical Display", and follow the instruction in console window to show the textual process plan in a window and graphical display in another window.

Figures 7 and 8 illustrate menus of the FBAPP software system in I-DEAS.

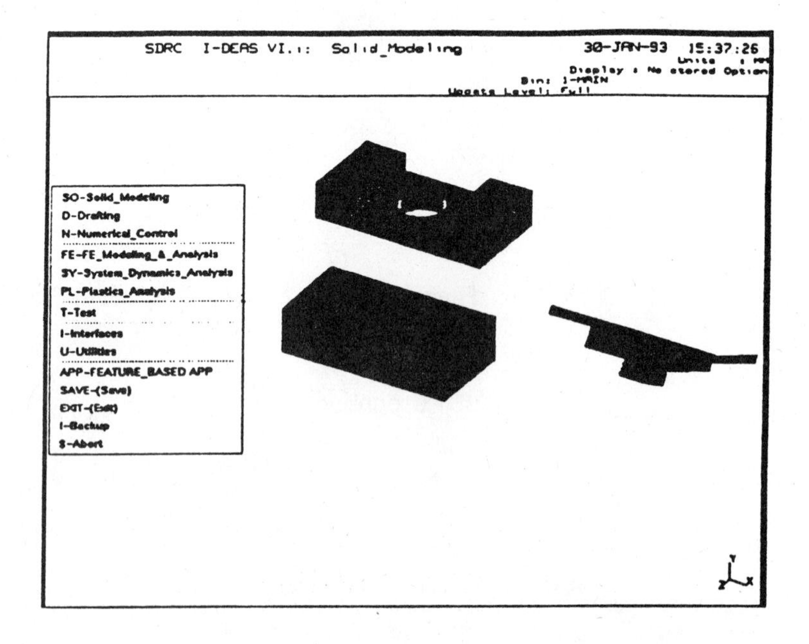

Figure 7 A Designed Part, a Blank and an ORV

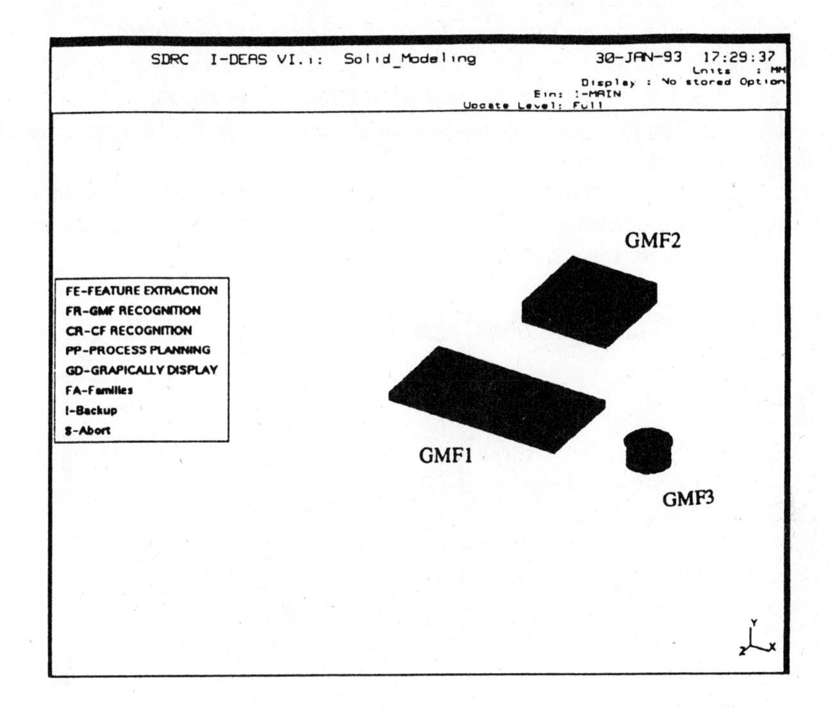

Figure 8 Representation of the Example GMFs

13.4.4 Example

A blank and a designed part are previously built and stored in I-DEAS universal files. By running the fe.prg, the blank and the designed parts are loaded into I-DEAS. The 3D designed part and blank can be manually arranged to a desire position. The overall removable volume can then be obtained by running Boolean operations. An illustration of this process is shown in Figure 7.

Following the instructions which appear in I-DEAS list window, the ORV can be decomposed into three GMFs. The three GMFs are stored into a series of universal files. Figure 8 illustrates the three GMFs for this example.

By clicking the feature recognition menu, the feature recognition information appears in a window as the following:

The feature is a rect_prism
The feature is a rect_prism
The feature is a right-circular-cylinder-----through-hole

By selecting the CF recognition, the system searches for any CMF. In this example no

CMF is recognized by the system (since no CMF is found in this example).

When process planning module is executed, it is required to provide dimension information for the blank and GMFs as well as tolerance information. After the required information is entered, the process plans are generated for each GMF. These process plans will be automatically stored in separate files.

By clicking the graphical display menu and following the instructions in the I-DEAS list window, the post-process planning can be performed. Figures 9 illustrates the post-process planing and graphic display for a slot.

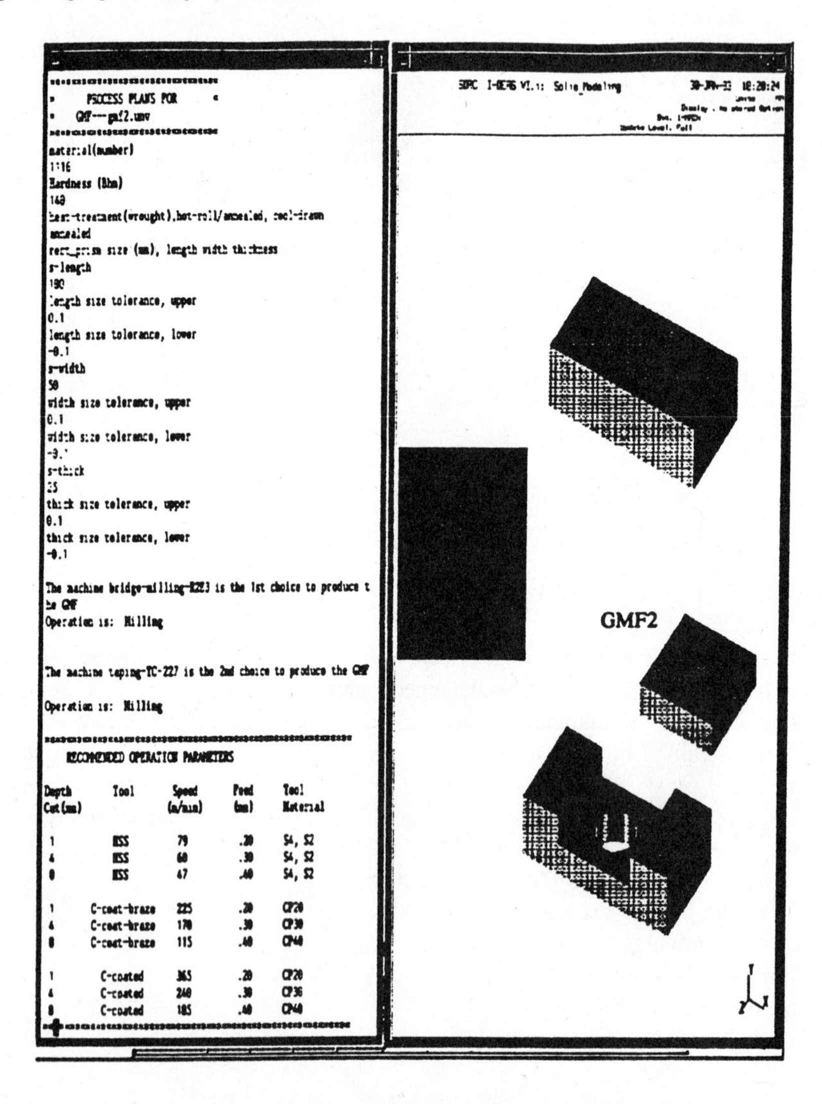

Figure 9 The Process Plans and the Graphical Display for GMF2

13.5 CONCLUSION

Feature-based automated process planning is a link between design and manufacturing. It plays an important role in concurrent design and manufacturing. The FBAPP prototype system presents an efficient way to avoid costly downstream traditional design and manufacturing process, and to make most product development decisions during the early stage of design. The techniques used for ORV generation, GMF extraction and recognition, GFMIS and the knowledge-based process planning make the FBAPP a promising system for the research in concurrent design and manufacturing.

13.6 REFERENCES

Alting, L. and Zhang, H-C., 1989, "Computer Aided Process Planning: the state-of-the-art survey" International Journal of Production Research, Vol.27,No.4,pp 553-585.

Armstrong, G. T., Carey, G. C. and Pennington, A. D., 1984, "Numerical Code Generation from Geometric Modeling System" Solid Modeling By Computer--from theory to application, Edited by M. S. Pickett and J. W. Boyse, Plenum Press.

Chang, T-C., 1990, Expert Process Planning for Manufacturing, Addison-Wesley Publishing Co, Reading, Massachusetts.

Chang, T-C., Wysk, R.A. and Wang, H.P., 1991, Computer-aided Manufacturing, Prentice Hall

Choi, B. K., Barash, M. M. and Anderson, D. C., 1984, "Automatic Recognition of Machined Surface From a 3D Solid Model" Computer-Aided Design, Vol.16/No.2, March.

Cunningham, J. J. and Dixon, J. R., 1988, "Design With Feature: The Origin of Feature", Proceedings of 1988 ASME International Computer in Engineering Conference Exhibition, Vol. 1, July 31-Aug.4, pp 237-243.

Cutkosky, M. R., Tenenbaum, J. M. and Muller, D., 1988 "Features in Process-Based Design", Proceedings of 1988 ASME International Computer in Engineering Conference Exhibition, Vol.1, July 31-Aug. 4, pp 557-562.

Cutkosky, M. R., Tenenbaum, J. M. and Brown, D. R., 1992, "Working With Multiple Representations in a Concurrent Design System" Journal of Mechanical Design, Vol.114, pp515-524.

Davies, B. J. and Darbyshire, I. L., 1984, "The use of expert systems in process-planning", Annals of the CIRP, 33(1)

Dong, J., 1993, "Design and Implementation a Feature-based Automated Process Planning System", Unpublished Ph. D. Dissertation, August, University of Louisville.

Dong, J and H.R. Parsaei, 1993. "Manufacturing Feature Extraction and Recognition", Computers and Industrial Engineering, Vol. 25, nos 1-4, pp. 325-328.

Dong, J. and H. R. Parsaei, 1993. "Feature-based Automated Process Planning System" The Proceedings of the 2nd IIE Research Conference, pp 11-15, Los Angeles, CA.

Dong, J., H. H. Jo, and H. R. Parsaei, 1992, "A Dynamic Feature-based Process Planning and Scheduling", Computers and Industrial Engineering, Vol. 23, pp 141-144, 1992.

Dong, J., E. Palmer, and H. R. Parsaei, 1992, "A Feature-based Automated Process Planning Prototype System", the Proceedings of the American Society of Naval Engineers, Product Engineering Symposium, Louisville, Kentucky.

Gomes, A. J. P. and Teixera, J. C. G., 1991, "Form Feature Modelling in a Hybrid CSG/BRep Scheme" Computer and Graphics, Vol 15. No. 2, pp 217-229.

Henderson, M. R. and Chang, G. J., 1988 "FRAPP: Automated Feature Recognition and Process Planning From Solid Model Data", the Proceedings of 1988 ASME international Computer in Engineering Conference and Exhibition, San Francisco, CA.

Lee Y. C. and Fu K. S., 1987, "Machine Understanding of CSG: Extraction and Unification of Manufacturing Features, IEEE Computer Graphics and Applications, 7(1), pp20-32.

Jakubowki, R., 1982, "Syntactic Characterization of Machined Part Shapes" Cybernetics and System, Vol. 13, pp1-24.

Joshi. S and Chang, T-C., 1988, "Graph-based Heuristic for Recognition of Machined features From a 3D Solid Model" Computer Aided Design, Vol.20/No.3, March, pp 1234-1255

Kim, Y. S., 1991, "Form Feature Recognition by Convex Decomposition" the Proceedings of 1991 ASME International Computer in Engineering Conference and Exposition, Aug. 18-22, Santa Clara, CA.

Kusiak, A., 1990, "Optimal Selection of Machinable Volumes" IIE Transactions, Vol.22/No.2, June, pp 234-250

Lee, Y. C. and Jea, K. F., 1987, "Machine Understanding of CSG: Extraction and Unification of Manufacturing Features" IEEE CG&A, pp20-32.

Lenau, T. and Alting, L., 1986, "XPLAN-an Expert Process Planning System" the 2nd International Expert Systems Conference, London, UK.

Matsushima, K., et al., 1985, "The integration CAD and CAM by Application of Artificial Intelligence Techniques" Annals of the CIRP, 24(1), pp329-332

Niebel, B. W.,1965, "Mechanical Process Selection for Planning New Designs" ASME paper, No. 737.

Nolen, J., 1989, Computer Automated Process Planning for World-Class Manufacturing, Marcel Dekker, Inc.

Nitschke, D. R., Chen, Y. M. and Miller, R. A., 1991, "A Feature Extraction Interface for CAD Part Models" the Proceedings of 1991 ASME International Computer in Engineering Conference and Exposition. Aug. 18-22, Santa Clara, CA.

Sakuria, H. and Gossarsd, D. C., 1988, "Shape Feature Recognition From 3D Solid Model" the Proceedings of 1988 ASME International Computer in Engineering Conference and Exhibition, San Francisco, CA.

Schenk, D. E., 1966, "Feasibility of Automated Process Planning" Ph.D Thesis, Purdue University, West Lafayette, Indiana, USA.

Shah, J. J., 1991, "Assessment of Features Technology" Computer-aided Design, 23(5), June.

SDRC, 1990, I-DEAS Solid Modeling Manuals, Structure Dynamics Research Co.

Tang, K. and Woo, T., 1991, "Algorithmic Aspects of Alternating Sum of Volumes. Part 1: Data Structure and Difference Operation" Computer Aided Design, Vol.23/No.5, June.

Waldman, H., 1983, "Process Planning at Sikorsky", CAPP, edited by J. Tulkoff(Computer and Automated Systems Association of SME).

Wang, W-T and Chang, T-C, 1990, "Feature Recognition for Automated Process Planning" the Proceedings of Manufacturing International, Part 2: Advances in Manufacturing System, March 25-28, pp 49-54, Atlanta, GA.

Wright, P. K. and Bourne, D. A., 1988, Manufacturing Intelligence, Addison-Wesely Publishing Company.

Wysk, R. A., 1977, "An Automated Process Planning and Selection Program: APPAS" Ph.D Thesis, Purdue University, West Lafayette, Indiana, USA.

14

CAD in Automatic Machine Programming

B.O. Nnaji, University of Massachusetts at Amherst

T.S. Kang, Auto-trol Technology Company

14.1 INTRODUCTION

Automatic machine programming requires that tasks which the machine must perform be specified to it in some task-level language. This method of specification implies that the machine must reason about its work environment in order to plan how to execute the task. One of the principal problems to be overcome in such planning is reasoning about the world of geometric entities whose shapes define the configuration of the world. Although the world information can be captured using a variety of approaches including vision, the understanding of the world implies understanding geometric shapes. Since the CAD system normally processes knowledge of parts geometric composition, obtaining and representing world information in the CAD system is the most viable approach.

By means of acquiring the world knowledge from the CAD system and generating these task level commands, a robot is able to automatically perform the task. The major drawback in the system so far developed (as well as in RAPT, and AUTOPASS) is that those systems have not been capable of interpreting geometry to recognize geometric features and inferring function which the features are intended to serve. Even for the most simple geometries -- the polyhedral objects, this has proven to be very cumbersome. The problem of inference of function from form (geometry) of form functional specifications is particularly difficult because it involves the regeneration of the designer's intent for the part or product. It invariably requires an experienced process planner to do this and amounts to a ''re-invention of the wheel."

Representing information to the machine (task specification) using the CAD system is in itself fraught with problems. This is largely due to the limited capacity of today's CAD systems. This problem is not limited to assembly but rather spans all of areas of applications of automatic process planning applications. Geometric data currently obtained from CAD systems are CAD system dependent. In addition, current CAD systems are incapable of carrying manufacturing process information other than geometry. What is needed is a new type of CAD system. A CAD system which is capable of capturing the designer's intent for the part or product and can transfer the product definition information to a process planner. This new CAD system should ideally be capable of translating the data so acquired at design stage into a neutral graphics medium in order to eliminate CAD system dependence of the geometric data. Function can be captured by the designer and transferred to the process planner, thus eliminating the huge stumbling block of today's automatic process planner -- the inference of function from form. In the next section, we describe the type of information which a desired CAD system must provide for the robot task planning system.

14.2 DESIRED CAD DATA

We have already mentioned that the future CAD system must produce both geometric and functional data which are relevant to the process planning function. In order to assure that future planning systems deal with data which is easy to manipulate, a study of the type of data appropriate for reasoning about geometry has been conducted [1]. The following observations were made: Since the essential geometric information in object reasoning is the boundary information, a good modeler must be able to present boundary information in the form of ***vertices, edges, loops,*** and ***faces***, from which features may be extracted. Also, the parameters used to represent a particular feature need to be explicitly represented. In addition, it seems a waste of very valuable time to design parts using the current approach of ***free design*** where models of parts are created and the process planner must ''tease out" the feature embedded on the part using some type of feature recognition system. It seems inevitable that at times a designer will necessarily be reduced to ''raw geometry," or that elaborations of features will take them outside classes whose manufacturing properties are explicitly present in a database [1]. The capability of recognizing instances of feature classes in an object description which does not have them explicitly present will be a continuing need. It is however obvious that a more reliable and efficient technique will be to employ the concepts of ***design with features*** [2,3]. This means the existence of a particular form feature has been directly expressed by the user of the system, or inferred as being part of a module so expressed. In contrast, the concept of ***free***

design merely requires the designer to describe with explicit geometry form of the features in a part without symbolic reference to the feature.

Although there are many available modeling schemes which can be used for geometric model representation on the CAD systems (e.g., sweep representation, primitive instances, spatial occupancy enumeration, cell decomposition) [4], we are concerned only with the boundary representation method because they possess the best attributes of all the representation schemes for geometric reasoning. The boundary representation is important because it is close to computer graphics, unambiguous, and available to computing algorithms. Disadvantages include: verbosity, complicated validity checking and difficulty in creating objects. CSG is useful in the creation process because it has features such as simple validity checking, unambiguity, conciseness, ease of object creation and high descriptive power. Its disadvantages include: inefficiency in producing line drawings, and inefficiency in graphic interaction operations. It is however possible to produce models using CSG and then transform the data to boundary data. These schemes essentially fit well in design with features, reasoning about geometry, transmutation of form into function, and the consequent product modeler; and result in a synergistic effect on the automatic process planning system.

It is well known that just as machine programming is currently machine dependent, product modeling is still CAD system dependent. In other words, there is no standard internal data representation across CAD system manufacturers. However, the advent of IGES, which is a neutral graphics data representation medium is the first step in this regard. For a CAD-based formalism to be truly beneficial to the manufacturing world, there must also be a standard CAD representation. IGES is one of the most popular neutral graphics system but is severely limited in the types of data which it can communicate. It can only manage with geometric information. Even this geometric information is quite restricted. New Standards which are intended to carry functional information as well as geometric information are under development at the time of this writing. The most prominent among them is PDES/STEP [5]. It is however possible for CAD data translated into a neutral form such as IGES to be restructured into a form understandable by a reasoning system [6]. This is useful in using such data for automatic process planning.

14.3 FEATURE, FEATURE REPRESENTATION AND CLASSIFICATION

Webster's dictionary defines a feature to be a prominent part or characteristic of an entity of interest. From an engineering point of view, a feature can be regarded as a portion of an object with a prominent characteristic. A universal definition of a feature has not yet been accepted. In this paper, a generalized feature definition is presented. We will then present a procedure for classification and representation of features for geometric reasoning in automated machine programming.

14.3.1 Feature Concept

A universally acceptable definition of a manufacturing feature is yet to be obtained. CAM-I described ''form feature" as a specific geometric configuration formed on the surfaces, edges, or corners of a workpiece intended to modify or to aid in achieving a given function [7].

This definition of ''form feature" has a very strong connection between ''form feature" and the material removal processes. Examples such as holes, grooves, slots, bosses, wall, rib, and pockets are the shapes or volumes associated with the surfaces of a part. However, considering the intrinsic nature of features in production industry, the definition of features should not focus on a particular process. Dixon defined feature as ''A feature is any geometric form or entity uniquely defined by its boundaries, or any uniquely defined geometric attribute of a part, that is meaningful to any life cycle issue" [8]. A more recent definition of a feature is, ''A feature is any named entity with attributes of both form and function" [9]. Nnaji and Kang defined feature as, '' A set of faces grouped together to perform a functional purpose" [10]. In PDES, form feature is defined in the conceptual model of feature information for the Integrated Product Information Model (IPIM) as a stereotypical portion of a shape or a portion of a shape that fits a pattern or stereotype [5]. The feature model is called ''Form Features Information Model" (FFIM). The FFIM is intended to be independent of product classes and user applications. However, the FFIM does not force user applications to implement specific form features nor application-independent nomenclature [5]. A feature can not just be standalone with specifications related only to''form." Other information such as shape-oriented(tolerances, surface finishes, dimensional items) and process(machining, assembly, sheet metal forming) information should also be included. There is a consensus among researchers in the field that features contain information with some bearing on both form and function. In this paper therefore, feature is defined as '' A set of geometric entities[1] (surfaces, edges, and vertices) together with specifications of the bounding relationships between them and which imply an engineering function on an object." [11] The characteristics of a generalized feature processing scheme are described as follows:

- a feature can be a combination of a group of relative geometric entities(face, edge, loop, vertex) or just a geometric entity itself,
- non-geometric and manufacturing information should be combined with the form of a feature itself,
- representation scheme and data structure of features should be pertinent to particular field of applications,
- a mapping function should be defined for an application-dependent system to map the form of features into a standardized format (PDES form feature representation, for example) and vice-versa (Figure 14-1) This will result in the possibility of exchange of product information from a CAD system to another or other advanced application systems.

In mechanical assembly or machining processes a feature is, usually, defined as a set of constituent faces [12,13,14,15,16]. The geometric information related to a feature is obviously a subset of the object although some geometric entities (such as the edges located on the feature boundary) should be ignored in accomplishing the recognition process. In addition to the geometric information, some non-geometric information associated with a feature is also essential for process planning.

---

[1]application-dependent

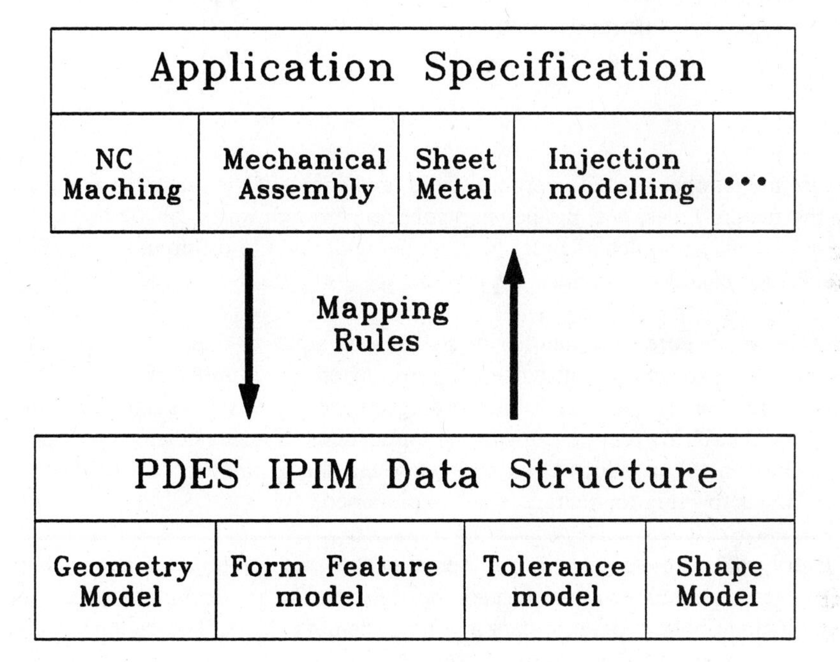

Figure 14-1: Mapping an application-dependent feature onto a standardized format.

14.3.2 The Euler Formula for Features

Features of a 3-D object can be considered as an open object. Open object is defined as an object which contains single faces and single edges. From a topological point of view, the graph of any feature is 2-manifold and can be embedded in a Möbius bend. The Möbius bend is a non-orientable surface and it is homomorphic to a disk. The Euler formula for features can be derived from the invariance of Euler characteristic of non-orientable surfaces. In this work, a face-edge incidence graph is used to represent features of a mechanical component. The Euler formula for the features with this representation is as follows:

$$v+f-e=s-h+r$$

where ***s, v, f, e*** are the number of shell, vertices, faces, and edges of the feature; ***r*** is the number of rings inside the feature faces; ***h*** is the genus(number of through holes for 3D objects) of the graph. Figure 14-2 shows examples of Euler formula being applied to different shapes. For each example the values for equation variables are given.

For a machining feature, the number of shells, *s*, is equal to one; while in sheet metal working *s* can be more than one (for instance, an embossment the number of shells is equal to two). The Euler formula for feature is useful for finding if the feature is a passage or non-passage and it is essential for classifying features for several applications. If a face-oriented representation (usually used in sheet metal working) is used to represent the geometric data of the part, the invariance of the Euler formula for feature is still maintained.

The primary goal in developing a representation scheme for object features should be completeness in order to pave the way for further applications. The product model representation of a part should include both geometric data and non-geometric data. The geometric data can be directly derived from a neutral graphics information such as the IGES through a geometric reasoning process. Some non-geometric information, such as rotational and reflective symmetries, of a part or a feature and pattern of features can also be obtained from topological and geometric reasoning based on graph and group theory. While others(e.g. tolerances) must be acquired by the designer through a product model interface.

Two types of representations are generally used to represent form features:

1. **Explicit Feature Representation** : features are represented using a group of faces associated with a coordinate frame and their relationships, this form of representation is shown in Figure 14-3;
2. **Implicit Feature Representation** : features are represented by a set of parameters and a homogeneous transformation matrix which specifies the location of a feature and associates a coordinate framwith it as shown in Figure 14-4.

The representation scheme for an explicit feature representation is lower in level when compared to an implicit feature representation scheme. The latter can always be obtained by a

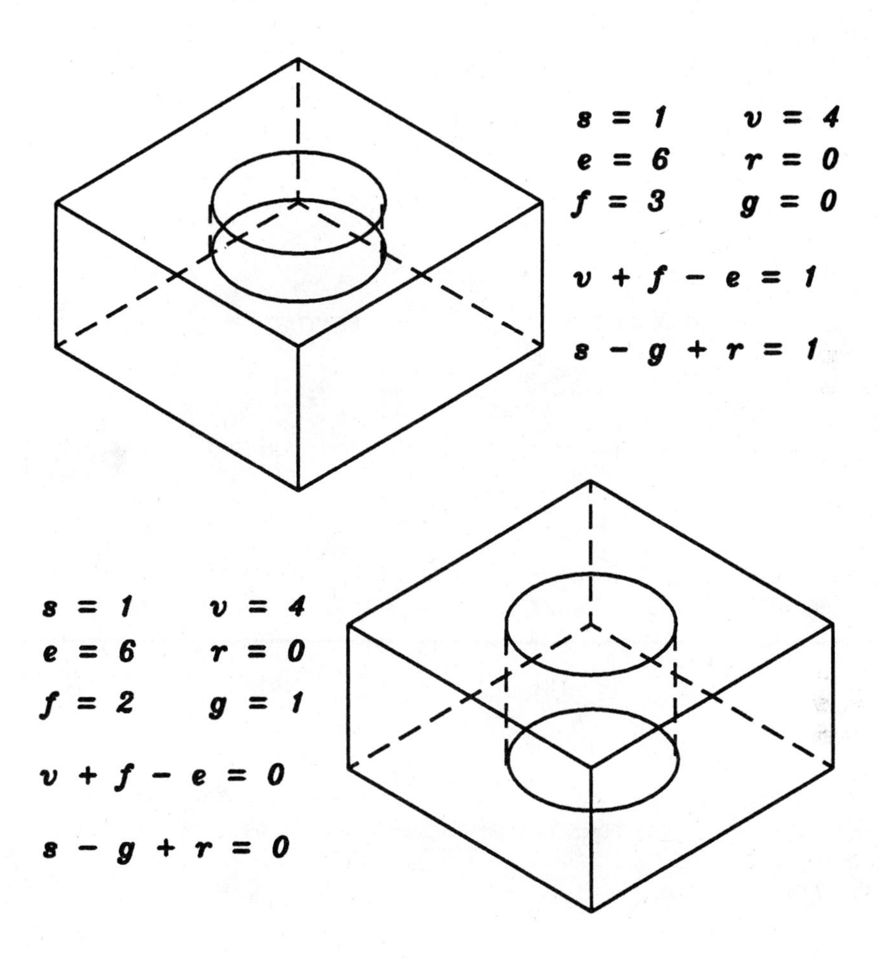

Figure 14-2: Examples of Euler formula being applied to different cases.

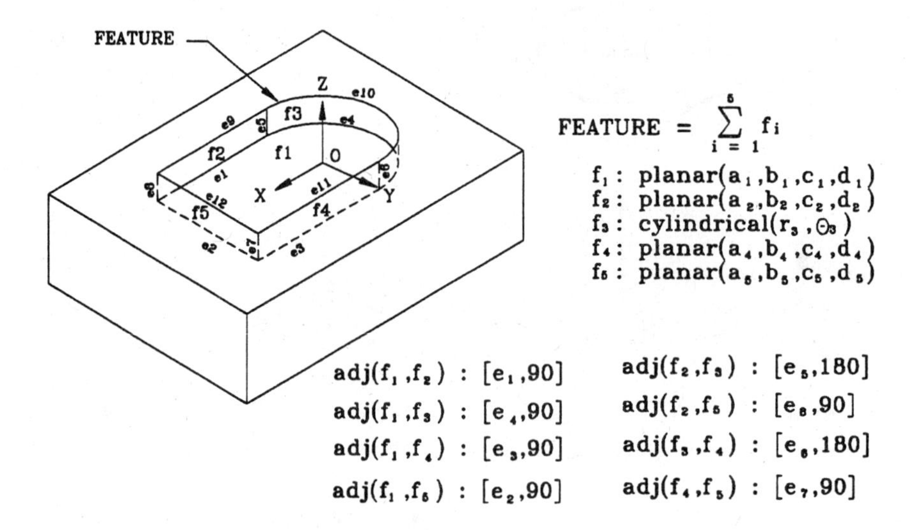

Figure 14-3: Explicit feature representation.

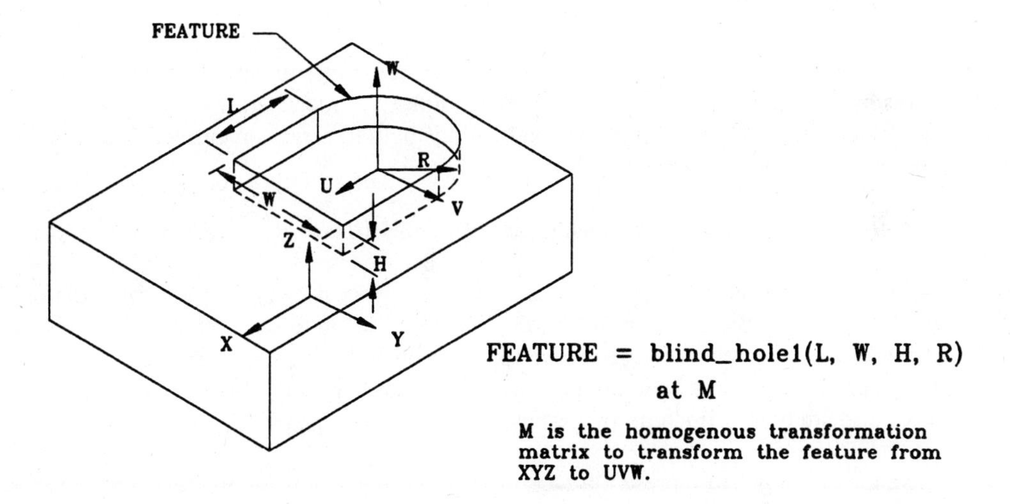

Figure 14-4: Implicit feature representation.

deduction process from the former. Implicit feature representation can be regarded as a ''macro view" of a feature and this representation may be useful in performing some operations in which the shape of a feature can be represented in a parametric way without ambiguity. This representation is used in feature-based design system for creating the part. However, the geometric information of features in this representation may not be complete if the boundary evaluation of the feature associated with the part is not followed. For example, a square hole in Figure 14-5 is difficult to represent using the implicit approach. In addition, without explicitly representing these features, the technological information, tolerances and surface finish, relative to faces cannot be specified. On the other hand, complex-shaped features (such as a thread on a cylindrical face) are hard to represent explicitly. In cases like these, implicit feature representation is suitable to represent the feature. In this paper, ''explicit feature representation" accompanied by parameters associated with implicit feature representation are used to represent features. This hybrid representation can reduce the complexity associated with the part creation process. For example, a threaded rod is hard and tedious both to create on current CAD systems and to interpret in a feature extraction system.

14.3.3 Representation of Feature Graph using Face-Edge Matrix

Using face-based feature graph to represent a feature is very convenient for a pictorial representation of the way different entities inside the feature relate to each other. However, for computer processing, a matrix representation for a graph is a useful and convenient way to represent a graph. From an algebraic point view, many known results of matrix algebra can be readily applied to study the structural properties of this kinds of graphs.

There are two ways to represent a face-based feature graph using matrix representation. These are face-face adjacency and the face-edge incidence matrices of a feature graph. Regardless of the existence of parallel links[2] and self-loops[3], both of adjacency and incidence matrices can be used to represent the feature without any ambiguity. However, the face-face adjacency graph cannot capture parallel links; while the face-edge incidence graph cannot capture the existence of a self-loop. The representations are described as follows: the adjacency matrix of a face-based feature graph, G, with n nodes and no parallel edges is an n by n symmetric binary matrix $A(G) = [\, x_{ij} \,]$ such that [11]

$$x_{ij} = x_{ji} = \begin{cases} 1 & \textit{there is an edge between } i^{th} \textit{ and } j^{th} \textit{ nodes,} \\ 0 & \textit{otherwise} \end{cases}$$

The incidence matrix of a face-based feature graph, G, with n nodes, m edges, and no self-loop is a m by n binary matrix, $I(G) = [\, y_{ij} \,]$, such that

[2] The number of intersecting edges between two faces is greater than one.

[3] A face is adjacent to itself.

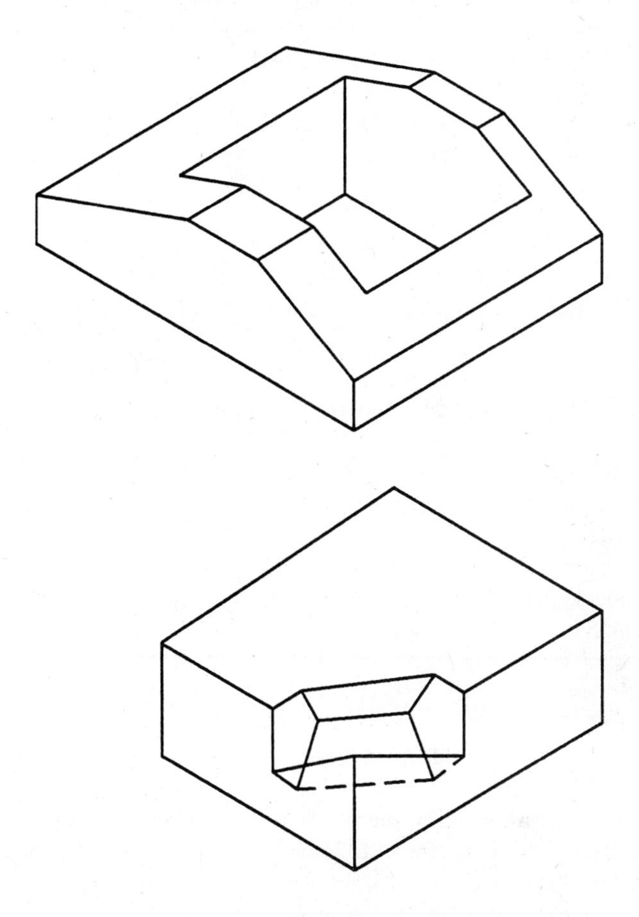

Figure 14-5: A square hole which is difcult to represent using implicit feature representation.

$$y_{ij} = \begin{cases} 1 & \textit{edge i is incident on node j}, \\ 0 & \textit{otherwise} \end{cases}$$

Since no self-loops are allowed in the feature graph and the parallel edges may exist in our representation mentioned in Section 14.2, face-edge incidence matrix of a feature graph is used to represent the configuration of a feature. It must be noted that elements in the adjacency and incidence graphs are not just carrying topological information. Geometric information such as the angle between two adjacent faces and the types of nodes and linkages is also included. Permutation of any rows or columns in an incidence matrix will result in a graph that is isomorphic to the original one. Two graphs with no self-loops are isomorphic if

$$I(G_1) = M \cdot I(G_2)$$

where $I(G_1)$ and $I(G_2)$ are the incidence matrices of $I(G_1)$ and $I(G_2)$. M is a row or column interchanging matrix.

The information structure of incidence matrix of the feature is as follows:

- feature edge-face incidence matrix
  1. incidence matrix mask : integer array,
  2. number of nodes : integer,
  3. number of linkages : integer,
  4. node sequence : face pointer chain,
  5. linkage sequence: edge pointer chain.

14.3.4 Feature Database

A feature database is used for the feature-based recognition and reasoning system. In this database, feature boundary loop ***FBL*** of a feature is ignored as shown in Figure 14-6. The characteristic of a feature is not affected by the boundary of the feature. The criteria to build a feature primitive in the feature database are as follows:

1. Faces and edges of a feature graph must be labeled.
2. The sequence of faces in the incidence matrix is determined using the starting face and edge and the loop direction of faces (or normal of the face).
3. The edge sequence in the incidence matrix is determined by the face sequence in the adjacency matrix and the loop direction of the faces.
4. The coordinate frame of the feature primitives should be determined.
5. The parameters of a feature primitive must be sufficient to represent the feature.

The methodology to determine the face and edge sequence for the incidence matrix of a

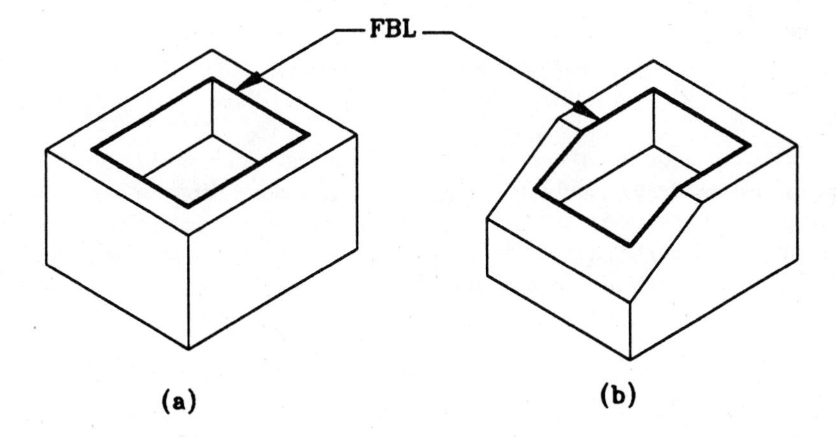

Figure 14-6: FBL of a feature should be ignored in the pattern matching process.

generic feature primitive is:

1. Choose a face, f_i, which has the highest degree number in face-based feature graph as the starting face.
2. Choose an adjacency edge, e_j, whose type appears the least number of times in the loop of f_i as the starting edge.
3. Start by traversing the faces in the feature graph using the backtracking search (depth first) according to the loop directions of the faces such that a face is visited only once.
4. Traverse the adjacent edges of the feature graph using the breadth-first search according the face sequence determined in (3) and the direction of the faces such that every edge is visited exactly twice except feature boundary edges which are only visited once.

For the feature recognition system, the incidence matrices of a feature graph is simply used for checking if a test feature and a feature primitive in feature database are topologically isomorphic. The result will be used to determine whether further geometric checking is necessary. The mapping relationships can be captured from topological checking, the geometric information needed to be carried by a feature primitive in order to perform the recognition process are as follows:

1. type of faces in a feature graph,
2. type of adjacent edges in the feature graph,
3. angle between any two adjacent faces (this information is contained in the data structure of edges).

For example, Figure 14-7 shows the information structure of two types of key_way primitives. The items shown in the incidence matrix indicate the hexadecimal values of each row of the matrices [11]. The face-edge incident relationships of the feature can be represented in a matrix form as follows:

face/edge	5	4	3	2	1	value
4	1	0	1	0	0	14
3	1	1	0	1	0	1a
2	0	1	0	0	1	09
1	0	0	1	1	1	07

Each row of the matrix is represented by an integer. For example, in the first row the value is <u>1</u> <u>0100</u> which is equal to 14 in hexadecimal notation. Using this representation will facilitate the topological checking process (especially in the recognition process) in a geometric

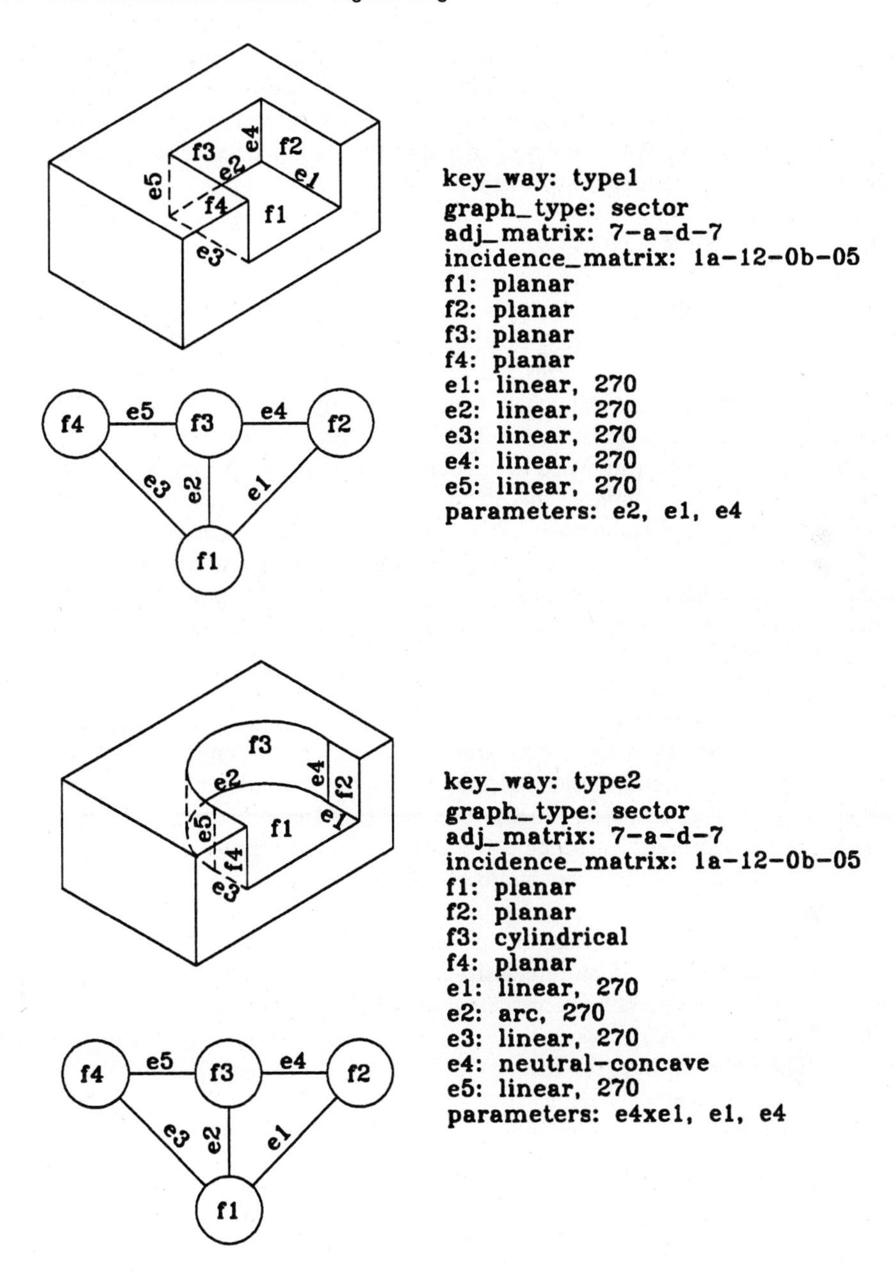

Figure 14-7: Examples of feature primitives in feature database.

reasoning system. In addition to the topological information, the incident edges also carry information which contains edge type and angle between adjacent faces(for example, e_1 in type 1 key_way shown in Figure 14-7 is a line segment and a common edge of f_1 and f_2, the solid angle between f_1 and f_2 is 270o; the type of feature faces are also included. The parameters needed to represent the feature are also carried.

14.4 FEATURE CLASSIFICATION

Since manufacturing features are application dependent, the classification of features needs to be dependent on the domain of application. Consequently, a ''protrusion" for machining process and a ''tab" in sheet metal fabrication would not be a feature. Machining and sheet metal fabrication are the applications associated with the material removal process. However, for mechanical assembly process a protrusion and tab can be essential features. The implication here is that features are significant only when the domain at hand and the task to be accomplished are clearly specified. Another example of this ''context-dependence" of features is an embossment. In sheet metal fabrication an embossment is a feature used to reinforce the material strength of the part. The same feature is completely devoid of any significance in assembly processes. The purpose of feature classification is to provide a storage and search procedure and medium for feature based design as well as reasoning systems. The classification scheme must capture topological and geometric variances and invariances among features in a hierarchical manner. The scheme to detect the variances and invariances for different areas of application must be developed and can be computerized. For product life-cycle considerations, the mapping of one or more features from an application to another is also an essential requirement. For example, two slot notches in sheet metal fabrication may imply the existence of a tab in the assembly process(Figure 14-8). This mapping process between two areas of application may warrant a change in the entire interpretation scheme for a manufacturing feature. The mapping process between two areas of application is beyond the scope of this paper.

In this paper, a feature is face-based and is represented by the face-edge incidence matrix. A generalized level for feature classification for mechanical assembly is developed as shown in Figure 14-9. The classification scheme is based on these prioritized criteria:

1. The Euler formula for features:
   From the value of genus h a feature is a passage or cannot be detected. The existence of genus implies there is a cutting through operation in the manufacturing process. A through hole is one of the most important properties of a manufacturing feature. If the number of genus is more than one then the feature cannot be produced only by one tooling operation. If the genus of a feature is h , then the number of feature boundary loop N_{FBL} of the feature is equal to $h + 1$. Each FBL implies a location that the tool can access to produce the feature (in machining and sheet metal working) or implies the insertion possibilities in an assembly process. Indeed, the number of operations, N_{OP}, needed to produce a feature is:

$$(N_{FBL} \ div \ 2) \leq N_{OP} \leq N_{FBL}$$

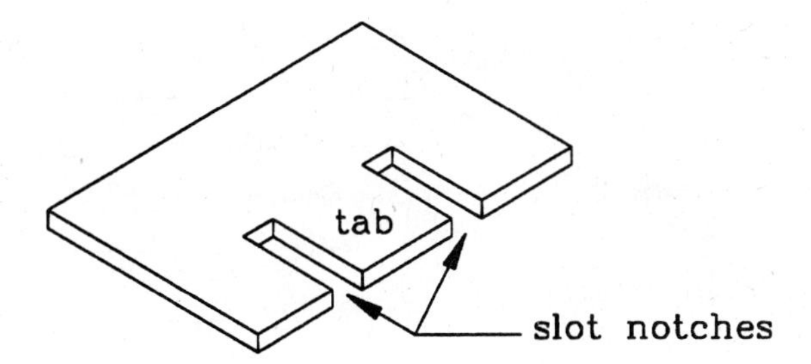

Figure 14-8: Mapping of features from an application to another.

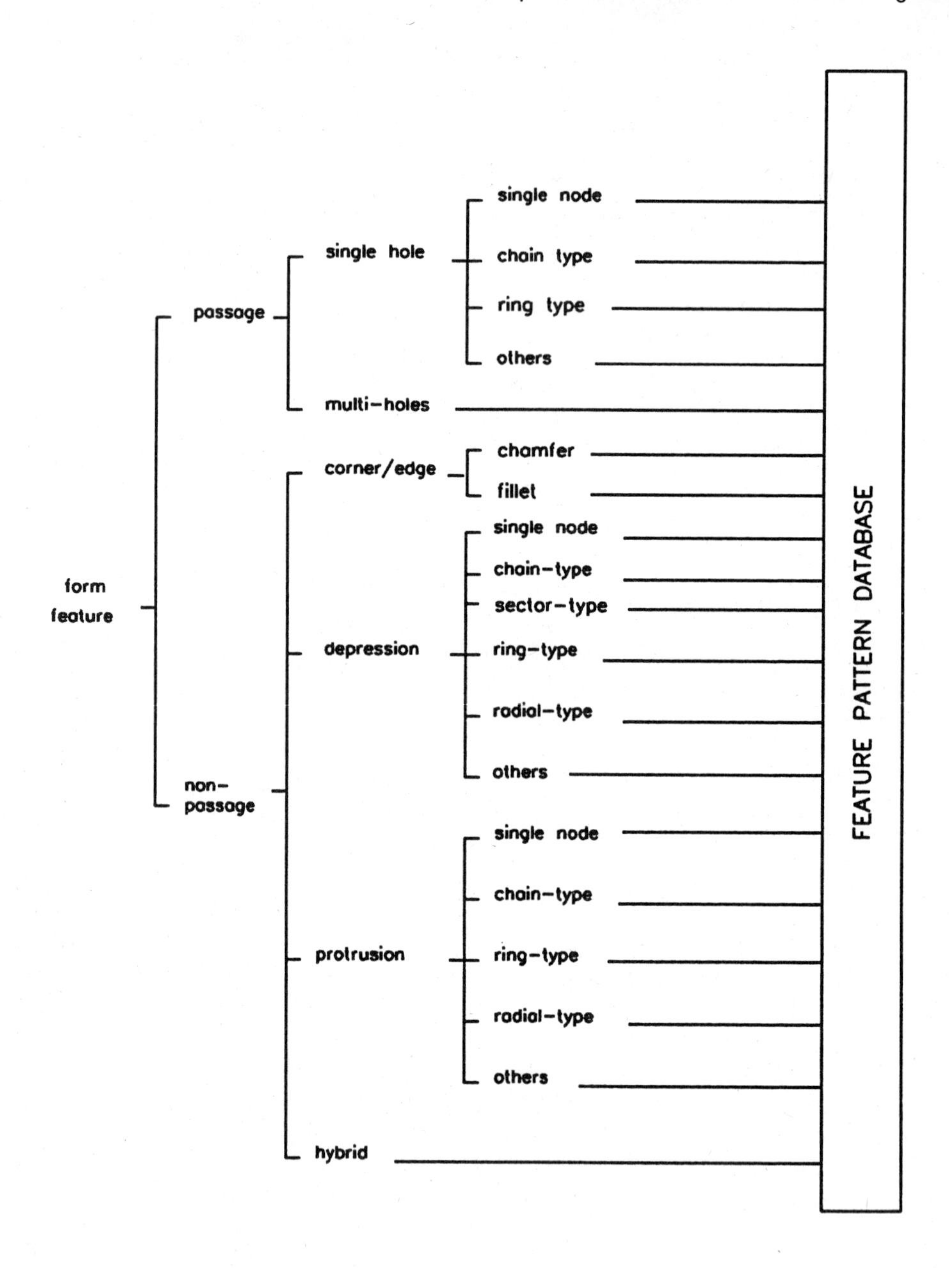

Figure 14-9: Generalized level feature classication for mechanical assembly.

where ***div*** is integer portion of the result of the division. Examples are shown in Figure 14-10.

2. Topological configuration of feature graph:
   Based on the configuration of the feature graph, the features whose graphs are homomorphic would be put into the same category. The type of configuration can be categorized into the following families (Figure 14-11):

   (a) single node
   (b) chain type
   (c) loop type
   (d) radial type
   (e) sector type

3. Type of edges in a feature boundary loop:
   The type of edge can be convex, concave, neutral convex, or neutral concave. For any edge located on the boundary of a feature, its edge type would be either convex or concave. Based the type of feature boundary edges, a feature is classified as being a protrusion, a depression or a hybrid type (Figure 14-12).

4. Solid angle of adjacent faces:
   This is used to find the group that preserves the angle between adjacent faces. Based on this, a square hole and a rectangular hole belong to the same family. This criteria is essential in the tool-matching process.

A face-oriented feature classification system for sheet metal fabrication, is shown in Figure 14-13. The criteria used are as follows:

1. Euler formula for the feature which is the same as that mentioned for mechanical assembly,
2. Cutting edges: in solid representation of an object, cutting edge is a convex edge that is connected to one or more thickness edges; in the face-oriented representation, a cutting edge is defined as an edge of a feature which is not the intersection of the feature faces,
3. Topological configuration of the feature,
4. Solid angle between the feature faces.

A careful observation of the properties of each application must be carried out for purposes of achieving unification and completeness(if possible) in a representation scheme.

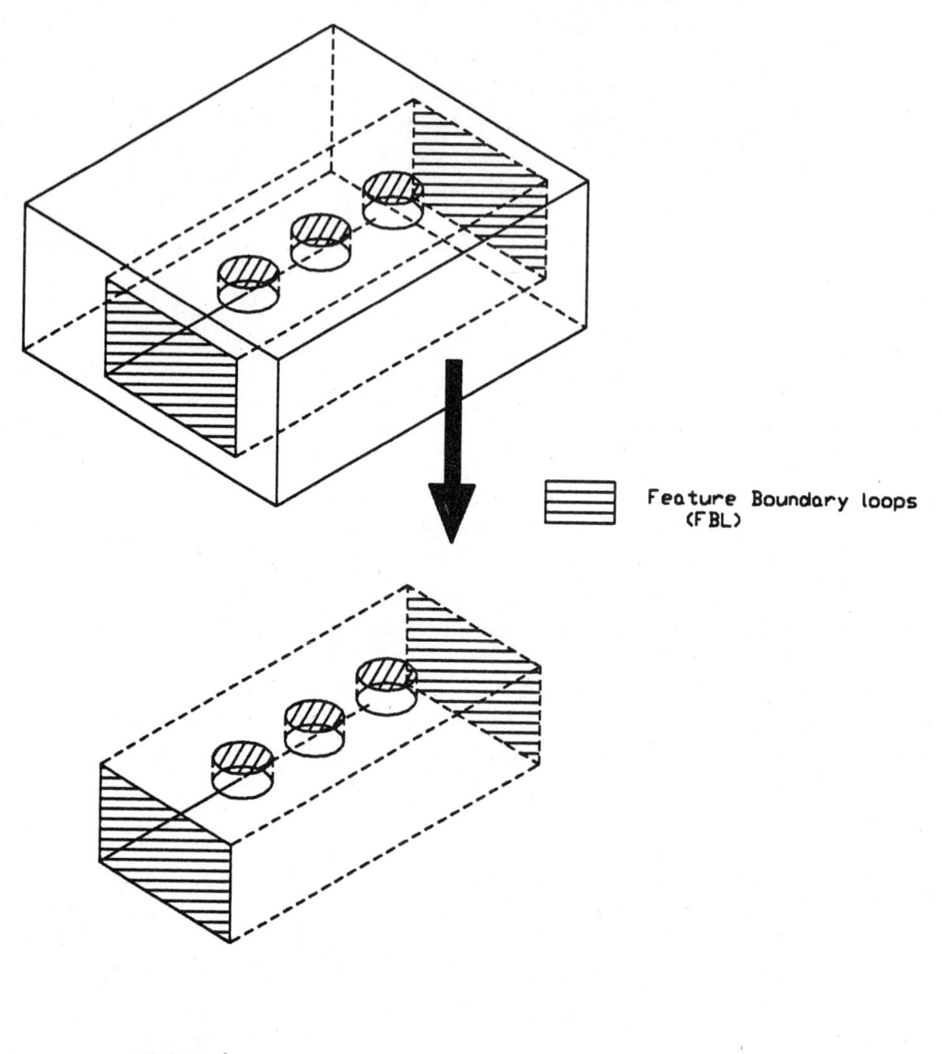

Figure 14-10: An example of genus for a manufacturing process.

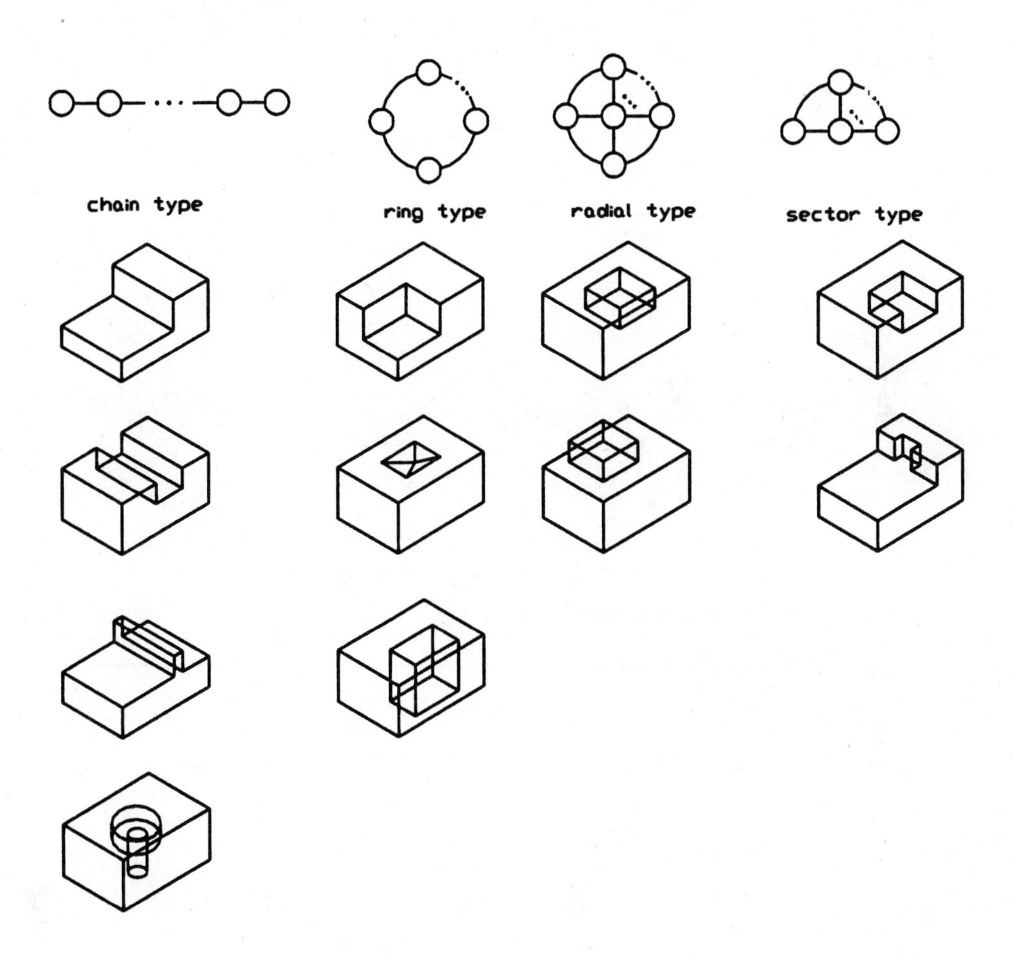

Figure 14-11: Types of feature topology.

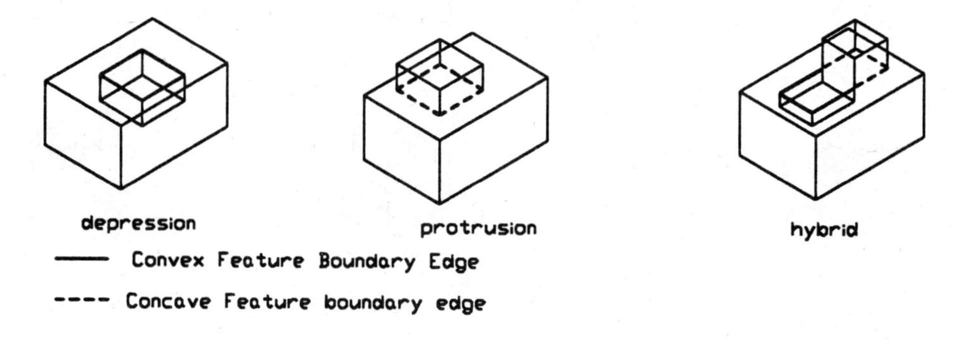

Figure 14-12: Feature boundary loops.

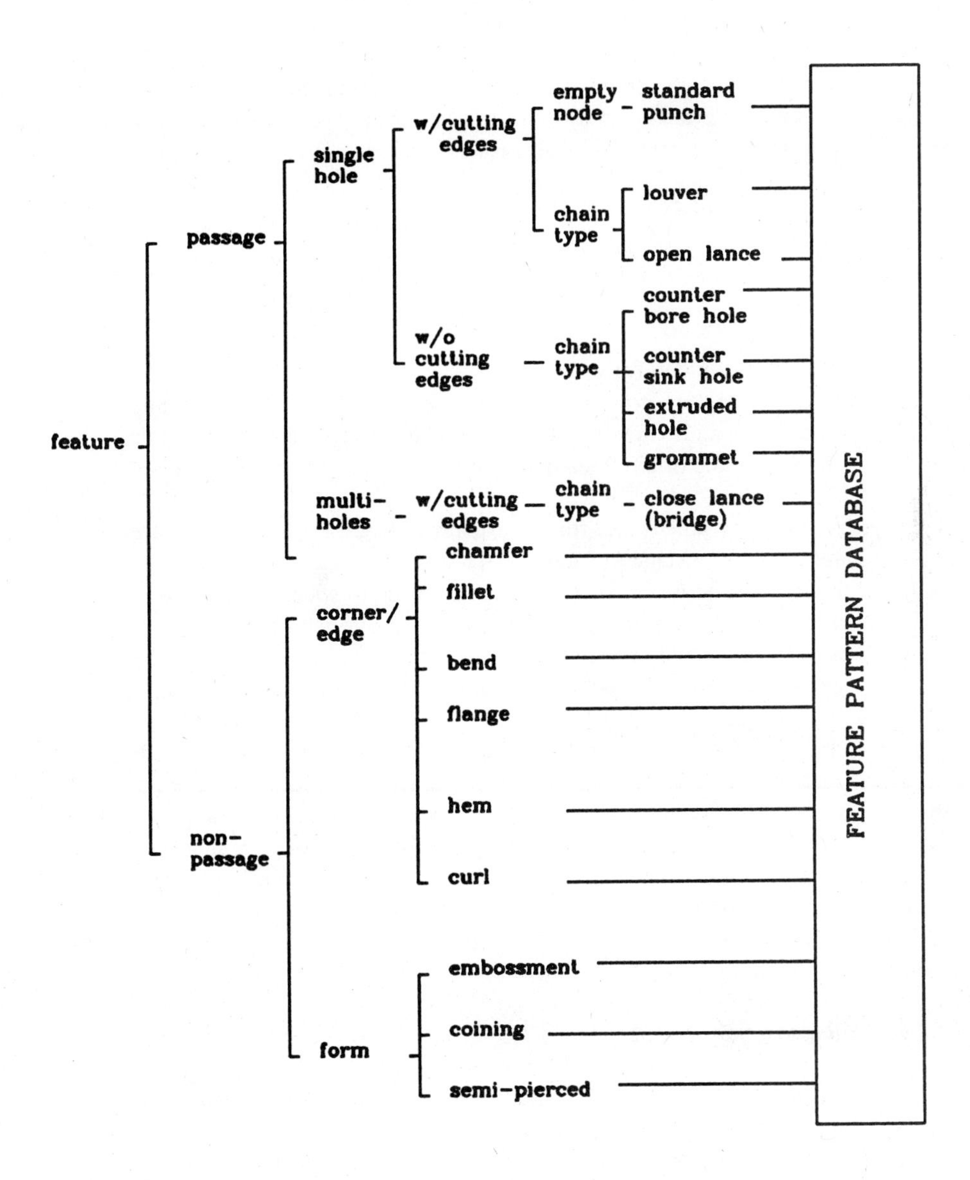

Figure 14-13: Feature classication for sheet metal application.

14.5 APPLICATIONS

In this section, we will show how the representation and classification of a feature discussed can be used in a feature reasoning system such as feature-based design system and feature recognition system. Figure 14-14 illustrates the logic for both systems.

14.5.1 Feature-based Design System

In order to manipulate features in a feature-based design system, the prerequisites are as follows:

1. A symbolic name for each feature.
2. A preset feature coordinate frame usually this coordinate frame coincides with the world coordinate frame. Based on this coordinate the original feature primitives in the feature database are defined.
3. A set of parameter set for each feature. Once the parameters are specified an instance of the feature will be created.
4. A homogeneous transformation matrix used to specify the location of the feature instance in 3-D space. This transformation matrix may be obtained by specifying the spatial relationships for the geometric entities among the feature instance and the existing part.

This symbolic feature representation can be obtained from the feature classification scheme mentioned above. The symbolic name of a feature is associated with a code in the classification system. This feature classification scheme provides a good management system for feature database when the number of feature primitives in the system is considerable. The proposed feature representation contains both explicit and implicit feature geometric information as well as non-geometric manufacturing information that would allow the designer to create the feature and pass the information to a process planner to determine the tooling process and produce the desired part.

14.5.2 Feature Recognition System

The most important principle to abide by in recognizing a feature is avoiding ambiguity. The criteria used for feature recognition process is based on the feature classification described early in this chapter. Then generic feature primitives or feature families which have different geometric characteristics under a specified category in feature classification must be developed. A feature may be decomposed during the recognition process. In general, the recognition process starts with topological checking. If necessary, this is followed by geometric checking. The schema for feature recognition process is shown in Figure 14-15.

The pattern matching process begins to match the extracted form feature with the generic feature primitives stored in the feature pattern database. If no feature primitives can match the extracted feature, the process would continue the process of decomposition of a feature into several partitions according to the rules set in the decomposition process and return to the pattern

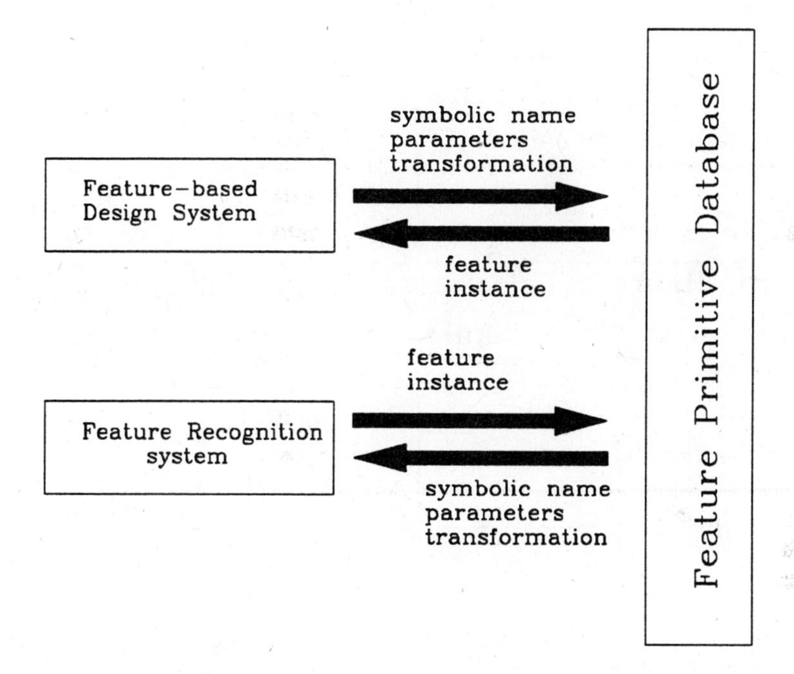

Figure 14-14: Feature-based design and recognition systems.

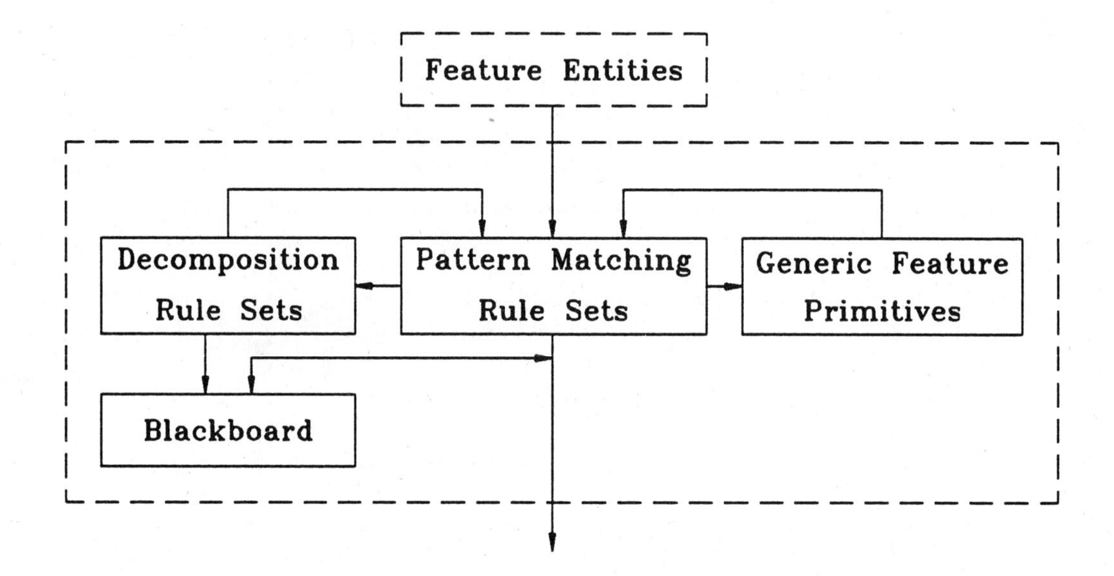

Figure 14-15: The schema of feature recognition process.

matching process to match these decomposed features individually. The pattern matching and decomposition processes would be executed repeatedly, until either all the decomposed features are recognized or any of the decomposed features cannot be decomposed any further. If a feature is not recognized, it would be put on a blackboard. The features on the blackboard will be eventually inserted into Generic Feature Primitive database manually [11].

14.5.3 Criteria Used for Feature Recognition

The feature recognition problem involves determining if there exists a one-to-one correspondence between the nodes of a face-based feature graph and a generic feature primitive stored in feature database. Based on the feature classification scheme discussed earlier, a feature can be classified into a category, according to the geometric and topological characteristics of the feature. The identical checking between the pattern and feature is then followed. A feature carries the information of both form and function, the form of a feature can be affected by the boundary condition while the function of the feature may not. Therefore, some information about feature entities should be ignored during the matching process. The feature boundary loop (***FBL***) of a feature is used to represent the relationship between the feature and the other part of an object. Using ***FBL*** as a criteria to do the feature recognition would be considered weak from a feature recognition standpoint. The length of a curve of a feature should also necessarily be ignored, since the size of the feature should be considered when the patterns are represented by primitives. The information of the feature entities employed in the recognition process would be addressed at the nodes and the entities that specified the relations between the nodes of the feature graph.

14.5.4 Pattern Matching

The recognition strategy used is "learning before recognition." The decision rules used to classify features into several categories are implemented to find the location of appropriate patterns quickly. The matching rules can be applied to locate the right primitive in feature pattern database for a specified testing form of a feature. Once a primitive is matched with the testing feature, the associated parameters and attributes of the primitive will be attached to this testing feature.

The matching rules are built according to the criteria described in Section 14.5.3. In order to detect that two face-based graphs are matched to each other, isomorphic checking must be performed. The isomorphism M maps a testing feature graph, T, onto a pre-set feature pattern graph, P, as shown in Figure 14-16.

Obviously, M is a bijective mapping and it should be coordinates independent. Not only must the adjacent topology of the feature graphs(***FG***s) of the testing feature and generic primitive be the same, but the information carried on each linkage of them should be identical. The entities of interest in M are the faces(nodes) and the adjacent edges(linkages). Since the information carried in faces is different, the nodes are not indistinguishable. Faces may be distinguished by their types. The angular dimension between any two faces should also be checked.

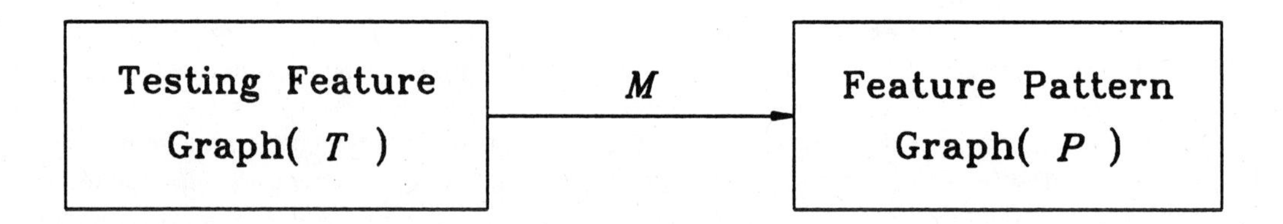

Figure 14-16: Mapping a testing feature graph onto a feature pattern graph.

The rules used in form feature classification are used as the decision rules to locate the feature primitive for a particular test feature. The procedure for the pattern matching process is described as follows:

1. Find the face sequence and edge sequence according to the procedure described in Section 14.3.4.
2. Check topological equivalence by comparing the corresponding values in the adjacency and incidence matrices of test feature and the generic feature primitive.
3. If the the test feature is isomorphic to the generic feature primitive, check the geometric contents of each node and linkage of the testing feature graph to its corresponding nodes and linkage in the generic feature graph; otherwise change the starting condition and go to (1).
4. If the geometric checking is passed then the feature is recognized; otherwise, change the starting conditions and go to (1).
5. If there is no match between the testing feature and the current feature primitive, pick the following feature primitive in the feature pattern database, repeat the whole process.

The geometric information is used to generate the face and edge sequences of the adjacency and incidence matrices for the topological checking. For a n nodes, m linkages feature graph, the complexity to check topologically identical is reduced from $n!$ to $2m$.

14.5.5 Feature Decomposition

After a feature, represented by a feature face set, is extracted using the procedure described above, it may be necessary to decompose this feature into two or more features if the feature is not recognized. Then all of the decomposed features should be tested individually.

The decomposition process will result in several independent co-level sub-features. The strategy used to decompose a feature is:

1. If the a feature graph is a chain-type graph in which each linkage of the graph has one endpoint in common. A chain-type graph is decomposable if two or more faces in the graph are co-surface. The sub-graphs can be found by extracting all the faces between every two co-surfaces. For any non-chain-type feature graph, if it is decomposable, there must exist one or more articulation nodes(faces). A node is said to be an articulation node, if the degree of this node is greater than one and the removal of a node from a connected graph will result in the generation of two or more disjoint sub-graphs.

2. Find the first level feature graph by extracting the faces from the envelope. Using the extracted feature faces as feature neighbouring faces (***FNF***) to find all the internal feature bounding loops, if any feature boundary loop exist then the sub-feature of the current level feature graph can be extracted by traversing the

current feature boundary loops.

3. Find the co-surface faces in feature graph. If they exist, extract all the faces in between as its sub-feature.

4. Find the articulation nodes of the feature graph(***FG***). The ***FG***, hence, can be separated into several disjoint sub-graphs by splitting the articulation nodes. These sub-graphs are 1-isomorphic to the original graph.

5. For each articulation node, ***A***, update the ***FNF*** by removing each face $F_i \in FNF$ such that ***A*** and F_i are disjoint. Then adding ***A*** to the updated ***FNF***.

$$FNF' = FNF \cup \{A\} - \{B\}$$

where $B \in FNF$, ***A*** and ***B*** are disjoint.

6. Consider every disjoint sub-graph as an individual graph.

   (a) If any ***IFBL*** is formed, then a sub-feature is generated. The face set of the sub-feature is a face set of this sub-graph subtracted from the articulation node.
   (b) If no sub-***FBL***s exist, then the feature is a major feature in this level. The feature face set is the product of the union of the sub-graph and the articulation face.

7. Find all the faces which have inner loops and extract all the connected faces as a sub-feature of the feature level in which the face belongs.
8. For each feature, find the feature type for the feature itself and its sub-features. Meanwhile, construct the feature graph for this feature.

An example of this is shown in Figure 14-17(a) and illustrates how this strategy can be applied. The first level feature faces can be found by traversing the first level feature boundary loop (***FBL***) which yield

$$FG_1 = \{ f_5, f_6, f_7, f_8, f_9, f_{10}, f_{11}, f_{12}, f_{13} \}$$

with

$$FNF_1 = \{ f_1, f_2, f_3, f_4 \}$$

Let the node FG_1 be the ***FNF***, a sub-***FBL*** then be found

$$FG_{1,1} = \{ f_{14}, f_{15}, f_{16} \}$$

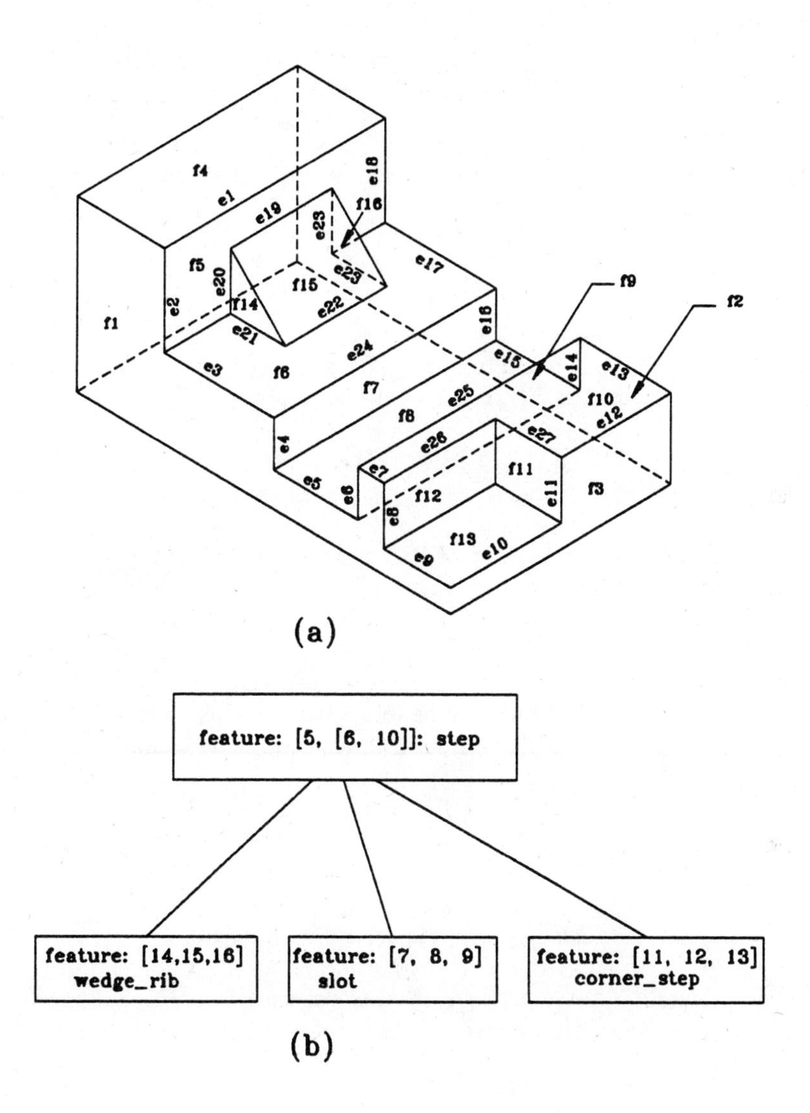

Figure 14-17: An example of feature decomposition.

From FG_1 , f_6 and f_{10} is co-surface, then the faces in between should be extracted as a sub-feature.

$$FG_{1,2} = \{ f_7, f_8, f_9 \}$$

After extracting this sub-feature, FG_1 should be updated to form FG_2 by removing $FG_{1,2}$ from FG_1:

$$FG_2 = \{ f_5, \{ f_6 , f_{10} \}, f_{11}, f_{12}, f_{13} \}$$

Now, an articulation node, f_{10}, is found from FG_1. By splitting f_{10} from FG_1 to yield two disjoint sub-graphs. These are defined as FG_3 and FG_4, respectively.

$$FG_3 = \{ f_5, \{ f_6, f_{10} \} \}, \qquad FG_4 = \{ f_{10}, f_{11}, f_{12}, f_{13} \}$$

Based on FNF the feature boundary loop (FBL) of FG_3 can be found:

$$FBL_{FG3} = \{ e_1, e_2, \{e_3, e_7\}, e_{26}, e_{27}, e_{12}, \{e_{13}, e_{17}\}, e_{18} \}$$

The feature boundary edges for FG_4 cannot be found based on the current FNF. Therefore, the articulation node, f_{10} needs to be remove from FG_4 and acts as an feature neighboring face.

$$FNF' = FNF \cup \{ f_{10} \} = \{ f_1, f_2, f_3, f_4, f_{10} \}$$

Based on FNF', for FG_4 - $\{ f_{10} \}$ there is a sub-FBL formed

$$FBL_{FG4} = \{ e_8, e_9, e_{11}, e_{26}, e_{27} \}$$

Therefore, a sub-feature $FG_{1,3} = \{ f_{11}, f_{12}, f_{13} \}$ is extracted from FG_4 . For FG_3, there is no sub-FBL formed under FNF', so the main feature can be found as:

$$FG_{1,0} = \{ f_5, f_6, f_{10} \}$$

The next step is to find the major features as well as the sub-features and build a hierarchical feature graph for this combination feature. By matching the generic feature in the database, the results are shown as follows:

$FG_{1,0}$: step,
$FG_{1,1}$: wedge-rib,
$FG_{1,2}$: groove,
$FG_{1,3}$: corner-step.

The feature graph of this feature can be built as shown in Figure 14-17(b).

14.6 CONCLUSION

Feature classification is of great importance in feature recognition as well as design with features and other geometric feature reasoning processes. Due to the fact that much of feature classification is domain specific, a generalized feature classification eludes geometric reasoning. In this paper, a classification scheme for mechanical assembly and sheet metal fabrication is presented. Several mathematical tools such as Euler's formula and graph theory were used to achieve a coherent and quick way of classifying features. Throughout the development, the role of unique and valid geometric data is emphasized. Classification schemes developed here are used for feature recognition as well as feature-based design systems for both mechanical assembly and sheet metal fabrication. We do not claim that the techniques discussed in this paper are all-encompassing. However, for domains of application that we have focused on, the results have been quite successful.

In this paper, we have shown the role of the computer-aided design system in providing necessary information for an automatic machine programming system. It is obvious that the for manufacturing operations such as assembly that the robotic system must possess the product data, and that this data should necessarily provide feature information. The reason for concentrating on features should be obvious. Assembly of components implies mating of features thus reducing the assembly reasoning on the level of mating yo feature reasoning. In addition, understanding the configuration of an object permits the possibility of reasoning about other aspects of the geometry for instance grasping certain faces at points that will provide for stable grasping and manipulation.

REFERENCES

1. Nnaji, B.O., and R.J. Popplestone. *A generalized shape and feature descriptor from models on a CAD/CAM system for automatic assembly: An approach to: form, function, design & manufacture.* In NSF Engineering Conference, Arizona, 1990.

2. Libardi, E.C., and J.R. Dixon. *Designing with features: Design and analysis of extrusions as an example.* In Spring National Design Engineering Conference and Show, pages 24-27, Chicago, March 1986.

3. Popplestone, R.J. *The Edingburgh designer system as a framework for robotics.* In IEEE International Conference on Robotics and Automation, pages 1972-1977, 1987.

4. Nnaji, B.O. *A framework for CAD-based geometric reasoning for robot assembly planning.* The International Journal of Production Research, Special Issue on Artificial Intelligence Applications in Manufacturing, Vol. 26, No. 5, pages 735-764, 1988.

5. National Institute of Standards and Technology. *Product Data Exchange Specification : The First Working Draft,* nistir 88-4004 edition, February 1988.

6. Nnaji, B.O., and J. Chu. *Ralph Static Planner : CAD based robotic assembly task planning for CSG-based objects.* International Journal of Intelligence Systems, 1989.

7. Pratt, M.J., and P.H. Wilson. *Requirements for support of form features in a solid modelling system.* Report R-85-ASPP-01, CAM-I, Arlington, Texas, June 1985.

8. Dixon, J.R., *et. al. Examples of symbolic representations of design geometric engineering with computers.* Expert Systems for Mechanical Design, Vol. 2, pages 1-10, 1987.

9. Dixon, J.R. *Research in designing with features.* In Design Theory 88, RPI, New York, June 1988.

10. Nnaji, B.O., and T.S. Kang. *Interpretation of CAD models through neutral geometric knowledge.* Artificial Intelligence for Engineering Design, Analysis and Manufacturing, pages 15-45, May 1990.

11. Kang, T.S. *Interpretation of CAD Models for an Automatic Machine Programming Planner.* PhD Thesis, University of Massachusetts at Amherst, 1991.

12. Shah, J.J. *Assessment of features technology.* Computer-Aided Design, Vol. 23, No. 5, pages 58-66, June 1991.

13. Joshi, S., and T.C. Chang. *Graph-based heuristics for recognition of machined features from a 3D solid model.* Computer-Aided Design, Vol. 20, No. 2, pages 58-66, March 1988.

14. Sakurai, H., and D.C. Gossard. *Recognizing shape features in solid models.* IEEE Computer Graphics & Applications, Vol. 10, No. 9, pages 22-32, September 1990.

15. Jakubowski, R. *Extraction of shape features for syntactic recognition of shape parts.* IEEE Transactions on Systems, Man & Cybernetics, Vol. 5, pages 642-651, September/October 1985.

16. Henderson, M.R. *Extraction of feature information from three-dimensional CAD data.* PhD Thesis, Purdue University, USA, 1984.

15

Fault Diagnosis of Large Manufacturing Processes

Wade O. Troxell, Colorado State University

A. Vincent Huffaker, Boston University

William F. Maulsby, Eastman Kodak Company

As manufactured products become more complex, the processes to create them and the fault diagnosis of the resulting processes become more complex as well. A framework to perform computer-assisted fault diagnosis of manufacturing processes is described. A proof-of-concept system is developed and demonstrated on a polypropylene-film manufacturing plant. The proof-of-concept system consists of three subsystems: (1) the process model, (2) an assumption-based truth maintenance system (ATMS), and (3) the inference engine. The process model is constructed using a recursive network modeling structure which allows the manufacturing process to be defined in an intuitive, hierarchical manner without committing to a global representation scheme. The process model assumes that the manufacturing process behavior is captured in terms of process principles. The actual manufacturing process is compared against the process model. Any contradictions between the process model and the actual process indicate the presence of one or more faults. The ATMS records these contradictions and computes a set of all possible resolutions of the contradictions that corresponds to the set of all possible fault diagnoses. The inference engine handles the interactions between the ATMS and the process model. The resulting program is able to successfully determine possible faults efficiently.

15.1 Introduction

Manufacturing processes are becoming more complex. Every new product has to be better than the previous products; every new product has to have the latest features and be of uncompromising quality. Some processes must be restricted and cannot have direct human intervention (e.g., chemical, nuclear, etc...). Some processes are continuous. They cannot be stopped; they just create bad output until the faults are fixed. Many processes are monitored by so many sensors that a human cannot assimilate all of the information.

At the same time, the world market is becoming more competitive. The ability to make quick, accurate diagnosis of process faults is becoming more important. Down time is becoming more expensive and critical. A company cannot accept less throughput because the products and the associated process are becoming more complex.

Improvement of manufacturing processes is a multi-faceted problem. Fault diagnosis is one area that can be improved. This chapter develops a framework for doing computer-assisted fault diagnosis of manufacturing processes.

15.2 Methods of Fault Diagnosis

Once one decides to perform computer-assisted fault diagnosis there are two approaches that can be followed:

-- model the diagnosis (rule-based), or

-- model the process (model-based).

If one decides to model the diagnosis, one will, in essence, create a ``trouble-shooting manual" or a rule-based fault decision tree. As an example, industry uses ``corrective action guideline" which list procedures to follow for given fault-driven situations. All the rules for computation-based fault diagnosis are defined as ``if {set of conditions} then {malfunction}". If a fault occurs, the procedure is to go through the rules, comparing the antecedent conditions with those in the rule-base until it finds a match and then execute the corresponding consequent conditions [1,2,3].

While this method seems easy and very intuitive, it does have numerous disadvantages:

1. *Hard to create.* Creating a rule-base with anymore than a trivial number of rules quickly becomes overwhelming. A rule-base is hard to construct because the rules must be constructed on a case-by-case basis. All the possible faulty conditions must be enumerated along with the corresponding corrective procedures to follow. The knowledge in the rule-base is limited to the list of possible faulty conditions initially declared and the knowledge of the expert who produced the corrective procedures [4].

2. *Hard to modify.* A rule-base is created for a specific process and there is very little carry-over of rules between processes. Any changes to the process (e.g., equipment upgrades) can arbitrarily affect any number of the rules. If a change occurs, one must decide the effects of the change on the overall behavior and diagnosis of the process. Any other process that is different from the original process would require the creation of a rule-base of its own with all the hardships that involves [4].
3. *Unwilling experts.* The same people who have the expert knowledge that this method requires are usually the same people who would feel threatened by giving their knowledge away.
4. *Increasing complexity beyond single faults.* Rule-based methods can only realistically diagnose single faults. In theory, of course, it is possible to diagnose multiple faults (e.g., ``if {set of conditions} then {several malfunctions}"), but for any given number of possible faults, n, there will need to be a diagnosis for every combination of those faults (2^n total possible combinations). Even if one limits the number of possible concurrent faults, the number of possible sets of faults still rises exponentially.

Model-based fault diagnosis is an alternative to rule-based modeling. One models the relationships between the components of the process and the characteristics of the output. Then, when the output becomes faulty, the program reasons back to the source of the problem [5,4,6].

Diagnostic reasoning based on a process model has several advantages.

1. *Easier to create.* A process model is easier to create than a rule-base because the structure and behavior of the process serve to guide the creation of the required knowledge [4].
2. *Easier to modify.* A process model is easier to modify than a rule-base because if the process changes, we can update the individual specifications for the process component that changes rather than having to decide the effects on the overall behavior of the process [4].
3. *Many sources of knowledge.* Although knowledge from experts is still desirable for some parts of the creation of the process model, knowledge can come from many other areas. For example, one can imagine a company shipping a part along with a model of the part. In addition, creating a process model does not require as much knowledge of the history of a process as does creating a rule-base.
4. *Multiple faults.* Model-based fault diagnosis does not automatically give us the ability to reason about multiple faults. Rather, by separating the representation of the process from the reasoning of that knowledge, it allows us to make the reasoning part of the framework as logically complete and as computationally efficient as needed.
5. *Novel faults.* ``It offers the possibility of dealing with novel faults....it takes the view that any discrepancy between observed and expected behavior is a bug, and it uses knowledge about the device structure and behavior to determine the possible sources of the bug. As a result, it is able to reason about bugs that are novel in the sense that they are not part of the `training set' and are manifested by symptoms not seen previously" [4].

15.3 Statement of the Problem

A manufacturing process contains a wealth of inherent information, including the structure and behavior of the process components. From a substantial process model created from the inherent structure available in a manufacturing process, it is possible to derive a knowledge base of beliefs and justifications that characterize the operation of the manufacturing process. When the observed behavior of the manufacturing process conflicts with the desired behavior, a fault has occurred and there will be a contradiction in the knowledge base. A truth maintenance system, in performing its job of maintaining the consistency of the knowledge base, will resolve any contradictions. These resolutions of the contradictions will be the sets of possible faulty process principles responsible for a faulty quality variable.

A framework for performing model-based fault diagnosis of manufacturing processes is described. The framework consists of three components: the process model, the inference engine, and the truth maintenance system.

Because the manufacturing process is a result of the *design process*, knowledge about its inherent structure can be captured and made into a *process model*. The structure can be divided into three components: *connectivity* (the layout of the process), *desired behavior* (the expected behavior of the process), and *observed behavior* (the actual behavior of the process). The observed behavior is measured by the *system inputs* (e.g., sensors and the meanings behind the sensor readings). The process model is created using *recursive network modeling* [7,8] which represents a process in a hierarchical manner. *Process components*, representing individual parts or subprocesses of the manufacturing process, are defined in increasingly more detailed layers. The bottom layer, or foundation, of the hierarchy consists of *process principles* that represent the fundamental, expressible behaviors of the manufacturing process. Because it is impossible to enumerate every possible causal relationship of a manufacturing process, the process model is described in terms of the effects of the process on *quality variables* (or important attributes of the final product).

If one or more faults occur in the output of the manufacturing process (as indicated by the system inputs), the inference engine computes which process principles, or sets of process principles, accounts for the fault(s). The inference engine can only use the process model as its basis for reasoning. The truth maintenance system maintains the *consistency of the ``beliefs"* or truths of the inference engine and allows the inference engine to reason with *multiple contexts* (or sets of justifying assumptions).

The proof-of-concept system is set up as follows:

1. The process model describes the connectivity, desired behavior, and observed behavior of the manufacturing process.
2. From the process model, the inference engine extracts information on the specific hierarchy of process components that affect the quality variable of interest and the system inputs which are affected by those process components. Then the inference translates that information into nodes and justifications in the assumption-based truth maintenance system (ATMS). The process principles are translated into assumptions. Each higher process component and each system input is translated into a node that is justified by one or more sets of assumptions (process principles).
3. The ATMS maintains the database of nodes and justifications that are passed to it from

the inference engine. If there is a contradiction, the ATMS quickly and efficiently computes the set(s) of underlying assumptions (process principles) that contribute to the contradiction. Each set of underlying assumptions is a possible fault diagnosis.

This chapter discusses each part of the fault diagnosis framework separately and how they work together. Section 15.4 discusses the general architecture of the framework. Sections 15.5 and 15.6 discuss how the process model is represented; Section 15.5 discusses how large-scale manufacturing processes can be described in general and Section 15.6 discusses how this framework specifically describes manufacturing processes. Section 15.7 discusses the reasoning for fault diagnosis and how specific possible faults are determined. Section 15.8 discusses a specific example of applying the framework. Section 15.9 concludes by summarizing the main points and discusses future directions of this research.

15.4 Architecture

Any model-based framework for fault diagnosis of manufacturing processes has two distinct components: the representation component and the reasoning component. The representation component simply acts as a knowledge base; it consists of the descriptions of the connectivity, the desired behavior, and the actual behavior of the manufacturing process. The reasoning component contains all the information needed to reason about possible faults from the knowledge available in the representation component.

Design versus Decomposition

There is a cyclical nature to the design and decomposition of a manufacturing process. See Figure 15-1. From the desired behaviors, *engineering design* creates a manufacturing process; from the manufacturing process, *top-down decomposition* of manufacturing process technology creates the desired behaviors.

In order to create the representation component, an existing manufacturing process must be decomposed into the appropriate underlying behaviors and the corresponding connections between them. This decomposition is possible because the manufacturing process is an artifact of the design process. The manufacturing process was created by listing the behaviors of the manufacturing process and then designing for them. As such, there is knowledge of the desired behaviors inherent in the structure of the manufacturing process.

Representation Component

The representation component of our framework is called the *process model*. The inherent structure in the manufacturing process is expressed in three sections (see Figure 15-2):

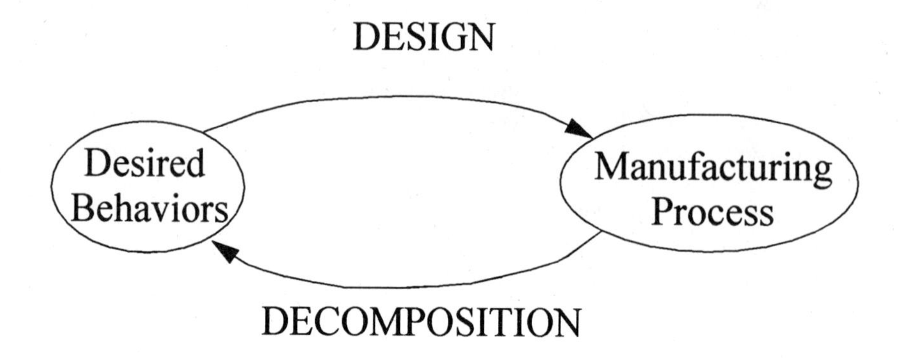

Figure 15-1: Design versus Decomposition

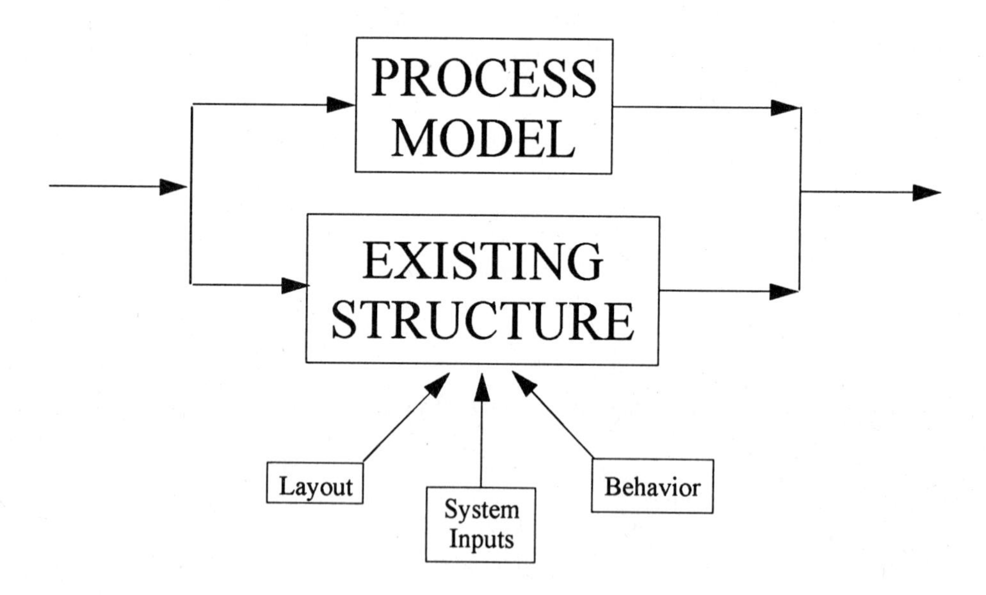

Figure 15-2: Relationship of existing structure of a manufacturing process and the process model.

1. *Connectivity* -- The connectivity of the process includes not only the overall connectivity, but also the connectivity at all relevant levels of detail of the process. This describes how the individual behaviors connect to create the overall behavior.
2. *Desired Behavior* -- This is the functional decomposition of the manufacturing process into process components and their expected behaviors. This component describes how the different process components affect the quality variables.
3. *Observed Behavior* -- This is the description of how well the manufacturing process is performing its desired behaviors. These observations come solely from the system inputs of the process.

Reasoning Component

The reasoning component of our framework is called the *problem-solver*. The problem-solver consists of two subsystems: the inference engine and the truth maintenance system (TMS) [21].

The inference engine is the part of the problem-solver that performs the reasoning of the fault diagnosis framework. After extracting the needed information from the process model, the inference engine applies a predetermined logic to that information to create new truths (beliefs). Because a manufacturing environment is very dynamic, information on the process will probably change. The inference engine needs to be able to keep track of all these changes and the effects that they have on the set of currently held beliefs. To do this, the inference engine utilizes a TMS.

A TMS is a formal method for maintaining a knowledge base of current beliefs and the justifications for those beliefs. For this project, we are using a specific type of TMS called an Assumption-based TMS (ATMS) [22]. The ATMS serves two purposes (1) it maintains the consistency of the beliefs of the inference engine, and (2) it allows the inference engine to reason in multiple contexts simultaneously.

ATMS versus JTMS for Fault Diagnosis

All truth maintenance systems perform basically the same general tasks. Consequently, any of them would work for this project. The ATMS was chosen for three reasons.

First, it finds all possible resolutions of any contradictions automatically whereas the JTMS only finds one resolution at a time. Returning all the resolutions at once is more desirable so the inference engine can analyze all the possible sets of faults and pick the most likely one(s) to present to the user.

Second, the ATMS maps nicely to the process modeling structure used in this project. In the ATMS, assumptions are the foundation upon which all nodes depend; in the process model, process principles are the foundation upon which all higher process components depend. It is easy to create a hierarchy where the process principles are declared as assumptions and the higher components are declared as nodes which are ultimately justified by underlying process principles.

Finally, the work involved in the ATMS is all done up front in the creation of the nodes and their environments. Once the environments have been determined, resolving any contradictions is trivial. The JTMS, however, only records the local dependencies, so creating the nodes is easy, but resolving any contradictions is difficult. Since the speed of the program is a concern for on-line fault diagnosis, the ATMS is more appealing. It is this feature that allows this framework to gracefully scale up to large-scale processes. The ATMS uses union and intersection operators to determine minimal sets and to propagate changes through the system. It is possible to choose the process representation and the union and intersection operators carefully to function very quickly.

In the proof-of-concept system presented later, the inference engine only reasons in one context; each belief only has one set of supporting justifications. This is specific to this problem, however, and may be different for other problems or domains. Depending on the representation of the problem, a belief may be justified by multiple sets of assumptions. In these cases, the ATMS automatically maintains all the contexts simultaneously.

15.5 Characterization of Large-Scale Manufacturing Processes

Modeling a manufacturing process can be very difficult because at any step along the process, many process principles are changing ``values", many of which are irrelevant. Imagine a piece of metal entering a cutting machine and being cut into two parts. Many ``values" are changed during this simple procedure including the heat of the metal, the area of the two parts, the perimeter of the two parts, the smoothness of the edges, the curvature of the parts, the heat of the surrounding air, and any number of other effects. What is important to understand is that in an actual manufacturing process, the important attributes and their values have already been defined. While it is impossible to explicitly model all possible interactions that occur during a manufacturing process, it is possible to model the effects on those pre-defined product quality variables. Consequently, the process model should only model how the process components affect these product quality parameters.

System Inputs

Any process model will necessarily have to be created according to an actual process. Most research projects on model-based fault diagnosis have created process models that take into account the actual connectivity of the process and its desired behavior but completely ignore the system inputs. However, if perfect information were available, then fault diagnosis would be trivial. The system inputs offer the only source of information that an automated fault diagnosis system has and the amount of information that they can provide is limited. The process model must include descriptions of the system inputs and the fault diagnosis system must use this knowledge as well as possible.

There are three types of detail that can come from sensed measurement: the actual numerical value, the units or meaning of that value, and the quality of that value. See Figure 15-3. Every measurement has an output that, by itself, is usually equated to simply a number lacking the necessary meaning. Because a number without an associated context (any units or meaning) is worthless, every measurement also has a corresponding context (e.g., volts, amps, number of aberrations, etc...). A *system input* is composed of a value and its context. In addition, some system inputs also have an output that corresponds to the quality of the attribute that they are measuring. While the quality of the sensed value may be declared deterministically, these system inputs are typically product quality variables monitored by statistical process control (SPC). See Figure 15-4.

An SPC system seeks to monitor the quality of the product in line with the process so that any statistically out-of-control conditions can be detected and quickly corrected at the source of the problem. For these inputs, the ranges of acceptable values have already been defined from empirically derived process averages. The quality knowledge simply states whether the actual value is within specifications, outside specifications, too high, or too low.

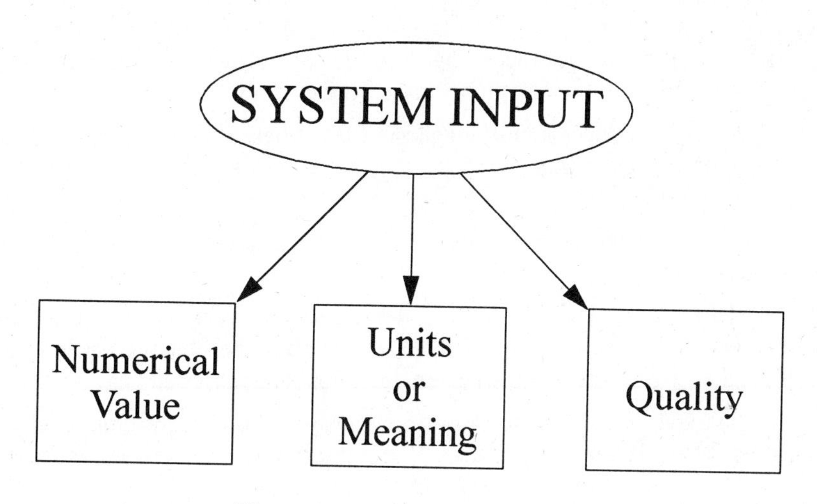

Figure 15-3: Available detail from system inputs.

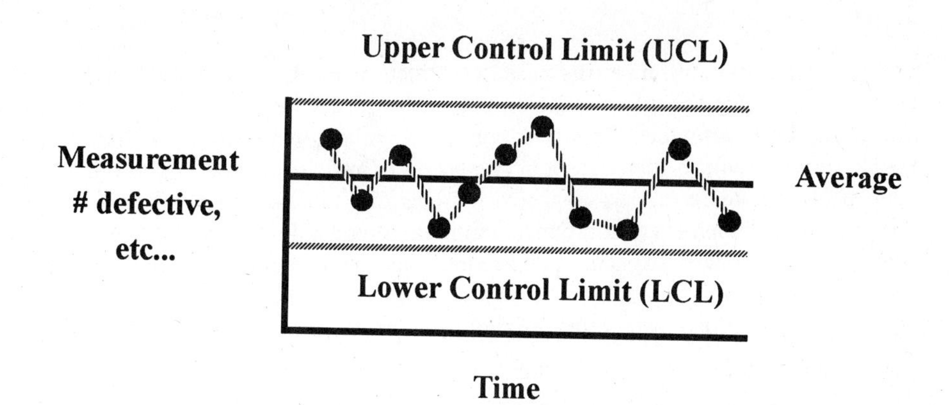

Figure 15-4: An example of an SPC control chart.

Comparison to Previous Efforts

There have been numerous papers detailing many approaches to model-based fault diagnosis.

Dishaw [9] introduces the framework, AESOP, for CMOS process diagnosis. AESOP derives ``causal relations in the knowledge base from underlying physical principles of process technology and device physics." The representation component of AESOP's architecture is a quantitative model of the device. As previously mentioned, while rule-based fault diagnosis may be easy to use, it is very difficult to create a rule-base. The reasoning component of AESOP creates a rule-base from the quantitative process model. AESOP works by exploring the ``effects of process excursions on electrical device characteristics of the CMOS process... Causal relations between process deviations and device characteristics [are] abstracted from the results of the simulations." The disadvantage of AESOP is that it requires a quantitative model which is very difficult to create for a manufacturing process because of the aforementioned problems of enumerating all the relationships within a manufacturing process.

Yamashita *et al.* [10] introduces a fault-diagnosis system based on qualitative reasoning. The representation component of the framework is a qualitative process model. The reasoning component performs four tasks.

1. *Rough estimation.* This step uses heuristics to remove unlikely elements from the list of possible fault candidates.
2. *Hypothesis and deduction.* During this step, the remaining possible causes are enumerated and checked to see if any one of them can explain the observed values of the devices. If so, that element becomes a candidate; if not, it is discarded.
3. *Analogy.* In this step, elements that were previously discarded are compared to the remaining candidates. If a previously discarded element is sufficiently similar to one of the remaining candidates, it becomes a candidate as well.
4. *Introduction of a time scale.* In this final step, the characteristic time-scales of the candidates are compared with the time-scale of the fault and all unlikely candidates are removed.

Davis [4] discusses fault diagnosis based on structure and behavior. In it, he describes how devices can be modeled according to their *functional organization* and their *physical organization.* Every component is modeled from both types of organization, creating two distinct sets of descriptions. He also introduces the concept of *categories of failure.* In performing fault diagnosis, the problem of complexity versus completeness becomes quite evident. If every possible interaction is enumerated, then every component might be responsible for a fault. But if we eliminate any interactions, then fault diagnosis may become impossible to perform. Davis gets around this by starting with the simplest possible category of failure (or *path of interaction*). If a consistent fault is not found, additional categories of failure are added until one is found.

Many of the ideas behind the framework presented in this chapter result from the work done by de Kleer *et al.* [11]. De Kleer and Williams present their general diagnostic engine (GDE) which utilizes an ATMS. Their research is noteworthy in several ways. First, ``failure candidates are represented and manipulated in terms of minimal sets of violated assumptions, resulting in an efficient diagnostic procedure." Because of that method of operation and its efficiency, GDE can handle the diagnoses of multiple faults. Finally, ``the diagnostic procedure is incremental, exploiting the iterative nature of diagnosis." The framework presented in this paper uses some of the same algorithms that GDE uses. Most of the advantages of GDE arise from its use of an ATMS. By also using an ATMS to maintain the consistency of the knowledge base, the framework presented here shares these advantages.

The aforementioned research [9,10,4,11] have had three noticeable characteristics that limit their use for performing fault diagnosis of manufacturing processes:

1. *Diagnosed digital circuits.* Circuits are described as containing many simple elements (electronic chips). The behavior of digital chips is quantized and the behavior of analog chips can almost always be calculated with well-defined, expressible mathematical equations. Furthermore, a circuit is described solely in terms of the voltages of the signals going through the chips. The behavior of processes, however, is much more difficult to describe. At any point along the process, many attributes of the product are being modified.
2. *Diagnosed single faults.* Although faults may tend to be independent in general, a fault diagnosis framework should not fail to properly diagnosis a process simply because more than one process fault occurs. In a tightly coupled process, faults tend to propagate through the system. In any critical application, the fault diagnosis system needs to be able to diagnose all the faults regardless of their number. Any attempts at diagnosing multiple faults are made by first checking all possible sets of one fault, then all possible sets of two faults, etc... This is an inefficient and very computationally expensive method of search.
3. *Ignored the system inputs.* The process models usually take advantage of the process topology and the process behavior, but they assume that any arbitrarily large amount of knowledge of the process is available. This is absolutely not the case. System inputs only provide a limited amount of information on the actual behavior of the process.

Previous efforts have contained portions of our desired framework, but never all of it. There have been papers on diagnosing processes, but they assume single faults. There have been papers on diagnosing multiple faults, but they only demonstrate their results on the simplest of digital circuits. Any fault diagnosis framework for manufacturing processes must both exploit the advantages and work around the disadvantages of this domain. The advantages include the inherent information available in the overall structure of the manufacturing process. The disadvantages include the limited information available from the system inputs. The fault diagnosis framework uses the available information in a manufacturing process and to get as much knowledge from the system inputs as possible.

15.6 Process Model

The foundation of this fault-diagnosis framework is the process model. There has to be a description of the desired behavior of the manufacturing process against which the real behavior is compared.

Methods of Process Modeling

In designing a process model, there are three levels of detail that can be used: causal, qualitative, and quantitative. Causal detail models the gross effects of the underlying parts (i.e., a certain process component effects a given quality variable); qualitative detail models the effects of change in the underlying parts (i.e., a change in a certain process component effects a change in a given quality variable); and quantitative detail models the effects of the state of the underlying parts (i.e., the state of a process component effects the given quality variable according to some mathematical function).

Causal detail is the simplest of the three types of detail. It expresses the underlying dependencies of a given quality variable. Causal detail is often represented using a cause-and-effect (or a ``fishbone'') diagram. See Figure 15-5. If there is a problem, the fishbone diagram for the affected attribute is traced through and possible factors are identified and checked until the problem is found. Such a diagram is constructed as follows [7]:

1. Identify the exact quality variable for which the tree will be created. The main backbone arrow of the diagram will represent this quality variable.
2. Enumerate the dominant process factors that affect the quality variable of interest. For each one, create a branch arrow pointing to the main backbone of the diagram.
3. For each branch created, identify the dominant process factors that affect it and create a branch arrow pointing to that branch. Repeat this process, creating more detailed branches until the desired level of detail has been achieved. The smallest branches of the diagram will represent the most detailed causes that can affect the quality variable of interest.

Qualitative detail expresses changes in the underlying dependencies of a quality variable and how those changes effect that quality variable. The relations between an attribute and an underlying factor are expressed using qualitative derivatives. Different systems use different names for the qualitative derivatives, but they all center around the three main descriptors of an attribute: decreasing, steady, or increasing. Using these derivative types, very general relations can be defined. Such relations are simple, yet can be a useful tool in fault diagnosis. If an attribute changes, one can reason back through the tree and find the possible factors which might have contributed to the change [12, 13, 14].

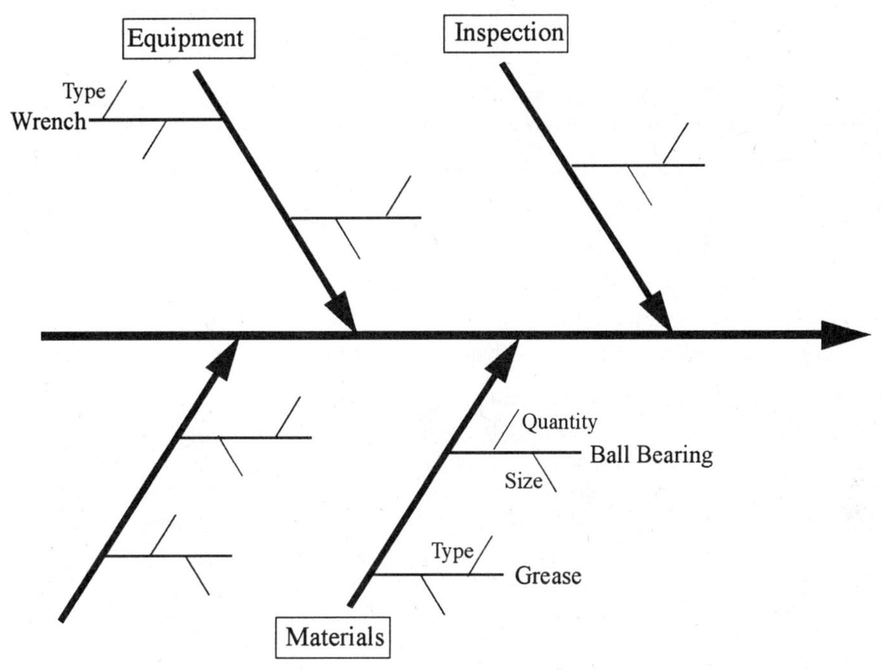

Figure 15-5: An example of a fishbone diagram relating probable causes to effects.

Qualitative detail can also be represented with a fishbone diagram. The backbone of the diagram represents a derivative of a quality variable. The branches of the diagram describe how changes in the dependencies creates the given derivative of the quality variable. A complete specification would require three different trees: one for each derivative (decreasing, steady, or increasing). From this viewpoint, causal detail can be described as a special case of qualitative detail where the derivative of the quality variable is steady.

Although causal and qualitative knowledge can be powerful tools for fault diagnosis, they still have two disadvantages. First, it is impossible to represent actual states of the process; only effects or changes in the process. Accordingly, the model cannot quantitatively reason about the exact states of the process components. Second, it is not actually modeling the process; it is modeling the qualitative effects of the process. If a given process can be expressed in more detailed terms, then using only causal knowledge quickly becomes too limiting [15].

Quantitative detail expresses the behavior of the process as a function of the states of the underlying factors. It attempts to numerically model the relationships between the quality variable and its underlying dependencies. Process models based on quantitative detail usually seek not only to describe the behavior of the process as a whole, but to describe all of the lower mechanisms that contribute to the overall behavior. At the lowest level, these mechanisms are termed first principles [6]. All higher behaviors can theoretically be expressed in terms of these first principles.

There are several problems with using quantitative detail as the only source of information for manufacturing processes. Stephanopoulos [16] lists four types of information that are not explicitly included in process models that are represented solely by systems of equations:

1. Underlying assumptions of the operational model,
2. Simplifications made by the modeler to limit the model's validity over a given range of conditions or to underscore the relative importance of sections of the model,
3. Original scope of the model (what the model was intended for), and
4. Many useful qualitative relationships.

Finally, most manufacturing environments are far too complex to model completely. This leads to attempts to model the uncertainty which is essentially modeling what one does not know which is at least as horrendous as modeling what one does know.

Manufacturing processes are less defined than any of the causal, qualitative, or quantitative types of detail. Parts of the process may be well understood and easily modeled mathematically. Other parts of the process may be less understood and may need to be expressed in terms of simple causal or qualitative knowledge. What is needed is a modeling structure that allows the user to define the model in the terms appropriate for a given process component.

Recursive Network Modeling

Small has introduced a modeling structure, System2, for constructing process models [7,8]. System2 is a recursive network model that separates the representation of process entities from the process principles that act on them. In System2, ``a process is composed of a collection of subprocesses that interact as a result of their connectivity. Each subprocess is itself an individual process that can be completely defined separately from the definition of the process that contains it." A process in System2 is defined recursively in a top-down fashion. The top-level consists of the highest possible abstraction of the process description in terms of process components and their connectivity. Subsequent layers contain the process descriptions of the underlying process components. Process definition continues until the desired level of detail has been achieved. The lowest levels will be process principles.

A process principle in System2 consists of a defined input-output relationship that is not dependent on any other process component. Process principles are the only process components that affect the quality variables. The higher levels simply declare how and when to apply the process principles. Process principles can be as detailed or as indefinite as needed and their level of detail can vary throughout the model. The designer is free to use whatever representation is convenient for the particular process component. A schematic of the recursive network model is given in figure 15-6. The vectors represent process entities and the square boxes represent process definitions.

System2 defines a process in terms of connectivity and behavior but allows the designer to specify the level of detail. This coincides very nicely with our need to represent the existing layout and behavior of a manufacturing plant but using whatever level of detail is available for a given process component. Consequently, System2 is a useful method for modeling manufacturing processes. It requires some modification, however, to scale to large systems. Three changes

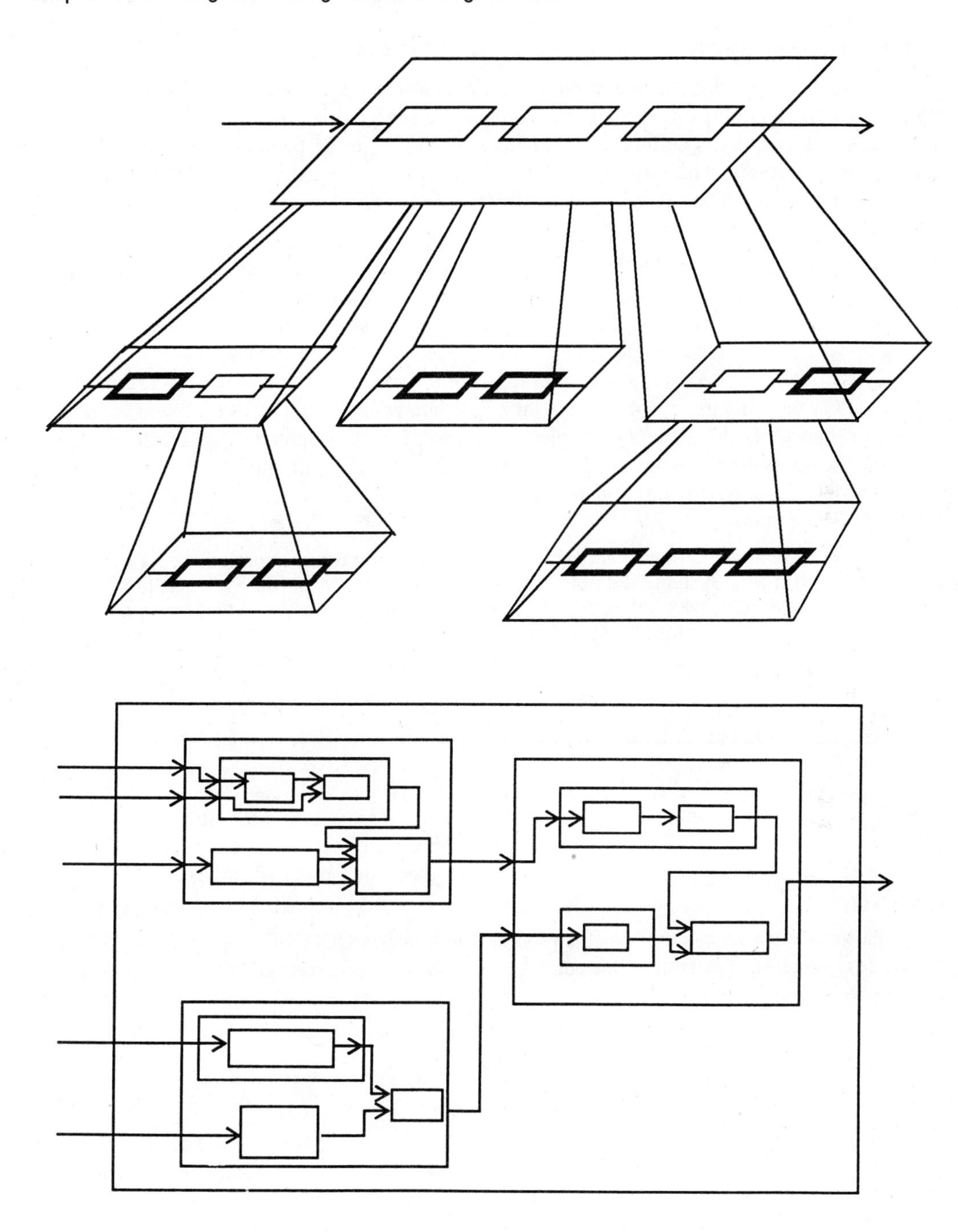

Figure 15-6: Two views of recursive network modeling.

have been added to improve the capability of modeling large processes.

System2 required every input and output port to be explicitly named. There is enough implicit information in the connectivity to eliminate the need for naming the ports. For example, assume that connectivity is defined with the function, (CONNECT from to). Given any defined process topology, it would be easy to find the beginning of the process, the end of the process, and the connectivity of all the parts in between. Explicitly naming the input and output ports of each part does not provide anymore information.

System2 did not separate the system inputs from the process itself. As a result, system inputs are declared as process components. Not only is this awkward, but in environments with a large number of system inputs, the fault diagnosis would waste considerable resources simply keeping up with them. System inputs simply monitor the process and serve as inputs into the process; they should not be declared as an integral part of the process. To this end, the current syntax incorporates system inputs in two ways. First, the numerical values of the system inputs are declared as inputs into the process principles (the meaning of the numerical value will be used in describing the behavior of the process principles in terms of the system inputs). Second, the quality parameters of the system input values are declared along with the process components that they are dependent upon.

A goal of recursive network modeling was to have a library of generic functions that each process was built from. System2 did not quite implement this concept. In System2, the first time a process was declared, it was fully defined. Then all subsequent copies of that process declared themselves to be a copy of the original process. The framework developed here separates the generic function library from the process declaration more fully than did System2.

Modeling Operators

The syntax is composed of five keywords: CONNECT, MODIFY-ATTRIBUTE, INSTANCE-OF, LOCAL-ATTRIB, and AFFECTED-BY. Together, they allow the designer to fully specify the three components of an existing manufacturing process: layout, behavior, and system inputs.

To specify the connectivity of the process, the function CONNECT is used. CONNECT takes three arguments: the name of the current level, the name of a part, and the name of a second part that the first part feeds into:

(CONNECT 'level-name 'from-name 'to-name).

The name of the top level is 'top. In lower levels, the level-name is simply the name of the process component whose internal connectivity is being specified.

The generic library of functions is made using the function, MODIFY-ATTRIBUTE. The inputs are the function name and how it affects the quality variables. For example, the definition of a function that increased the thickness looks like this:

(MODIFY-ATTRIBUTE 'lib-function-name 'thickness 'amount)

Because of the recursive network structure, the exact expression used to describe the effect of the function on a quality variable is up to the designer. It could be a mathematical formula expressing the amount as a function of the states of certain system inputs; it might also be a qualitative expression stating how the function effects the attribute in general; or it might be simply be a 'nil since the function modifies the attribute and there is no information available on how the function effects it. This is where the meaning of the numerical values of the system inputs are to be used (i.e., to describe what the system inputs mean to the quality of the quality variables) [17].

The next two functions are used to declare how the process principles depend on the generic library functions. The first function is INSTANCE-OF that takes as inputs the process principle and the library function of which it is a copy. An example of this call is:

(INSTANCE-OF 'process-component 'lib-function-name).

There will probably be local differences between process components even if they are instances of the same generic function. They might have different system inputs or different specifications. The last function defined here acts as a parameter list into the library function. LOCAL-ATTRIB takes as inputs the name of the process principle and a parameter list. An example of this is:

```
(LOCAL-ATTRIB 'process component '((parameter-name1 parameter value)
                                    (parameter-name2 parameter value)
                                    (parameter-name3 parameter value))).
```

The numerical value of the system input will be used as an input into a process principle.

Finally, there needs to be a way to declare the quality knowledge of the system inputs. This is done with the function, AFFECTED-BY that takes as inputs the name of the system input and the process components which affect it. The parts and subprocesses can be defined at any level of detail. An example of this call is:

```
(AFFECTED-BY 'system-input '(process-component1
                              process-component2
                              process-component3))
```

15.7 Fault Diagnosis Problem Solver

The process model and the system inputs are the sole sources of information that this framework will use for performing fault diagnosis. They contain descriptions of the process as well as information on the goodness of the actual process.

A function of the inference engine is to serve as the interface between the process model and the system inputs and the ATMS. The inference engine does this by translating the process model and the system inputs into nodes and justifications in the ATMS. The ATMS detects any contradictions that may have occurred. Finally, when the ATMS returns the set of possible faults to the contradictions, the inference engine selects the most appropriate fault(s) to present to the user.

Formalization of Fault Diagnosis

The fault diagnosis framework presented here is patterned after the framework presented by de Kleer and Williams [11]. They discuss a method of diagnosis that allows for diagnosing multiple faults. The algorithms used are formalized in [6]:

Definition 1: A *conflict set* for {SD,COMPONENTS, OBS} is a set $\{c_1, \ldots, c_k\} \subseteq$ COMPONENTS such that

$$SD \cup OBS \cup \{\neg AB(c_1), \ldots, \neg AB(c_k)\}$$

[where SD = System Description, OBS = Observed behavior, and AB = abnormal]

A conflict set for (SD, COMPONENTS, OBS) is *minimal* iff no proper subset of it is a conflict set for (SD, COMPONENTS, OBS).

Proposition 1: $\Delta \subseteq$ COMPONENTS is a diagnosis for (SD, COMPONENTS, OBS) iff Δ is a minimal set such that COMPONENTS $- \Delta$ is not a conflict set for (SD, COMPONENTS, OBS).

Definition 2: Suppose C is a collection of sets. A *hitting set* for C is a set $H \subseteq \bigcup_{S \in C} S$ such that $H \cap S \neq \{\ \}$ for each $S \in C$. A hitting set for C is *minimal* iff no proper subset of it is a hitting set for C.

The following is our principal characterization of diagnoses, and will provide the basis for computing diagnoses:

Theorem 1: $\Delta \subseteq$ COMPONENTS is a diagnosis for (SD, COMPONENTS, OBS) iff Δ is a *minimal hitting set* for the collection of conflict sets for (SD, COMPONENTS, OBS).

Briefly restated, a conflict set is a set of components that are not consistent with the declared system description and desired behavior. A hitting set of a collection of sets, C, is a set, H, such that H contains part of each set in C. A diagnosis is a set of components which is a minimal hitting set for the collection of conflict sets.

The ATMS maintains minimal sets of justifications for each belief. If the inference engine declares a belief as being contradictory, the corresponding sets of justifications are considered to be conflict sets and are added to the nogood-table of the ATMS. The nogood-table, like the environments of the nodes, is also maintained so that the sets in it are minimal sets. A possible diagnosis is any set of process principles that compose a hitting set of the sets of minimal conflict sets in the nogood-table of the ATMS [18, 11, 19].

Logic for Fault Diagnosis

The fault diagnosis representation is held separate from the process model. The process model contains all the information on the expected behavior of the process. Different programs will access the information available in the process model as needed (e.g,. fault diagnosis, performance prediction, etc...). See Figure 15-7.

As mentioned previously, the process model can contain any or all of three types of detail: causal, qualitative, or quantitative. The designer must decide how the fault diagnosis program will use the information available in the process model. The designer is free to use any, all, or none of the information in the process model.

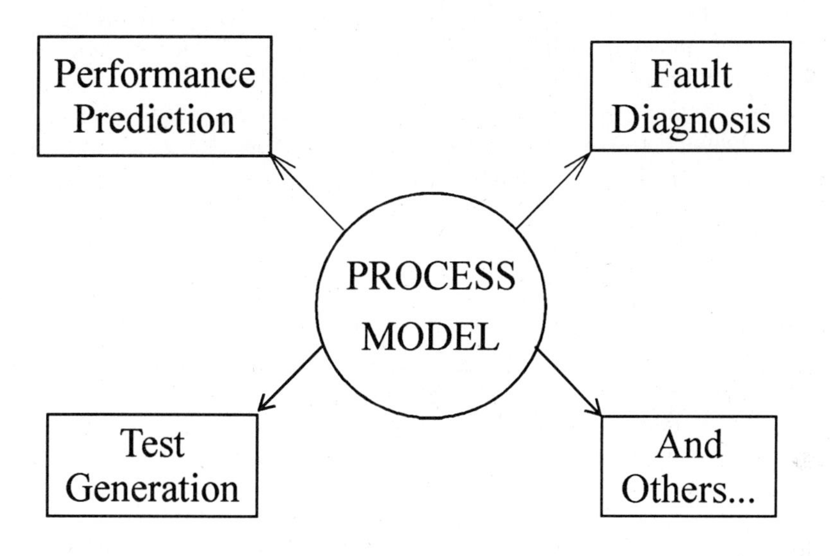

Figure 15-7: Relationship between the process model and other programs.

The ATMS does not inherently know how to find the possible faults. The ATMS simply maintains a database of beliefs and justifications. There must be a methodology or logic for deciding that a particular part is a possible fault. This logic may or may not be explicitly defined in the inference engine (depending on how the inference engine is set up). Instead, the logic will be represented in how the inference engine creates the nodes and justifications. Consequently, the logic must be fully understood before the inference engine can properly create them.

One of the simpler logics is to only use the causal detail in the process model. A part is a suspect if it affects the quality variable in any way. If a system input returns a bad value, then all process principles that contributed to the value of that system input are suspects. The goal of the inference engine is now to maintain the dependencies of the system inputs. The next step up in complexity involves using the qualitative detail in the process model (e.g., too much, too little, just right).

Creating the Nodes and Justifications

The ATMS resolves contradictions between the process model and the system inputs. The two implied assumptions are that any discrepancy in the system inputs will result in one or more contradictions and that the resolution of the contradiction(s) is meaningful. These assumptions are not implemented automatically; they must be carefully planned for in the inference engine.

In order for meaningful contradictions to occur, the ATMS will keep track of which process principles affect which system inputs. Then when a system input returns a bad value, the ATMS knows which process principles might be at fault. From the information available in the process model, it is simple to construct a dependency network like the one shown in Figure 15-8. The nodes represent process components. Believing a node corresponds to believing that process principle is functioning correctly. Such a network contains descriptions of the processes and the dependent system inputs. Each system input is justified by the set of process principles that ultimately support it. If the system input returns a bad value, the ATMS will know which process principles that have caused it.

Fault Diagnosis

As the framework is set up, a bad value from a system input will contradict in a meaningful way. Moreover, this specific method of fault diagnosis comes with a free feature: if more than one system input returns a bad value, the ATMS will automatically take care of finding the minimal conflict sets that justify all the faulty system inputs. So, we can easily take advantage of any extra knowledge that comes in.

15.8 Example Process Model

To demonstrate proof-of-concept on a process model, a polypropylene-film manufacturing process operation was selected. This plant is characteristic of plastic-film manufacturing plants.

This type of plant was chosen for three reasons:

1. There is a need for quick fault diagnosis because the process is continuous and cannot be stopped.
2. There is a need to diagnose multiple faults because the process is tightly coupled and faults may propagate through the system.
3. The general method for creating plastic film is well understood and documented.

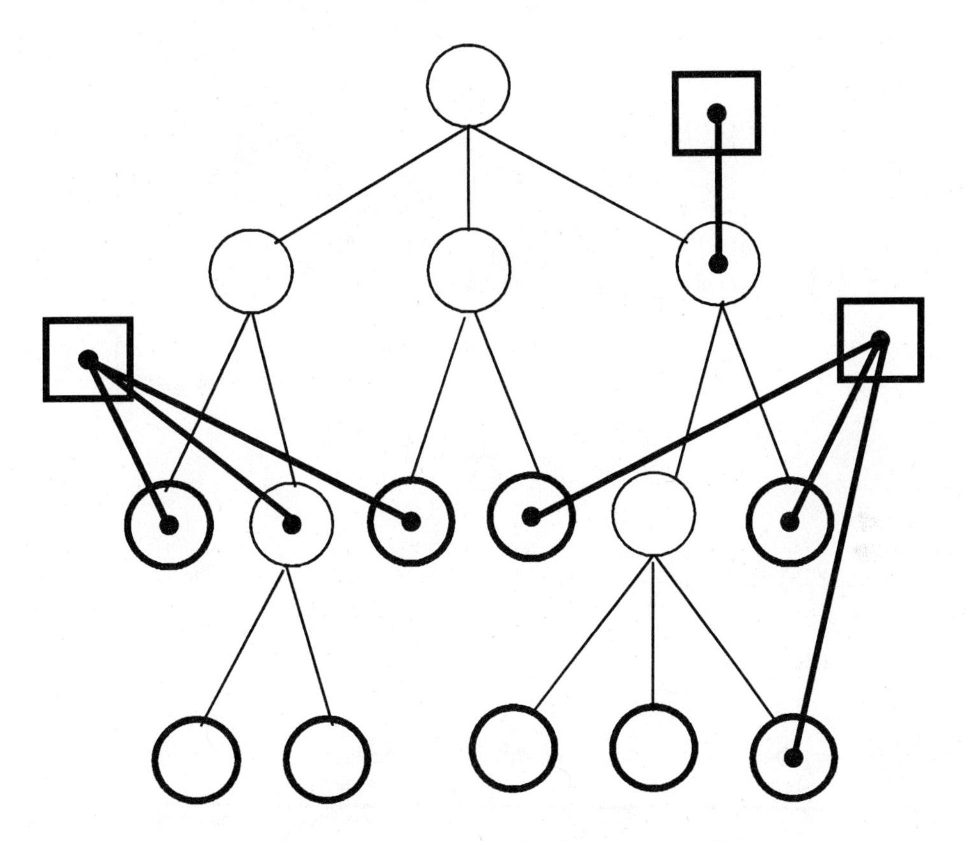

Circles = Process Components as declared in the process model
Bold Circles = Process Principles
Bold Squares = System Inputs (connected to the dependent process components)

Figure 15-8: Dependency tree used for fault diagnostic model.

The process model that is used in this example problem is derived from books on plastic film manufacturing. Figure 15-9 shows an outline of the process for making polypropylene film [20]. First, the plastic is melted and placed on the casting wheel. Second, the plastic is stretched lengthwise. Next, the plastic is stretched width-wise. Finally, the plastic is cooled and wound up on rollers. Before each stretch, the plastic is heated to facilitate stretching. After each stretch, the plastic is partially cooled to set it.

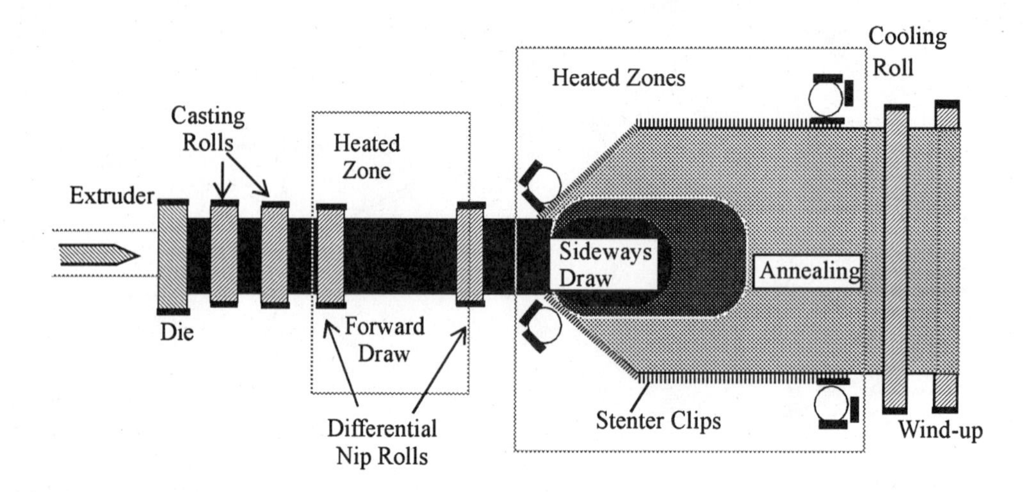

Figure 15-9: Outline of a manufacturing process to make polypropylene film. Adapted from [20].

Example Results

When the program is run, a quality variable of interest is defined and then only process components that effect that quality variable are used. There are several quality variables of interest in a film manufacturing process. In this case, we chose to focus on the thickness of the film. Consequently, only process components which effect thickness are used. Other quality variables include scratches, flatness, or other film blemishes.

Every different quality variable has its own dependency tree of process components which affect it. Any given manufacturing process model will have many corresponding dependency trees that characterize how the process affects the quality variables.

A polypropylene film manufacturing process model is used to test the fault diagnosis framework. The process model is read in and the information is stored as property lists. The causal detail from the process model is then translated into nodes and justifications in the ATMS. The process principles are declared as assumptions. The higher process components are declared as nodes that are justified by their immediate sets of dependent process components. The resulting dependency tree derived from the process model is shown in Figure 15-10. The specific model results in 20 nodes and 4 justifications in the ATMS plus those created for the SPC input nodes.

The bottom nodes of the dependency tree are the process principles. The information in the process principles state how they modify the quality variable of interest (specifically, thickness). For example, consider the process principle, Air-Heat-1, in the specific model below. It is included in the dependency tree because it effects the thickness of the film. From the process

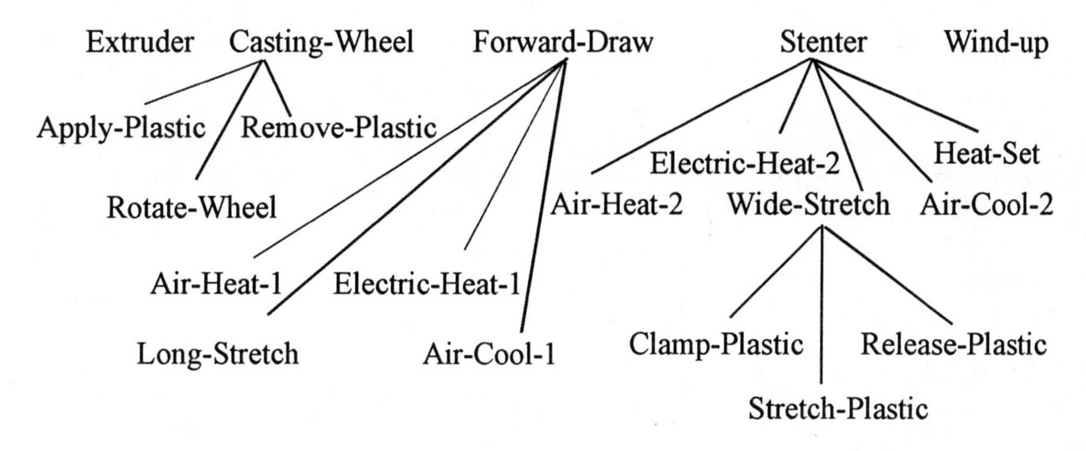

Figure 15-10: Resulting dependency tree to process fault diagnosis.

model, it can be seen that its effect on the thickness of the film is expressed as a numerical function dependent on the pressure of the air, the temperature of the air, the temperature of the air around the film, the number and diameter of the air jets, and the volume of the enclosed area. After declaring all the process components, selected nodes representing SPC inputs are made contradictory (i.e., simulating a fault). When this happens, the ATMS determines the underlying process principles of the system input and places them in the nogood-table of the ATMS.

Because the nogood-table is kept minimal, additional information can be usefully integrated. For example, if the Stenter in the example model were to be faulty, the justifying process principles (Air-Heat-2, Electric-Heat-2, Clamp-Plastic, Stretch-Plastic, Release-Plastic, Air-Cool-2, and Heat-Set) would be added to the nogood-table. If later on, the Wide-Stretch process component were to be faulty, its justifying process principles (Clamp-Plastic, Stretch-Plastic, and Release-Plastic) would also be added to the nogood-table. But since the second set is a subset of the first set, the ATMS would only keep the second set. After all the faulty SPC inputs have been declared, the resulting list of suspect process principles is found by simply observing at the nogood-table of the ATMS.

The ATMS does all its work up front in the creation of the nodes. As a result, the creation of the nodes takes considerable more time than the resolution of any contradictions. As an example, this program took 24 seconds to load the model and create the appropriate nodes in the ATMS (17 seconds were required to create the nodes). After loading the model, the program took only 0.5 seconds to resolve any contradiction.

15.9 Conclusions

A framework for performing computer-assisted fault diagnosis of large-scale manufacturing processes is discussed and applied to a proof-of-concept system. There were three important aspects of this project: we performed model-based fault diagnosis, we performed it on manufacturing processes, and the resulting framework will gracefully scale up to larger manufacturing processes.

Model-Based Fault Diagnosis

The framework was designed to perform fault diagnosis from the process model, as opposed to an expert-designed rule-base. From a process model incorporating the connectivity, the desired behavior, and the observed behavior of a process, we were able to create a knowledge base of that information. A fault was indicated by a contradiction between the desired behavior and the observed behavior. By resolving those contradictions, the ATMS returned a set of possible suspect process principles that might account for the fault.

Fault Diagnosis of Manufacturing Processes

Existing manufacturing plants already have an abundance of knowledge inherent in the structure of the plant. This includes the layout of the plant, the expected behavior of the components of the plant, the system inputs which monitor the behavior of the plant, and a list of attributes (or quality variables) which are important to monitor. This information was captured using a recursive network modeling structure. With it, a process is broken down into a hierarchical network of process components. The lowest levels of the hierarchy consist of the process principles. The process principles are the only process components that effect the quality variables. How a specific process principle is declared as effecting the quality variable can be as detailed or as indefinite as needed and the level of detail can vary throughout the process model. Because of its ability to represent the behavior of a process in varying levels of detail and its ability to represent the connectivity of a process, recursive network modeling is ideally suited for modeling manufacturing processes.

Fault Diagnosis of Large-Scale Manufacturing Processes

From the viewpoint of the user, using the hierarchical process declaration in the recursive network modeling structure will allow large processes to be easily broken into more manageable parts. From a computational viewpoint, the ATMS is a fast and efficient method for resolving contradictions that can realistically handle a large number of process components.

Future Work

This proof-of-concept provides a base to focus future efforts in diagnosis of system faults. It is expected that future interactions will more fully develop these concepts of fault diagnosis within an existing large manufacturing process. As mentioned previously, the process model can contain three types of detail: causal, qualitative, and quantitative. The fault diagnosis framework presented here, however, only uses the causal detail. It seems apparent that one of the next steps should be to modify the framework to utilize qualitative and/or quantitative process detail. Using qualitative detail may be as simple as maintaining three different knowledge bases for the three different qualitative derivatives. Using quantitative detail will be more difficult and will probably require a completely different type of problem representation. Using all types of available detail will be much harder.

Usually when automated fault diagnosis has been performed, only one type of process detail has been used. Most research in fault diagnosis creates an artificial association between the process detail and the fault diagnosis reasoning; they only use the type of process detail their reasoning program can understand. There is a great lack of understanding as to how to incorporate more than one type of process detail. This theory will have to be developed more before attempting to incorporate more types of detail. In the final analysis, more work is needed in unifying theories of model-based fault diagnosis.

Acknowledgments

This project was funded, in part, by the Colorado Advance Technology Institute through a Colorado Advanced Software Institute research grant. We want to recognize the Kodak Colorado Division for their cooperation on this project.

References

1. Chandrasekaren, B., and Sanjay Mittal. ``Deep versus compiled knowledge approaches to diagnostic problem-solving." *International Journal of Man-Machine Studies*, 19:425--436, 1983.

2. Dhar, Vasant, and Harry E. Pople. ``Rule-based versus structure-based models for explaining and generating expert behavior." *Communications of the ACM*, 30:542--555, June 1987.

3. Sudduth, A.L. ``Knowledge-based systems in process fault diagnosis." *Nuclear Engineering and Design*, 113:195--209, 1989.

4. Davis, Randall. ``Diagnostic reasoning based on structure and behavior." *Artificial Intelligence*, 24:347--410, 1984.

5. Davis, Randall. ``Reasoning from first principles in electronic trouble-shooting." *International Journal of Man-Machine Studies*, 19:403--423, 1983.

6. Reiter, Raymond. ``A theory of diagnosis from first principles." *Artificial Intelligence*, 32:57--95, April 1987.

7. Small, Daniel H. *A structured approach to process representation.* Master's thesis, Colorado State University, 1988.

8. Troxell, Wade O., Daniel H. Small, William Maulsby, and Eric J. Krengel. ``Automating the identification of perturbations within large manufacturing processes: A structured approach to process representation." Technical report, CIAI-TR-89-03, January 1989.

9. Dishaw, Patrick J. ``Aesop: A simulation-based knowledge system for CMOS process diagnosis." *IEEE Transactions on Semiconductor Manufacturing*, 2:94--103, August 1989.

10. Yamashita, Y., *et al.* ``A fault-diagnosis system based on qualitative reasoning." *International Chemical Engineering*, pages 151--159, January 1990.

11. de Kleer, Johan, and Brian C. Williams. ``Diagnosing multiple faults." *Artificial Intelligence*, 32:97--130, 1987.

12. Forbus, Kenneth D. ``Qualitative process theory." *Artificial Intelligence*, 24:85--167, 1984.

13. Forbus, Kenneth D. ``Qpe: Using assumption-based truth maintenance for qualitative simulation." *International Journal for Artificial Intelligence*, pages 200--215, 1988.

14. Whitehead, J. Douglass, and John W. Roach. ``Hoist: A second-generation expert system based on qualitative physics." *AI Magazine*, pages 108--119, Fall 1990.

15. Shrager, J., D. Jorday, T. Moran, G. Kiczales, and D. Russell. ``Issues in the pragmatics of qualitative modeling: Lessons learned from a xerographics project." *Communications of the ACM*, 30:1036--1047, December 1987.

16. Stephanopoulos, G., *et al.* ``MODEL.LA. a modeling language for process engineering -- I. the formal framework." *Computers in Chemical Engineering*, 14(8):813--846, 1990.

17. Roffell, Brian, and Patrick Chin. *Computer Control in Process Industries.* Lewis Publishers, 1987.

18. Arlabosse, Francois, Jean-Bart Bruno, and Nathalie Porte. ``An efficient problem solving architecture using ATMS." *AI Communications*, 1(4):6--15, December 1988.

19. Struss, P. *Extensions to ATMS-based diagnosis*. Elsevier, 1988. Edited by J. S. Gero.

20. Rigby, H. A. *Production, Properties, and Packaging Applications of Polypropylene*. Modern Packaging Films, Butterworths, 1967. Pages 56-78.

21. Doyle, John. ``A truth maintenance system." *Artificial Intelligence*, 12:231--272, 1979.

22. de Kleer, Johan. ``An assumption-based TMS." *Artificial Intelligence*, 28:127--162, 1986.

AUTHOR INDEX

SUBJECT INDEX